LASER SCANNING

International Society for Photogrammetry and Remote Sensing (ISPRS) Book Series

ISSN: 1572–3348

Book Series Editor

Zhilin Li
Department of Land Surveying and Geo-Informatics, The Hong King Polytechnic University, Hong Kong, P.R. China

information from imagery

Laser Scanning

An Emerging Technology in Structural Engineering

Editors

Belén Riveiro

School of Industrial Engineering, University of Vigo, Vigo, Spain

Roderik Lindenbergh

Department of Geoscience and Remote Sensing, Delft University of Technology, Delft, The Netherlands

CRC Press
Taylor & Francis Group
Boca Raton London New York Leiden

CRC Press is an imprint of the
Taylor & Francis Group, an **informa** business
A BALKEMA BOOK

CRC Press/Balkema is an imprint of the Taylor & Francis Group, an informa business

Typeset by Apex CoVantage, LLC

Library of Congress Cataloging-in-Publication Data
Applied for

Published by: CRC Press/Balkema
Schipholweg 107c, 2316 XC Leiden, The Netherlands
e-mail: Pub.NL@taylorandfrancis.com
www.crcpress.com – www.taylorandfrancis.com

ISBN: 978-1-138-49604-0 (Hbk)
ISBN: 978-1-351-01886-9 (eBook)
DOI: https://doi.org/10.1201/9781351018869

Printed and bound by CPI Group (UK) Ltd, Croydon, CR0 4YY

Laser Scanning – Riveiro and Lindenbergh
© 2020 ISPRS, ISBN 978-1-138-49604-0

Contents

Preface

Laser scanning technology has significantly evolved in the last years; its ability to collect large amounts of accurate geometric 3D data at a relatively low cost has motivated the adoption of this geomatic technology for many different applications. The growing need to digitally record our environment in 3D, together with the developments in artificial intelligence, particularly relevant for the automated interpretation of digital images, motivated the adoption of laser scanning as a fundamental tool for geometrical recording. This is especially relevant in the construction and civil engineering domains, among others, where laser scanners have been included as routine measuring technology for projects requiring geometrical control, 3D modelling, or building/infrastructure management, deformation of structural health monitoring, etc. As a consequence of its expansion in other domains outside of geomatics, new needs appeared, namely: more efficient acquisition systems, automated object identification and parameterization, etc.

Since terrestrial laser scanners have become more accessible to researchers and professionals working in the industrial sector, there has been an explosion of articles reporting successful applications of the technology, proposing new acquisition systems, even new sensors, and more intensely, automatic data processing procedures. Even though access to the huge number of papers that are being published every day becomes easier, there exists a need to compile in a single document those aspects of laser scanning technology that have evolved in the last years. Additionally, such a document can serve to update the reference books for students, not only in the geomatic domain but most importantly, for those students and professionals from other domains such as civil engineering, construction, or architecture (among others), who are introducing this powerful geomatic technique as a basic tool for surveying, measurement, 3D modelling, and other subsequent operations.

In this context, this book provides an overview on the evolution of laser scanning technology and its noticeable impact in the structural engineering domain. It aims to provide an up-to-date synthesis of the state of the art of the technology for the reverse engineering of built constructions.

Data processing of large point clouds collected with laser scanning systems has in parallel fashion experienced an important advance, and thus, an intense activity in the development of automated data processing algorithms has been noticed. The emergence of these algorithms is related to an intense demand of engineers and technicians from outside the geomatics domain. Different applications have in common that they all demand an efficient conversion of the raw laser scan point cloud data to 3D models, but all with purpose-dependent requirements. This book provides an overview of state-of-the-art algorithms, different best practices, and most recent processing tools such as the emergent application of deep learning to dense point clouds.

Finally, the book is completed with a set of chapters presenting some relevant practical applications of laser scanning point clouds to structural engineering such as routine bridge inspection, documentation and automated inspection of historical constructions, exploitation of dense point clouds to determine stresses in metallic structural members, construction site monitoring, and conversion of point clouds into BIM models for a more efficient management of existing constructions.

Laser Scanning – Riveiro and Lindenbergh
© 2020 ISPRS, ISBN 978-1-138-49604-0

Editors

Belén Riveiro was born in Pontevedra (Spain) in 1983. Since 2012, she is an Associate Professor at the Department of Materials Engineering, Applied Mechanics and Construction at the University of Vigo (Spain). She holds an MSc in Construction Engineering (2015), an MESc in Forestry Engineering (2006) and a PhD in Environmental Engineering (2011). She has been an Associate Professor in Geomatics at Newcastle University (UK) in 2011, and was a postdoctoral research fellow at the University of Minho (2012 and 2014), the University of Cambridge (2015), and at Delft University of Technology (2016). Her research is focused on the application of remote sensing technologies in structural engineering for the automated modelling, inspection and material characterization using inverse analysis procedures. At a larger scale, she uses Lidar for the modelling of terrestrial transport infrastructure and urban spaces. She has been principal investigator in several research and innovation projects (at national and international levels) and coordinator of an European action focused on the resilience of transport infrastructure to extreme events (within the H2020 Framework Program). She has published more than 80 research papers in SCI-JCR journals (h-index 22) and more than 60 communications to international conferences. She is co-author of nine patents and has supervised six PhD theses. She is the Secretary of the ISPRS WGII/10 3D Mapping for Environmental & Infrastructure Monitoring for the period 2016–2020. In 2017 she was awarded by the Spanish Royal Academy of Engineering with a medal of the Juan López de Peñalver Prize within the Awards for Young Researchers.

Roderik Lindenbergh is Associate Professor in Laser Scanning at the Department. of Geoscience and Remote Sensing of Delft University of Technology. He studied mathematics at the University of Amsterdam and studied six months at RWTH Aachen for his MSc project on computational group theory. After eight months at QM&W College, London, he worked on his PhD in geometry and topology at Utrecht University, where he obtained his PhD on limits of Voronoi diagrams in 2002. From then he has been working at the current Dept. of Geoscience and Remote Sensing of TU Delft. His main research interests are properties, processing methodology, and applications of laser scan point clouds and digital terrain analysis using satellite data. He was elected best lecturer 2017 of the Faculty of Civil Engineering and Geosciences. He was Chair of Laser Scanning 2017, a workshop organized at the Geospatial Week in Wuhan, China. He is associate editor of the ISPRS journal on photogrammetry and remote sensing, and he is the 2016–2020 Chair of ISPRS WG II/10 on 3D mapping for environmental and infrastructural monitoring.

Laser Scanning – Riveiro and Lindenbergh
© 2020 ISPRS, ISBN 978-1-138-49604-0

Contributors

Fabrizio Banfi
Department of Architecture, Built Environment and Construction Engineering
Politecnico di Milan, Milan, Italy
Email: fabrizio.banfi@polimi.it

Luigi Barazzetti
Department of Architecture, Built Environment and Construction Engineering
Politecnico di Milan, Milan, Italy
Email: luigi.barazzetti@polimi.it

Roland Billen
Geomatics Unit, University of Liège, Liège, Belgium
Email: rbillen@uliege.be

Manuel Cabaleiro
Department of Materials Engineering, Applied Mechanics and Construction
University of Vigo, Vigo, Spain
Email: mcabaleiro@uvigo.es

Pierre Charbonnier
Cerema, Equipe Projet ENDSUM, Strasbourg, France
Email: Pierre.Charbonnier@cerema.fr

Borja Conde
Department of Materials Engineering, Applied Mechanics and Construction
University of Vigo, Vigo, Spain
Email: bconde@uvigo.es

Susana Del Pozo
Department of Cartographic and Land Engineering
University of Salamanca, Ávila, Spain
Email: s.p.aguilera@usal.es

Philippe Foucher
Cerema, Equipe Projet ENDSUM, Strasbourg, France
Email: Philippe.Foucher@cerema.fr

Diego González-Aguilera
Department of Cartographic and Land Engineering
University of Salamanca, Ávila, Spain
Email: daguilera@usal.es

Pierre Grussenmeyer
Photogrammetry and Geomatics Group, ICube Laboratory UMR 7357, INSA, Strasbourg, France
Email: pierre.grussenmeyer@insa-strasbourg.fr

Samuel Guillemin
Photogrammetry and Geomatics Group, ICube Laboratory UMR 7357, INSA, Strasbourg, France
Email: samuel.guillemin@insa-strasbourg.fr

Harri Kaartinen
Department of Remote Sensing and Photogrammetry
Finnish Geospatial Research Institute FGI, Masala, Finland
Department of Geography and Geology, University of Turku, Finland
Email: harri.kaartinen@nls.fi

Sanna Kaasalainen
Department of Navigation and Positioning, Finnish Geospatial Research Institute – FGI, Masala, Finland
Email: sanna.kaasalainen@nls.fi

Matthieu Koehl
Photogrammetry and Geomatics Group, ICube Laboratory UMR 7357, INSA, Strasbourg, France
Email: mathieu.koehl@insa-strasbourg.fr

Antero Kukko
Department of Remote Sensing and Photogrammetry
Finnish Geospatial Research Institute FGI
Masala, Finland
Department of Built Environment
Aalto University, Espoo, Finland
Email: antero.kukko@nls.fi

Debra F. Laefer
Center for Urban Science & Department of Civil and Urban Engineering
New York University, New York, US
Email: debra.laefer@nyu.edu

Derek D. Lichti
Department of Geomatics Engineering
The University of Calgary, Calgary, Alberta, Canada
Email: ddlichti@ucalgary.ca

Roderik Lindenbergh
Department of Geoscience and Remote Sensing
Delft University of Technology, Delft, The Netherlands
Email: R.C.Lindenbergh@tudelft.nl

Cserep Mate
Department of Geoscience and Remote Sensing
Delft University of Technology, Delft, The Netherlands
Email: mcserep@caesar.elte.hu

Fabio Menna
3D Optical Metrology Unit
Bruno Kessler Foundation, Trento, Italy
Email: fmenna@fbk.eu

Emmanuel Moisan
Cerema, Equipe Projet ENDSUM, Strasbourg, France
Email: emmanuel.moisan@insa-strasbourg.fr

Ángel Luis Muñoz-Nieto
Department of Cartographic and Land Engineering
University of Salamanca, Ávila, Spain
Email: almuni@usal.es

Adriaan van Natijne
Department of Geoscience and Remote Sensing
Delft University of Technology, Delft, The Netherlands
Email: A.L.vanNatijne@tudelft.nl

Erica Nocerino
Laboratoire des Sciences de l'Information et des Systèmes (LSIS laboratory), Aix-Marseille Université, France
Theoretical Physics, ETH Zürich, Switzerland
Email: erica.NOCERINO@univ-amu.fr

Abdul Nurunnabi
Department of Geoscience and Remote Sensing
Delft University of Technology, Delft, The Netherlands
Email: pyal1471@yahoo.com

Florent Poux
Geomatics Unit, University of Liège, Liège, Belgium
Email: fpoux@uliege.be

Luís F. Ramos
ISISE, Department of Civil Engineering, University of Minho, Guimarães, Portugal
Email: lramos@civil.uminho.pt

Belén Riveiro
Department of Materials Engineering, Applied Mechanics and Construction
University of Vigo, Vigo, Spain
Email: belenriveiro@uvigo.es

Pablo Rodríguez-Gonzálvez
Department of Mining Technology, Topography and Structures
Universidad de León, Ponferrada, Spain
Email: p.rodriguez@unileon.es

Luis J. Sánchez-Aparicio
Department of Cartographic and Land Engineering
University of Salamanca, Ávila, Spain
Email: luisj@usal.es

Ana Sánchez Rodríguez
Department of Materials Engineering, Applied Mechanics and Construction
University of Vigo, Vigo, Spain
Email: anasanchez@uvigo.es

Mario Soilán Rodríguez
Department of Materials Engineering, Applied Mechanics and Construction
University of Vigo, Vigo, Spain
Email: msoilan@uvigo.es

Sylvie Soudarissanane
Department of Geoscience and Remote Sensing
Delft University of Technology, Delft, The Netherlands
Email: sylvie.soudarissanane@gmail.com

Linh Truong-Hong
Department of Geoscience & Remote Sensing
Delft University of Technology, Delft, The Netherlands
Email: L.Truong@tudelft.nl

Cheng Wang
School of Information Science and Engineering
Xiamen University, Xiamen city, Fujian Province, China
Email: cwang@xmu.edu.cn

Jinhu Wang
Department of Geoscience and Remote Sensing
Delft University of Technology, Delft, The Netherlands
Email: jinhu.wang@hotmail.com

Martin Weinmann
Institute of Photogrammetry and Remote Sensing
Karlsruhe Institute of Technology (KIT), Karlsruhe, Germany
Email: martin.weinmann@kit.edu

Chenglu Wen
School of Information Science and Engineering
Xiamen University, Xiamen city, Fujian Province, China
Email: clwen@xmu.edu.cn

Daniel Wujanz
technet GmbH
Am Lehnshof 8, 13467 Berlin, Germany
Email: daniel.wujanz@technet-gmbh.com

Laser Scanning – Riveiro and Lindenbergh
© 2020 ISPRS, ISBN 978-1-138-49604-0

Chapter 1

Introduction

Belén Riveiro, Ana Sánchez-Rodríguez and Mario Soilán

Terrestrial mobile mapping technologies have changed the vision on how to conduct routine surveying and inspection of constructions, which for many years was done by visual inspection methods complemented with rudimentary measurements. With the growing of digital technology, the inspection procedures also evolved thanks to the use of special machines, which in many cases relied on remote and nondestructive techniques to measure and visually detect damages by an experienced operator. Within these new approaches, laser scanning is probably one of the technologies that has impacted the construction and civil engineering fields in a greater manner due to its ability to massively, accurately and efficiently register the 3D geometry of existing buildings and structures.

Vosselman and Maas (2010) reported in the precedent book about *Airborne and Terrestrial Laser Scanning* that laser scanning was (at that date) a relatively young technique, but the authors already foresaw its impact as a 3D measuring technique for many different applications. In their book, the fundamental principles and pioneer methods were already presented, so with the aim of avoiding overlapping contents, the present book is focused to update the state of the art of laser scanning by focusing in the advances suffered by the technique in the last years, with special application to the structural engineering domain.

In the precedent publications, researchers and professionals were reporting works and investigations based on existing commercial laser scanners, which were accompanied by specific software packages that permitted the processing of point clouds in a quite straightforward manner (scan registration, point cloud visualization and measurements, ending with 3D modelling tools). However, in the last ten years, there has been a significant evolution of static stations to mobile scanning systems that can be mounted in many different platforms such as vans, boats, backpack solutions, indoor trolleys, etc. Similarly, talking about data processing, many authors had already proposed automatic or semiautomatic methods applying different approaches that include heuristic rules, or more advanced machine learning techniques, where deep learning applied to 3D point clouds represents one of the hot topics in the state of the art.

This book has been structured in a set of chapters that cover those topics where laser scanning has evolved more importantly in the last years, together with those particular applications to structural and construction engineering that have motivated the definitive adoption of laser scanning as a basic but powerful technique within these sectors.

Thus, the book starts with some chapters devoted to data acquisition in laser scanning, namely new sensors and innovative platforms.

Chapter 2 introduces one of the main innovations regarding sensors, referring to multispectral laser scanning. Even though very few systems exist providing simultaneous 3D point clouds and spectral measurements, research on multiwavelength laser returns has been reported in the literature. For that reason, this book starts with a chapter summarizing these first experiences on the use of multispectral TLS and its potential to efficiently and robustly solve problems in science and industry.

In continuation, a set of chapters focusing on new mobile scanning systems is presented. Thus, Chapter 3 starts presenting the advances in kinematic scanning solutions, with a chapter presenting multiplatform approaches to the application of mobile laser scanning in various surveying tasks in outdoor environments (3D data for terrain modeling, industrial scenes, construction sites and infrastructure, as well as urban mapping). The chapter outlines the idea of multiplatform mobile laser scanning particularly implementing solutions such as vehicle-, trolley- and ATV-operated data acquisition, and boat-mounted equipment. Chapter 4 continues focusing in portable mobile mapping systems, whose main distinctive feature regarding the conventional mobile mapping systems is the ability to efficiently acquire 3D information on the move in narrow, indoor and confined environments. The chapter includes a taxonomy of the available platforms together with a description of hardware components and data processing algorithms. Chapter 5 continues focusing on indoor modeling, thus presenting recent advances of laser scanning for this purpose. The chapter focuses on two essential parts: first, the calibration for multibeam laser scanners and indoor building modeling using backpacked laser scanning point clouds; second, an approach for semantic labeling of laser scanning point cloud data into the walls, ceiling, floor and other objects. The method ends with the extraction of line structures from the labeled points to achieve an initial description of the building line framework.

Chapter 6 focuses on reviewing some fundamental concepts for point cloud quality control. These includes reference measures to inform technicians when inspecting geometric changes in the dimensions of structural members, namely: random and systematic errors, geometric network design and error propagation, etc. The chapter ends with a real work where these concepts are illustrated, particularly the effect of propagation of random errors.

Chapter 7 addresses key aspects in data processing: semantic segmentation of dense point clouds, as this has been a field where laser scanning reported more investigations and publications. This chapter revises the fundamentals for the three main approaches followed for point cloud segmentation: standard classification techniques based on manually defined features for segmentation at point level, strategies for point cloud segmentation at object level, and the most recent approaches for point cloud segmentation based on deep learning. The chapter also presents some commonly used benchmark datasets for evaluating the performance of a semantic segmentation approach.

The chapters regarding point cloud processing end with Chapter 9, where the Smart Point Cloud concept is introduced, as this can be understood as a broad framework for the semantic enrichment and structuration of point clouds for intelligent agents and decision-making systems. The approach presented in the chapter provides a conceptual data model to structure 3D point data, semantics and topology proficiently. A multimodal infrastructure that integrates the aforementioned data model integrates knowledge extraction, knowledge integration and knowledge representation for automatic agent-based decision-making over enriched point cloud data.

The last section of the book presents a series of chapters presenting different applications where laser scanning has been demonstrated to be a very useful tool in the structural and construction engineering domains. Thus, Chapter 10 presents a successful methodology to integrate laser and sonar data and thus being able to reconstruct semi-immersed structures. Chapter 11 presents experiences where point clouds acquired with laser scanners can contribute to the structural diagnosis of historic constructions. Thus the geometry and the spectral backscattered reflectance data acquired by terrestrial laser scanners can be used to map and evaluate structural and chemical damages of historical constructions in a semiautomatic and robust way. Chapter 12 focuses on the application of laser scanning to bridge inspection. A review of the state of the art is presented, where the potential of laser scanning for bridge inspection is presented including geometric modelling, structural deformation, and damage detection. Chapter 13 reviews applications where laser scanning has contributed to structural analysis including geometric modelling for the creation of structural models and structural behavior simulation. In the second part of the chapter, it is demonstrated how laser scanning data can be used to solve inverse problems (estimation of the value of influential material properties), or even contribute in forensic of existing structures. Chapter 14 focuses on monitoring of construction sites and how laser scanning can contribute as an efficient and accurate tool. After evaluating some aspects relevant for the monitoring of construction sites, the focus is set then on data acquisition and referencing of point clouds, which represents a vital prerequisite for monitoring.

The last chapter addresses building information modelling (BIM) and how dense point clouds and their smart processing can contribute to automatically elaborate these BIM models. The authors present the SCAN to BIM approach. The authors propose a modelling management information system (MMIS) which is able to convert point clouds into as-built BIM (AB-BIM), solving some issues related to generative modelling, and in a more productive manner. The authors in their approach considered the interoperability constrains in the models generated, so the implementation of automatic verification systems is proposed to check metric accuracy between the original point clouds and the 3D objects.

REFERENCE

Vosselman, G. & Maas, H.G. (2010) *Airborne and Terrestrial Laser Scanning*. CRC Press, Scotland, UK.

Laser Scanning – Riveiro and Lindenbergh
© 2020 ISPRS, ISBN 978-1-138-49604-0

Chapter 2

Multispectral terrestrial lidar

State of the art and challenges

Sanna Kaasalainen

ABSTRACT: The development of multispectral terrestrial laser scanning (TLS) is still at the very beginning, with only four instruments worldwide providing simultaneous three-dimensional (3D) point cloud and spectral measurement. Research on multiwavelength laser returns has been carried out by more groups, but there are still only about ten research instruments published and no commercial availability. This chapter summarizes the experiences from all these studies to provide an overview of the state of the art and future developments needed to bring the multispectral TLS technology into the next level. Although the current number of applications is sparse, they already show that multispectral lidar technology has potential to disrupt many fields of science and industry due to its robustness and the level of detail available.

1 INTRODUCTION

Multispectral laser scanning enables target identification and analysis by combining both spectral and spatial features. Multiwavelength airborne laser scanners are already on the market, such as the Optech Titan 3-channel airborne laser scanner, and an increasing number of applications are being published (e.g. Wichmann *et al.*, 2015; Matikainen *et al.*, 2017; Axelsson *et al.*, 2018). Conversely, multispectral terrestrial laser scanners (TLS) only exist as prototypes built by a few individual research groups. This chapter summarizes the current status of multispectral TLS and discusses the challenges and future prospects of this emerging technology.

Today, there are a few multispectral TLS and active imaging projects, ranging from two-dimensional (2D) active spectral imaging, sometimes with a separate range measurement (Wei *et al.*, 2012; Manninen *et al.*, 2014), to instruments producing full-scale three-dimensional (3D) point cloud data (Douglas *et al.*, 2012; Powers and Davis, 2012; Hakala et al., 2012; Danson *et al.*, 2014). There are also pointwise systems, where only the intensity is measured for spectral analysis, and waveform lidars for vertical profiling (Rall and Knox, 2004; Du *et al.*, 2016). Alternatively, a point cloud is constructed by combining data from separate monochromatic laser scanners (Gong *et al.*, 2015; Hartzell *et al.*, 2014; Elsherif *et al.*, 2018). While all these provide important input for analyzing the multispectral laser returns, the main focus of this chapter is on the simultaneous capture of spectra and target 3D geometry.

The most obvious advantage of multispectral lidar has already been shown in the pioneering studies (cf. Wichmann *et al.*, 2015): no time gaps or registration errors exist between different colors (or spectral channels) or between geometric (topography) and spectral features, which improves the results compared to the traditional approach of combining point clouds with hyperspectral imagery (e.g. Guo *et al.*, 2011). Also, the target identification capability of active hyperspectral imaging enables a wide variety of applications. But the prospect of mapping spectral properties in 3D over the target and in the same spatial resolution as conventional lidars has not yet been fully utilized or even understood. Many of the studies on multispectral laser scanning have so far focused on spectral performance without including the spatial interpretation of the point clouds. For example, Du *et al.* (2016) present leaf nitrogen retrieval with a 32-channel lidar, but the topographic aspect is not considered. A similar approach was presented in Li *et al.* (2014), who also focused on leaf spectral indices rather than point clouds. However, one-shot acquisition of point cloud and spectral data will crucially increase the level of detail and information content available from the measurements. There is also the prospect of getting non-destructive and large-scale data on the properties that have so far been measurable with destructive means only (Hakala *et al.*, 2015), but this idea still needs comprehensive studies with improved instruments, as well as systematic laboratory reference to be established as a method.

Radiometric calibration of lidar intensity has been studied for more than a decade by now. A comprehensive review on radiometric calibration methods, along with the basic physical concepts, is provided in Kashani *et al.* (2015). They classify the levels of intensity processing according to the accuracy and information quality of the calibrated laser returns, starting from the digital numbers (raw intensity) available directly from the detector (level 0). Level 1 comprises the range or incidence angle correction, while in level 2, scaling or normalization, such as histogram correction, is carried out. Full radiometric calibration (level 3) results in values comparable to target reflectance and requires the use of reference targets with known reflectance. The level 3 process is also termed absolute radiometric calibration (Briese *et al.*, 2012).

Kashani *et al.* (2015) also identified some important knowledge gaps, such as difficulty in comparison with intensities obtained with different lidar instruments or variation between different wavelengths. That, plus the fact that lidar intensity is still somewhat underutilized, especially for TLS (Li *et al.*, 2016; Schofield *et al.*, 2016), has been one of the drivers for the development of multiple-wavelength laser scanners. Radiometric calibration has sometimes been seen as a preliminary step toward multispectral lidar (Matikainen *et al.*, 2017). Using multiple channels also enables the use of spectral ratios, which may in some cases help overcome the problem that the laser backscatter intensity does not directly represent the hemispherical reflectance and hence the target physical properties. Simultaneous data acquisition at different wavelengths enables the comparison of spectral differences or trends, which can sometimes be done without absolute radiometric calibration using only the range-corrected or normalized return intensity (Kashani *et al.*, 2015; Li *et al.*, 2016). This kind of approach usually works when changes in target properties are observed within the same experiment or with the same instrument. However, even the relative measurements are prone to errors caused by inaccuracy in sampling or calibration of the laser beam at different wavelengths. It has also turned out that the radiometric calibration is instrument specific (Calders *et al.*, 2017). In any case, rigorous radiometric calibration is necessary to be able to interpret physically the reflectance measurements available from multispectral TLS. It is also critical for combining and comparing results from different experiments.

This chapter aims to provide a comprehensive overview on the status of multispectral TLS so far. It also discusses some important challenges related to multispectral laser scanning in particular, although many of these problems are familiar to monochromatic TLS as well. Even though the number of currently existing multispectral TLS instruments is sparse, those few available already indicate the vast potential of this new technology. The chapter is organized as follows: Section 2 reviews the state of the art in multispectral TLS and identifies some difficulties commonly met in instrument development and data analysis. Applications are summarized in Section 3, while Section 4 highlights some future aspects related to the technology.

2 MULTISPECTRAL TERRESTRIAL LIDAR: STATE OF THE ART AND CHALLENGES

Even though the idea of active hyperspectral sensing has been available for almost two decades (Johnson *et al.*, 1999), the first multi-wavelength terrestrial laser scanner instruments were presented by Douglas *et al.* (2012), Powers and Davis (2012), Hakala *et al.* (2012), and Gaulton *et al.* (2013). Research efforts for multispectral TLS technologies are slowly increasing. Some applications have been previously summarized by Eitel *et al.* (2016) and Hancock *et al.* (2017). Much of the research has focused on vegetation (Wichmann *et al.*, 2015).

2.1 *State of the art*

Dual- or multiwavelength 3D TLS point clouds have been published so far by four projects. The Dual-Wavelength Echidna Lidar (DWEL; Douglas *et al.*, 2012) and the Salford Advanced Laser Canopy Analyser (SALCA; Danson *et al.*, 2014) are dual-wavelength instruments. The Finnish Geospatial Research Institute Hyperspectral Lidar (FGI HSL; Hakala *et al.*, 2012), and the spectral LADAR by Powers and Davis (2012) have multiple channels. All these are full-waveform digitizing scanning lidars. In addition to these four, there are multispectral lidar studies that have mainly focused on spectral or waveform analysis of the laser return. Table 2.1 lists all these research efforts, followed by a more detailed description on each project in this section.

2.1.1 Multispectral TLS providing 3D topography and reflectance information

Dual-Wavelength Echidna Lidar (DWEL) is a portable two-channel scanning lidar developed at the Boston University (Douglas *et al.*, 2012; Howe *et al.*, 2015). Two coaxial near-infrared (NIR, 1064 nm) and shortwave infrared (SWIR, 1548 nm) lasers (with beam diameter of 7 mm), which are ideal for leaf separation and are synchronized with an external trigger (Douglas *et al.*, 2015). An additional green laser is applied to be able to see the scan path. The returning pulses are collected with a telescope. Full waveforms are captured with two indium gallium arsenide (InGaAs) photodiode detectors, for which the beams are separated with a beam splitter. The measurement range is about 70 meters. A radiometric calibration scheme, based on reference panels, has recently been outlined to relate the measured reflectance to target radiative and structural properties (Li *et al.*, 2016). In

Table 2.1 Summary of multispectral lidar projects. The top four provide one-shot multispectral point clouds, whereas the others focus on spectral analysis of pointwise or two-dimensional depth images.

Project	*Laser Source*	*Channels (nm)*	*Detector/Sampling*	*Application*
DWEL (Douglas *et al.*, 2012)	Two coaxial lasers, 5.1 ns, 20 kHz	1064, 1548	InGaAs photodiodes, 2 GHz digitizer	Separating leaves vs. bark
SALCA (Danson *etal.*, 2014)	Two asynchronous lasers, 1–3 ns, 5 kHz	1063, 1545	Single digitizer 1GHz	VI's related to moisture
FGI HSL (Hakala *et al.*, 2012)	Supercontinuum, 1 ns, 5 kHz	8 channels: 450–1050	Spectrograph, APD array, 1 GHz digitizer	Target identification, VI's
Spectral LADAR (Powers and Davis, 2012)	Supercontinuum, 1.8–4.0 ns, 25 kHz	25 channels: 1080–1620	Spectrograph, InGaAs APD, 5 GHz digitizer	Material type classification
MSCL (Woodhouse *et al.*, 2011)	Tunable Nd:YAG, 4.75ns, 20Hz	531, 550, 690, 780	Silicon photodiode + oscilloscope, 5 GHz	VI's such as PRI, NDVI
GECO (Eitel *et al.*, 2014a)	Two laser diodes, 1 kHz	532, 658	Red/green filters, two photodiodes	Chlorophyll observation
HSL Beijing (Li *et al.*, 2014)	Supercontinuum, 1–2 ns, 20–40 kHz	4–32 channels: 409–914	APD arrays, 5 GHz digitizer	Leaf biochemical contents
TCSPC Lidar (Wallace *et al.*, 2014)	Supercontinuum, < 50 ps, 2 MHz	531, 570, 670, 780	4 single photon APDs + TCSCP modules	Conifer needle NDVI
MSL Wuhan (Gong *et al.*, 2015)	4 synchronous diode lasers, 800 Hz	556, 670, 700, 780	4 PMTs, oscilloscope, range finder	Object classification
HL System (Du *et al.*, 2016)	Supercontinuum, 1–2-ns, 20–40 k Hz	32 channels: 538–910	Grating spectrometer, APD array	Rice leaf nitrogen from SVM

Note: VI = Vegetation Index, APD = Avalanche Photodiode, PRI = photochemical reflectance index, NDVI = normalized difference vegetation index, TCSPC = Time-Correlated Single-Photon Counting, PMT = Photo Multiplier Tube, SVM = support vector machine.

addition to improving the quantitative results on vegetation structure from spectral responses, the calibration scheme also enables the comparison and collaborative use of different instruments.

Salford Advanced Laser Canopy Analyser (SALCA; Gaulton *et al.*, 2013; Danson *et al.*, 2014) uses two sequentially emitting lasers. Full waveforms at NIR (1063 nm) and SWIR wavelengths (1545 nm) are detected and digitized with a single detector. These wavelengths have proven ideal in, e.g., leaf-bark separation. Because of the asynchronous pulses, there is a small (1%) offset in the footprints. Calibration is done by means of reference panels. The maximum range is 105 m, and the footprint sizes are 8.0 mm (1063 nm) and 9.2 mm (1545 nm) at 10 m. The capability of the SALCA lidar in leaf water content retrieval has been demonstrated with normalized difference (vegetation) indices observed for three different species (Gaulton *et al.*, 2013). Recently, a radiometric calibration scheme based on neural networks was presented for SALCA (Schofield *et al.*, 2016). As the radiometric response to range, reflectance, and laser temperature was found to be complex, a neural network solution proved a robust tool for reflectance calibration. An effect of internal temperature on intensity was detected, and since it was more pronounced in SWIR, this might pose another challenge to multispectral lidar measurement and analysis in general.

The FGI Hyperspectral lidar (HSL; Hakala *et al.*, 2012) is a multiwavelength full-waveform laser scanner, with a supercontinuum laser light source producing 1 ns pulses at 420–2400 nm (the spot size being 5 mm at 4 m for 543 nm). The optical system consists of a parabolic mirror, which collects the returned pulse energy into an optical fiber connected to a spectrograph. The spectral dispersion is detected with an avalanche photodiode (APD) array (with 450–1050 nm response range). Currently, 8 of 16 available APD spectral channels can be digitized, but the APD array can be moved with respect to the spectrograph to adjust the wavelengths to be detected. In the 2012 breadboard prototype, the digitizer operated on 1 GHz sampling rate, but increasing the digitization rate is investigated as part of an ongoing development work toward an operational field instrument and improved target characterization (Kaasalainen *et al.*, 2018b). The ongoing development and intensity calibration efforts also aim at increasing the detector sensitivity and extending the measurement range from currently achieved 5–50 meters. Here too, external reference panels are used in the calibration.

The spectral laser detection and ranging (LADAR) laboratory demonstrator prototype developed in Maryland by Powers and Davis (2012) is also based on supercontinuum technology. In addition to the supercontinuum source, the system comprises transmitter and receiver mirrors and optics, a mechanically tuned spectrometer, APD array driving a transimpedance amplifier (TIA), and a 5 GSa/s digitizer. Similarly to the FGI HSL, the time-of-flight measurement is triggered by picking up a fraction of the outgoing beam with a beam splitter. The measured range has been tested up to 40 m. The instrument was designed with military imaging applications in mind, and therefore its capability of detecting obscured objects from behind a camouflage was demonstrated. The near-infrared wavelengths of operation also ensure eye safety.

2.1.2 Lidar projects focusing on spectral analysis

There are also lidar studies, where mainly the intensity or the spectrum of the returning laser pulse is analyzed from a fixed point on the target surface. An instrument closest to a one-shot multi-wavelength lidar was introduced by Gong *et al.* (2015). Their approach was based on Wei *et al.* (2012), where a four-wavelength synthesized beam was produced with four semiconductor laser diodes (555, 670, 700, and 780 nm). Spatial data were obtained using a simultaneous laser range finder for distance measurement, which means that the spectral and spatial information would not come from the same source. The technical solution was described to be equivalent to four monochromatic lidars, but the combination of a single-wavelength point cloud and a multispectral image could be used for classification similarly to multispectral TLS. The study focused on classification algorithms for target identification, e.g. separating green and dry leaves and inorganic materials.

The Edinburgh University Multispectral Canopy Lidar (MSCL) is based on tuneable laser operation in four wavelengths: 531, 550, 690, and 780 nm (Woodhouse *et al.*, 2011). No point clouds were presented, but tree vertical structure was analyzed from digitized waveforms, together with intensity information for analyzing bark, twigs, and leaves. The full-waveform configuration allowed the acquisition of canopy profile and NDVI with respect to tree height. The NDVI profiles could be related to chlorophyll concentration (Morsdorf *et al.*, 2009). Four wavelengths were also recorded with a time-correlated single-photon counting (TCSPC) technique by Wallace *et al.* (2014), which would also allow the collection of 3D depth image data. Supercontinuum laser was used as a source, but the output was filtered. The single photon counting approach was based on an earlier design by Buller *et al.* (2005). The TCSPC lidar was also used to produce underwater 3D depth images at a single wavelength (Maccarone *et al.*, 2015).

Crop foliar nitrogen retrieval was carried out with the Green Economic Chlorophyll Observation (GECO) sensor, consisting of two laser diodes (532 and 658 nm; Eitel *et al.*, 2014a). Only the returning laser intensity was analyzed, and range data were not utilized. A horizontal path lidar for chlorophyll estimation in two wavelengths was also presented by Rall and Knox (2004).

Supercontinuum laser has been tested for target spectral analysis by Du *et al.* (2016) and Sun *et al.* (2017), who measured rice leaf nitrogen at 32 channels ranging from 538 to 910 nm. In this experiment, the laser was pointed to the target at a fixed incidence angle. A grating spectrometer was used for creating the dispersion of the returned supercontinuum laser pulse. Nitrogen content was retrieved and classified using the spectral data only. A similar approach with full-waveform digitizing capacity was developed by Li *et al.* (2014) for measuring leaf biochemical contents, e.g. nitrogen. Point clouds were not presented either, but range measurement was tested in a later study (Niu *et al.*, 2015). They also investigated the ranging accuracy but presented no 3D point clouds either.

2.2 *Challenges*

As multispectral TLS instruments are not available commercially, and because of the novelty of the technologies, there are many challenges related to hardware design, e.g. beam alignment in multilaser systems, range accuracy, detector sensitivity (especially signal to noise), etc. (cf. Danson *et al.*, 2014; Howe *et al.*, 2015). Many of these are related to the fact that the laser power is dispersed into multiple channels. Also, much of the analysis is based on comparison of intensity values or trends, which emphasizes the need for consistent spectral performance to be able to separate reliably the hits from different targets. This causes extra work with data pre-processing, calibration, and information extraction (cf. Danson *et al.*, 2018).

Laser stability is vital for spectral performance. Temperature was found to affect the intensity of SALCA returns (Danson *et al.*, 2018). For DWEL, the drifts in laser power are monitored with a reference panel sampling the outgoing pulses during each mirror rotation (Li *et al.*, 2016). In the case of supercontinuum laser used in the FGI HSL, fluctuations in laser stability are normalized by dividing each pulse with the trigger pulse, which is separated from each transmitted pulse with a beam sampler. Initial tests have indicated a reasonable spectral stability, but the stability is being further assessed in the ongoing calibration study (Kaasalainen *et al.*, 2018b). Powers and Davis (2012) also use the transmitted pulse to normalize the backscattered signal.

Another important issue is the considerable data volume, especially when the number of channels and scanning resolution are increased. This introduces more demands for data processing, storage, and transfer, especially if on-site pre-processing is needed for real-time operation. Therefore, even if the addition of channels would be technically straightforward, for example by just adding another digitizer component, dealing with the increased data volume would require considerable changes in the processing and calibration procedures. Optimizing the processing steps is one of the future aims, and the solutions are likely to be application specific. The challenges of processing, storing, and analyzing vast amounts of data produced by full-waveform lidar in general have been addressed in many studies (e.g. Pfeifer *et al.*, 2014).

It is already known from single-wavelength TLS that full-waveform echo digitizing will improve not only the range but also target detection capacity (Ullrich and Pfennigbauer, 2011). A lot of effort has been put into sampling the laser waveforms. As a preprocessing step, sampling is usually implemented by the manufacturer in commercial scanners (cf. Calders *et al.* (2017) who discuss the pulse sampling for RIEGL VZ-400 terrestrial lidar), but as multispectral lidars are research instruments, the sampling has to be individually solved in each case. Narrow-width pulses can be expected from low signal returns (Danson *et al.*, 2018). This may not cause a major problem in, e.g. vegetation mapping, where strong differences between NIR and visible are easily observed. But for targets for which the spectral differences are small (or multiple hits must be detected) or those with great variation in intensity, the inaccuracy of weak signals may be an issue, especially when the digitizing rate is the same order as the pulse width (e.g. sampling 1 ns pulses at 1 GHz; Kaasalainen *et al.*, 2018b). Pulse width is also crucial for range resolution, which can also be improved in the full-waveform case with signal processing approaches (Powers and Davis, 2012).

There are other effects, such as ringing artifacts (Hakala *et al.*, 2012; Li *et al.*, 2016; Danson *et al.*, 2018) and telescopic and saturation, which occur especially when scanning from near to far ranges (Li *et al.*, 2016). There is another calibration challenge coming from far-range measurements, where the signal to noise ratio falls and hence decreases the accuracy. Specifications, such as scanning resolution versus beam divergence, also need to be optimized (Li *et al.*, 2018). Niu *et al.* (2015) discovered a synchronization inaccuracy between wavelengths for a supercontinuum laser, which may affect the range accuracy. While solutions can be found for all these problems, there is still a great need for comprehensive data collection to be able to utilize better the combined spectral and point cloud information accurately and robustly. This is particularly true because so few multispectral TLS instruments are available, and most applications are still to be developed. New applications are likely to call for further hardware and software development.

3 APPLICATIONS

3.1 *Vegetation*

While the potential of TLS in measuring vegetation structure can be utilized with monochromatic scanners, the added value from simultaneous spectral data is considerable. Woodhouse *et al.* (2011) identified some potential vegetation applications for multispectral lidar, such as mapping species composition, identifying healthy versus stressed canopies, monitoring the plant photosynthetic capabilities, or diagnostic monitoring via pigment concentrations. Many of these have since been tested with newly developed instruments. Vegetation targets are usually challenging, since they represent complex, multitarget situations, where each laser pulse is likely to hit more than one surface.

3.1.1 Monitoring the photosynthetic activity and health

Vegetation photosynthetic capacity is an important research topic because it is directly related to the role of forests as carbon sinks, which is important in understanding the dynamics of climate change and the global carbon cycle (Gaulton *et al.*, 2013; Wallace *et al.*, 2014). Vegetation health, productivity, and stress level are related to the leaf biophysical parameters, such as the amount of chlorophyll in plants. Chlorophyll can be monitored by measuring the changes in the so-called vegetation spectral red-edge and estimating the chlorophyll levels with various vegetation indices using values near the red-edge domain (Rall and Knox, 2004). Therefore, there is a strong research interest toward retrieving vegetation spectral indices with multispectral TLS. These indices were previously studied by means of passive remote sensing (e.g. Kalacska *et al.*, 2015; Jay *et al.*, 2017). Nevalainen *et al.* (2014) provide a summary of 27 published vegetation indices and tested them for pine chlorophyll retrieval with the FGI HSL. The modified simple ratio (MSR) and the modified chlorophyll absorption ratio index (MCARI), where reflectance values at 705 and 750 nm (i.e. those near the spectral red edge) were utilized, were found to be most sensitive for chlorophyll estimation. Similar indices were later monitored throughout the growing season with the FGI HSL to demonstrate a non-destructive time series of pine chlorophyll content, validated with laboratory analysis (Hakala *et al.*, 2015).

In addition to forests, crop monitoring has also been explored. Nitrogen concentration in oat samples was derived from the FGI HSL data by means of the chlorophyll absorption ratio index (CARI), for which laser returns at reflectance 700 nm, 670 nm, and 550 nm were selected (Nevalainen *et al.*, 2013). Support vector machine regression was tested for nitrogen concentration in rice crops by Du *et al.* (2016). Here the number of wavelengths played a crucial role.

Vegetation moisture content is an important indicator of tree health, drought stress, and fire risk. Moisture content has been mapped from SALCA point clouds using SALCA normalized ratio index (SNRI) and other indices, where laser return intensities at 1064 nm and 1545 nm were compared (Gaulton *et al.*, 2013; Hancock *et al.*, 2017). A relationship was found between equivalent water thickness (EWT) and SALCA reflectance at both channels, but the relationship was strongest with EWT and spectral indices. The suitability of SALCA wavelengths for leaf moisture estimation was also demonstrated with a leaf reflectance model.

3.1.2 Identifying tree parts or tree species

DWEL and SALCA have also been applied in the separation of leaf (foliage) and woody material (such as bark) to be able to study forest structure and function (Douglas *et al.*, 2015; Danson *et al.*, 2018). The recognition is based on the stronger leaf absorption at 1548 nm compared to stems, resulting in a difference between the intensities from two NIR channels, which can then be observed for each point. This improves substantially the separation capability, since for monochromatic laser scanners, even the waveform is not enough to separate hits from leaves with partial hits from edges of trunks or branches (Douglas *et al.*, 2015). Leaves can also be separated by point cloud classification using vegetation indices such as NDVI for each point (Woodhouse *et al.*, 2011; Nevalainen *et al.*, 2014).

Tree species classification has been demonstrated for spruce and pine with the FGI HSL (Vauhkonen *et al.*, 2013) based on classification features that combined range and reflectance properties. The returns from inside the foliage turned out to be essential with respect to the classification accuracy. The results are somewhat similar to those obtained with a multispectral airborne laser scanning at three wavelengths with Optech Titan X (Axelsson *et al.*, 2018), where combined spectral and range information proved most efficient in classification.

3.2 *Object classification*

3.2.1 Separating different materials

Multispectral TLS has also been used to separate inorganic materials from organic ones or each other (see Figure 2.1 for an example). Initial tests of the FGI HSL have been carried out to explore its potential for other applications than vegetation. Monitoring target moisture has been tested for cardboard and wooden objects in indoor conditions with spectral indices related to water absorption band at 970 nm (Kaasalainen *et al.*, 2017). Moisture in targets can be detected from these indices, while it can be automatically localized from point cloud information. Preliminary results for snow surfaces have also been obtained, mainly to explore the capability of the HSL to distinguish pollution or to investigate the possible wavelength effects on the incidence angle on snow (Anttila *et al.*, 2016). The incidence angle behavior was found to be similar for all HSL wavelengths (540–1000 nm) for a sample of melting (wet) snow. An ongoing research effort focuses on mineral identification in mines, where ores are distinguished from their spectral properties and mapped from 3D point clouds of mine tunnels. In comparison with strong signatures in the vegetation red-edge domain, mineral detection has proven somewhat more challenging, as the differences in intensity are more subtle and hence pose more challenges to the accuracy of the radiometric calibration (Kaasalainen *et al.*, 2018b).

Gong *et al.* (2015) used both spectral and spatial information to classify and distinguish different objects with the Wuhan MSL: white wall, ceramic pots, *Cactaceae*, carton, plastic foam block, and healthy and dead leaves, with a support vector machine (SVM) supervised classification method. The point clouds were obtained by combining a multispectral image captured with four lasers and

Figure 2.1. Example of target identification from a multispectral point cloud: camouflage and artificial targets can be separated from vegetation based on both spectral and shape recognition. The same applies for tree trunks and leaves. Point cloud obtained with the FGI HSL (Hakala *et al.*, 2012; Nevalainen *et al.*, 2014; Puttonen *et al.*, 2015).

Source: Image by Olli Nevalainen, FGI.

range information from a traditional lidar. Comparison of results to those from single-wavelength lidar or traditional passive remote sensing only indicated that a higher classification was achieved with simultaneous multispectral and spatial data.

3.2.2 Defense and security applications

There has been a long interest in active spectral imaging for military applications. Besides target classification, detection of concealed targets, e.g. those behind a camouflage nets have been an object of interest (Johnson *et al.*, 1999; Nischan *et al.*, 2003; Powers and Davis, 2012). Other important features in defense- and security-related imaging are the eye safety and the capability of long-range measurement, even up to 1 km (Manninen *et al.*, 2014). 3D spectral point clouds have been utilized by Powers and Davis (2012), who demonstrated the advantages over passive spectral sensing in the detection of featureless, flat surfaces and objects obscured by camouflage netting with a spectral LADAR operating in NIR. The separation was based on the range difference and spectral features, using a K-means algorithm for classifying spectral vectors.

Puttonen *et al.* (2015) applied the FGI HSL for identifying artificial targets (camouflage net, LECA brick, plastic chair) and separating them from vegetation. It was found that while the differences in spectral responses were significant and allowed classification, spatial aggregation of individually classified points increased the classification accuracy. This would mean that point cloud segmentation is necessary for a robust solution. It was also shown that spectral data enhance the detection of hidden targets (those masked with camouflage nets) using a combination of spectral indices (cf. Johnson *et al.* (1999), who introduced this idea along with the concept of active hyperspectral imaging).

4 FUTURE ASPECTS

Many of the results presented in this chapter are preliminary and first of their kind and call for extensive future research, both to establish the method and to show its full value in different scientific and commercial applications. However, they all point out the potential in robust and automatic target identification and monitoring. Considering the robotics and automation megatrends related to, e.g. intelligent transport, smart agriculture, or the internet of things (IoT), all these developments call for robust sensor-based environment perception.

There is an ongoing trend in TLS toward portability, rapid data capture, and also cost efficiency. While the data quality may not be as high as with a high-performance lidar, the practicality of cost-effective, lightweight TLS instruments is likely to increase their usage and thus allow larger areas to be investigated in the first place (Paynter *et al.*, 2016 and references therein). These devices have the potential to complement and extend the spatial range of observations acquired with high-performance ones. There will be a pressure for lower cost in the multispectral TLS applications as well, but the technology should first be well established, and currently this is only possible with high-performance research instruments.

Further research is still needed for full spatial interpretation of multi/hyperspectral TLS data, especially to establish a nondestructive means for, e.g. mapping vegetation pigment concentrations in 3D for diagnostics (Eitel *et al.*, 2014b, Hakala *et al.*, 2015; Sun *et al.*, 2017) or automatic localization of moisture in built environment (Kaasalainen *et al.*, 2017). In addition to the technical issues described above, a few significant future aspects are discussed in more detail.

4.1 *Data analysis methods*

Producing and analyzing multichannel point clouds is a demanding task: not only the instrumentation is challenging, but software needs to be capable of handling vast amounts of information produced by different types of digitization equipment. This information has to be processed and calibrated with methods customized for each instrument and finally converted into a form readable for data analysis software (which will also have to be custom made, as standard software for multispectral lidar is nonexistent) and formats suitable for other users. The LASer (LAS) format already allows the addition of extra attributes to each point in a point cloud (e.g. RIEGL, 2012). Because of vast information content, automatizing these processes still requires a major effort, and optimization, especially if real-time mapping is aimed at (such as that in simultaneous localization and mapping (SLAM)). All the recent studies call for more work on the data analysis front (Douglas *et al.*, 2015; Danson *et al.*, 2018; Kaasalainen *et al.*, 2018b).

Even though manual delineation of objects could be possible using the spectrum of even a single point, combined use of both spatial and spectral information will be necessary for the solutions to be automatic and robust (cf. Li *et al.*, 2018). The need of information is application specific. Puttonen *et al.* (2015) also found that, because of the complexity of the targets such as trees, an individual laser return can have almost any intensity, and hence it is difficult to identify a source of a single return. It was also shown that more than one spectral index was needed to separate some camouflage objects from organic ones. Kaasalainen *et al.* (2018a) found that for vegetation, leaf angles affect the spectrum of some plants, and a physical correction of this inaccuracy is not realistic. This may introduce uncertainty in the vegetation indices.

While spectral (vegetation) indices have played a major role in multispectral lidar studies so far, other detection methods, such as spectral unmixing algorithms (Powers and Davis, 2012; Altmann *et al.*, 2015) can be applied. This calls for increasing the number of channels (from two or four), which is most practical by the use of supercontinuum lasers. Du *et al.* (2016) suggested that adding more channels improves the classification accuracy in the SVM regression.

4.2 *Measurement geometry*

Effects of measurement geometry, in this case, the incidence angle must be better understood. They have recently been shown to affect the spectral indices measured with laser scanners (Eitel *et al.*, 2014a, Hancock *et al.*, 2017). It is not possible to correct these for leaf canopies since the leaf incidence angle is mostly not known. Therefore, an empirical correction scheme may turn out to be the only alternative for vegetation (Kaasalainen *et al.*, 2018a) but in any case, the variation in incidence angles is likely to limit the accuracy of leaf spectral data (Hancock *et al.*, 2017). Further work is also needed to find out whether the incidence angle behavior would be different for relative (such as NDVI) or absolute vegetation indices (MCARI; cf. Nevalainen *et al.*, 2014). Gaulton *et al.* (2013) and Shi *et al.* (2015) did not observe the incidence angle effect but discussed the use of relative indices to cancel out or reduce it. Some targets do not exhibit any difference between wavelengths, such as a sample of wet snow (Anttila *et al.*, 2016). It is clear that more testing and further experiments are needed.

4.3 *Eye safety versus number of channels*

Near-infrared lasers have provided useful in moisture estimation (Gaulton *et al.*, 2013; Manninen *et al.*, 2014), and it is possible to build systems that meet the eye safety requirements. Overall, the eye safety problem is easier to tackle with instruments operating at discrete laser wavelengths, because they are easier to detect, as all laser power is concentrated on narrow wavelength bands. The downside is the inaccuracy in the co-alignment of the laser beams (Li *et al.*, 2018). Then again, supercontinuum lasers provide the only way to increase the number of channels or make a hyperspectral implementation. For visible wavelengths, filtering is one option, especially if the number of channels can be limited. For experiments carried out in laboratory or indoor environments (such as tunnels), this problem may be solved with traditional laser safety measures (limiting the access, safety goggles, etc.). Furthermore, the need for laser power in supercontinuum applications may decrease along with improving detectors as the technical readiness level increases from prototype level, or the power can be otherwise lowered in close-range experiments. In any case, hyperspectral TLS utilizing supercontinuum lasers may never be a multipurpose instrument, but it should be tailored for the purpose to optimize the laser safety vs. wavelength range. Technical solutions enabling versatility will be crucial.

5 SUMMARY

Multispectral lidars represent the next generation of terrestrial laser scanning. This chapter has reviewed the current state of the art and discussed the prospects and challenges related to instrumentation, usage, and data interpretation from multispectral TLS. Once these challenges have been tackled, multispectral TLS will provide comprehensive environment perception and target identification in an entirely new level of robustness, detail, and accuracy. It has already been shown with the applications demonstrated so far that multispectral TLS has the potential of disrupting many fields of science and industry. As it seems that the technical implementation will strongly depend on the application, commercial availability may have to wait until the performance of the technology has been better established. This is an object of active ongoing research, with more and more results coming up in the near future.

REFERENCES

Altmann, Y., Wallace, A. & McLaughlin, S. (2015) Spectral unmixing of multispectral Lidar signals. *IEEE Transactions on Signal Processing*, 63(20), 5525–5534. https://doi.org/10.1109/TSP.2015.2457401.

Anttila, K., Hakala, T., Kaasalainen, S., Kaartinen, H., Nevalainen, O., Krooks, A., . . . Jaakkola, A. (2016) Calibrating laser scanner data from snow surfaces: Correction of intensitys. *Cold Regions Science and Technology*, 121, 52–59. https://doi.org/10.1016/j.coldregions.2015.10.005.

Axelsson, A., Lindberg, E. & Olsson, H. (2018) Exploring multispectral ALS data for tree species classification. *Remote Sensing*, 10(2), 183. https://doi.org/10.3390/rs10020183.

Briese, C., Pfennigbauer, M., Lehner, H., Ullrich, A., Wagner, W. & Pfeifer, N. (2012) Radiometric calibration of multi-wavelength airborne laser scanning data. *ISPRS Annals of Photogrammetry, Remote Sensing and Spatial Information Sciences*, I-7, 335–340. https://doi.org/10.5194/isprsannals-I-7-335-2012.

Buller, G.S., Harkins, R.D., McCarthy, A., Hiskett, P.A., MacKinnon, G.R., Smith, G.R., . . . Rarity, J.G. (2005) Multiple wavelength time-of-flight sensor based on time-correlated single-photon counting. *Review of Scientific Instruments*, 76(8), 83112. https://doi.org/10.1063/1.2001672.

Calders, K., Disney, M.I., Armston, J., Burt, A., Brede, B., Origo, N., . . . Nightingale, J. (2017) Evaluation of the range accuracy and the radiometric calibration of multiple terrestrial laser scanning instruments for data interoperability. *IEEE Transactions on Geoscience and Remote Sensing*, 55(5), 2716–2724. https://doi.org/10.1109/tgrs.2017.2652721.

Danson, F.M., Gaulton, R., Armitage, R.P., Disney, M., Gunawan, O., Lewis, P., . . . Ramirez, A.F. (2014) Developing a dual-wavelength full-waveform terrestrial laser scanner to characterize forest canopy structure. *Agricultural and Forest Meteorology*, 198–199, 7–14. https://doi.org/10.1016/j.agrformet.2014.07.007.

Danson, F.M., Schofield, L.A. & Sasse, F. (2018) Spectral and spatial information from a novel dual-wavelength full waveform terrestrial laser scanner for forest ecology. *Interface Focus*, 8(2), 20170049. http://doi.org/10.1098/rsfs.2017.0049.

Douglas, E.S., Martel, J., Li, Z., Howe, G., Hewawasam, K., Marshall, R.A., . . . Chakrabarti, S. (2015) Finding leaves in the forest: The dual-wavelength Echidna Lidar. *IEEE Geoscience and Remote Sensing Letters*, 12(4), 776–780. https://doi.org/10.1109/lgrs.2014.2361812.

Douglas, E.S., Strahler, A., Martel, J., Cook, T., Mendillo, C., Marshall, R., . . . Lovell, J. (2012) DWEL: A dual-wavelength Echidna Lidar for ground-based forest scanning (pp.4998–5001). *IEEE International Geoscience and Remote Sensing Symposium (IGARSS '12)*. https://doi.org/10.1109/igarss.2012.6352489.

Du, L., Gong, W., Shi, S., Yang, J., Sun, J., Zhu, B. & Song, S. (2016) Estimation of rice leaf nitrogen contents based on hyperspectral Lidar. *International Journal of Applied Earth Observation and Geoinformation*, 44, 136–143. https://doi.org/10.1016/j.jag.2015.08.008.

Eitel, J.U.H., Höfle, B., Vierling, L.A., Abellán, A., Asner, G.P., Deems, J.S., . . . Vierling, K.T. (2016) Beyond 3-D: The new spectrum of Lidar applications for earth and ecological sciences. *Remote Sensing of Environment*, 186, 372–392. https://doi.org/10.1016/j.rse.2016.08.018.

Eitel, J.U.H., Magney, T.S., Vierling, L.A., Brown, T.T. & Huggins, D.R. (2014b). LiDAR based biomass and crop nitrogen estimates for rapid, non-destructive assessment of wheat nitrogen status. *Field Crops Research*, 159, 21–32. https://doi.org/10.1016/j.fcr.2014.01.008.

Eitel, J.U.H., Magney, T.S., Vierling, L.A. & Dittmar, G. (2014a). Assessment of crop foliar nitrogen using a novel dual-wavelength laser system and implications for conducting laser-based plant physiology. *ISPRS Journal of Photogrammetry and Remote Sensing*, 97, 229–240. https://doi.org/10.1016/j.isprsjprs.2014.09.009.

Elsherif, A., Gaulton, R. & Mills, J. (2018) Estimation of vegetation water content at leaf and canopy level using dual-wavelength commercial terrestrial laser scanners. *Interface Focus*, 8(2), 20170041. https://doi.org/10.1098/rsfs.2017.0041.

Gaulton, R., Danson, F.M., Ramirez, F.A. & Gunawan, O. (2013) The potential of dual-wavelength laser scanning for estimating vegetation moisture content. *Remote Sensing of Environment*, 132, 32–39. https://doi.org/10.1016/j.rse.2013.01.001.

Gong, W., Sun, J., Shi, S., Yang, J., Du, L., Zhu, B. & Song, S. (2015) Investigating the potential of using the spatial and spectral information of multispectral LiDAR for object classification. *Sensors*, 15(9), 21989–22002. https://doi.org/10.3390/s150921989.

Guo, L., Chehata, N., Mallet, C. & Boukir, S. (2011) Relevance of airborne lidar and multispectral image data for urban scene classification using random forests. *ISPRS Journal of Photogrammetry and Remote Sensing*, 66(1), 56–66. https://doi.org/10.1016/j.isprsjprs.2010.08.007.

Hakala, T., Nevalainen, O., Kaasalainen, S. & Mäkipää, R. (2015) Technical note: Multispectral Lidar time series of pine canopy chlorophyll content. *Biogeosciences*, 12(5), 1629–1634. https://doi.org/10.5194/bg-12-1629-2015.

Hakala, T., Suomalainen, J., Kaasalainen, S. & Chen, Y. (2012) Full waveform hyperspectral LiDAR for terrestrial laser scanning. *Optics Express*, 20(7), 7119. https://doi.org/10.1364/OE.20.007119.

Hancock, S., Gaulton, R. & Danson, F.M. (2017) Angular reflectance of leaves with a dual-wavelength terrestrial lidar and its implications for leaf-bark separation and leaf moisture estimation. *IEEE Transactions on Geoscience and Remote Sensing*, 55(6), 3084–3090. https://doi.org/10.1109/TGRS.2017.2652140.

Hartzell, P., Glennie, C., Biber, K. & Khan, S. (2014) Application of multispectral LiDAR to automated virtual outcrop geology. *ISPRS Journal of Photogrammetry and Remote Sensing*, 88, 147–155. https://doi.org/10.1016/j.isprsjprs.2013.12.004.

Howe, G.A., Hewawasam, K., Douglas, E.S., Martel, J., Li, Z., Strahler, A., . . . Chakrabarti, S. (2015) Capabilities and performance of dual-wavelength Echidna ® Lidar. *Journal of Applied Remote Sensing*, 9(1), 95979. https://doi.org/10.1117/1.JRS.9.095979.

Jay, S., Maupas, F., Bendoula, R. & Gorretta, N. (2017) Retrieving LAI, chlorophyll and nitrogen contents in sugar beet crops from multi-angular optical remote sensing: Comparison of vegetation indices and PROSAIL inversion for field phenotyping. *Field Crops Research*, 210, 33–46. https://doi.org/10.1016/j.fcr.2017.05.005.

Johnson, B., Joseph, R., Nischan, M.L., Newbury, A.B., Kerekes, J.P., Barclay, H.T., . . . Zayhowski, J.J. (1999) Compact active hyperspectral imaging system for the detection of concealed targets. In: Dubey, A.C., Harvey, J.F., Broach, J.T. & Dugan, R.E. (eds) *Detection and Remediation Technologies for Mines and Minelike Targets IV*. p. 144. https://doi.org/10.1117/12.357002

Kaasalainen, S., Åkerblom, M., Nevalainen, O., Hakala, T. & Kaasalainen, M. (2018a) Incidence angle dependency of leaf vegetation indices from hyperspectral Lidar measurements. *Interface Focus*, 8(2), 20170033. https://doi.org/10.1098/rsfs.2017.0033.

Kaasalainen, S., Malkamäki, T., Ilinca, J. & Ruotsalainen, R. (2018b) Multispectral terrestrial laser scanning: New developments and applications. *IEEE International Geoscience and Remote Sensing Symposium (IGARSS '18)*. IEEE. https://doi.org/10.1109/IGARSS.2018.8517590.

Kaasalainen, S., Ruotsalainen, L., Kirkko-Jaakkola, M., Nevalainen, O. & Hakala, T. (2017) Towards multispectral, multi-sensor indoor positioning and target identification. *Electronics Letters*, 53(15), 1008–1011. https://doi.org/10.1049/el.2017.1473.

Kalacska, M., Lalonde, M. & Moore, T.R. (2015) Estimation of foliar chlorophyll and nitrogen content in an ombrotrophic bog from hyperspectral data: Scaling from leaf to image. *Remote Sensing of Environment*, 169, 270–279. https://doi.org/10.1016/j.rse.2015.08.012.

Kashani, A., Olsen, M., Parrish, C. & Wilson, N. (2015) A review of LIDAR radiometric processing: From ad hoc intensity correction to rigorous radiometric calibration. *Sensors*, 15(11), 28099–28128. https://doi.org/10.3390/s151128099.

Li, W., Sun, G., Niu, Z., Gao, S. & Qiao, H. (2014) Estimation of leaf biochemical content using a novel hyperspectral full-waveform LiDAR system. *Remote Sensing Letters*, 5(8), 693–702. https://doi.org/10.1080/2150704X.2014.960608.

Li, Z., Jupp, D., Strahler, A., Schaaf, C., Howe, G., Hewawasam, K., . . . Schaefer, M. (2016) Radiometric calibration of a dual-wavelength, full-waveform terrestrial lidar. *Sensors*, 16(3), 313. https://doi.org/10.3390/s16030313.

Li, Z., Schaefer, M., Strahler, A., Schaaf, C. & Jupp, D. (2018) On the utilization of novel spectral laser scanning for three-dimensional classification of vegetation elements. *Interface Focus*, 8(2), 20170039. https://doi.org/10.1098/rsfs.2017.0039.

Maccarone, A., McCarthy, A., Ren, X., Warburton, R.E., Wallace, A.M., Moffat, J., . . . Buller, G.S. (2015) Underwater depth imaging using time-correlated single-photon counting. *Optics Express*, 23(26), 33911. https://doi.org/10.1364/OE.23.033911.

Manninen, A., Kääriäinen, T., Parviainen, T., Buchter, S., Heiliö, M. & Laurila, T. (2014) Long distance active hyperspectral sensing using high-power near-infrared supercontinuum light source. *Optics Express*, 22(6), 7172. https://doi.org/10.1364/oe.22.007172.

Matikainen, L., Karila, K., Hyyppä, J., Litkey, P., Puttonen, E. & Ahokas, E. (2017) Object-based analysis of multispectral airborne laser scanner data for land cover classification and map updating. *ISPRS Journal of Photogrammetry and Remote Sensing*, 128, 298–313. https://doi.org/10.1016/j.isprsjprs.2017.04.005.

Morsdorf, F., Nichol, C., Malthus, T. & Woodhouse, I.H. (2009) Assessing forest structural and physiological information content of multi-spectral LiDAR waveforms by radiative transfer modelling. *Remote Sensing of Environment*, 113(10), 2152–2163. https://doi.org/10.1016/j.rse.2009.05.019.

Nevalainen, O., Hakala, T., Suomalainen, J. & Kaasalainen, S. (2013) Nitrogen concentration estimation with hyperspectral LiDAR. *ISPRS Annals of Photogrammetry, Remote Sensing and Spatial Information Sciences*, II-5/W2, 205–210. https://doi.org/10.5194/isprsannals-ii-5-w2-205-2013.

Nevalainen, O., Hakala, T., Suomalainen, J., Mäkipää, R., Peltoniemi, M., Krooks, A. & Kaasalainen, S. (2014) Fast and nondestructive method for leaf level chlorophyll estimation using hyperspectral LiDAR. *Agricultural and Forest Meteorology*, 198–199, 250–258. https://doi.org/10.1016/j.agrformet.2014.08.018.

Nischan, M., Joseph, R., Libby, J. & Kerekes, J. (2003) Active spectral imaging. *Lincoln Laboratory Journal*, 14, 131–144. Available from: www.ll.mit.edu/publications/journal/pdf/vol14_no1/14_1activespectral.pdf.

Niu, Z., Xu, Z., Sun, G., Huang, W., Wang, L., Feng, M., . . . Gao, S. (2015) Design of a new multispectral waveform LiDAR instrument to monitor vegetation. *IEEE Geoscience and Remote Sensing Letters*, 12(7), 1506–1510. https://doi.org/10.1109/LGRS.2015.2410788.

Paynter, I., Saenz, E., Genest, D., Peri, F., Erb, A., Li, Z., . . . Schaaf, C. (2016) Observing ecosystems with lightweight, rapid-scanning terrestrial lidar scanners. *Remote Sensing in Ecology and Conservation*, 2(4), 174–189. https://doi.org/10.1002/rse2.26.

Pfeifer, N., Mandlburger, G., Otepka, J. & Karel, W. (2014) OPALS – A framework for Airborne Laser Scanning data analysis. *Computers, Environment and Urban Systems*, 45, 125–136. https://doi.org/10.1016/j.compenvurbsys.2013.11.002.

Powers, M.A. & Davis, C.C. (2012) Spectral LADAR: Active range-resolved three-dimensional imaging spectroscopy. *Applied Optics*, 51(10), 1468. https://doi.org/10.1364/AO.51.001468.

Puttonen, E., Hakala, T., Nevalainen, O., Kaasalainen, S., Krooks, A., Karjalainen, M. & Anttila, K. (2015) Artificial target detection with a hyperspectral LiDAR over 26-h measurement. *Optical Engineering*, 54(1), 13105. https://doi.org/10.1117/1.oe.54.1.013105.

Rall, J.A.R. & Knox, R.G. (2004) Spectral ratio biospheric Lidar. *IEEE International Geoscience and Remote Sensing Symposium, 2004 (IGARSS '04). Proceedings, 2004*. IEEE. https://doi.org/10.1109/igarss.2004.1370726.

RIEGL. (2012) *LAS Extrabytes Implementation in RIEGL Software*. RIEGL Laser Measurement Systems GmbH. Available from: www.cs.unc.edu/~isenburg/lastools/las14/Whitepaper_LAS_extrabytes_implementation_in_Riegl_software.pdf [accessed 15th Mar 2018].

Schofield, L.A., Danson, F.M., Entwistle, N.S., Gaulton, R. & Hancock, S. (2016) Radiometric calibration of a dual-wavelength terrestrial laser scanner using neural networks. *Remote Sensing Letters*, 7(4), 299–308. https://doi.org/10.1080/2150704X.2015.1134843.

Shi, S., Song, S., Gong, W., Du, L., Zhu, B. & Huang, X. (2015) Improving backscatter intensity calibration for multispectral LiDAR. *IEEE Geoscience and Remote Sensing Letters*, 12(7), 1421–1425. https://doi.org/10.1109/lgrs.2015.2405573.

Sun, J., Shi, S., Gong, W., Yang, J., Du, L., Song, S., . . . Zhang, Z. (2017). Evaluation of hyperspectral LiDAR for monitoring rice leaf nitrogen by comparison with multispectral LiDAR and passive spectrometer. *Scientific Reports*, 7, 40362. https://doi.org/10.1038/srep40362.

Ullrich, A. & Pfennigbauer, M. (2011) Echo digitization and waveform analysis in airborne and terrestrial laser scanning. In: Fritsch, D. (ed) *Photogrammetric Week '11*. Wichmann/VDE Verlag, Belin, Germany and Offenbach, Germany, 2011. Available from: www.ifp.uni-stuttgart.de/publications/phowo11/220 Ullrich.pdf

Vauhkonen, J., Hakala, T., Suomalainen, J., Kaasalainen, S., Nevalainen, O., Vastaranta, M., . . . Hyyppa, J. (2013) Classification of spruce and pine trees using active hyperspectral LiDAR. *IEEE Geoscience and Remote Sensing Letters*, 10(5), 1138–1141. https://doi.org/10.1109/lgrs.2012.2232278.

Wallace, A.M., McCarthy, A., Nichol, C.J., Ximing Ren, Morak, S., Martinez-Ramirez, D., . . . Buller, G.S. (2014) Design and evaluation of multispectral LiDAR for the recovery of arboreal parameters. *IEEE Transactions on Geoscience and Remote Sensing*, 52(8), 4942–4954. https://doi.org/10.1109/TGRS.2013.2285942.

Wei, G., Shalei, S., Bo, Z., Shuo, S., Faquan, L. & Xuewu, C. (2012) Multi-wavelength canopy LiDAR for remote sensing of vegetation: Design and system performance. *ISPRS Journal of Photogrammetry and Remote Sensing*, 69, 1–9. https://doi.org/10.1016/j.isprsjprs.2012.02.001.

Wichmann, V., Bremer, M., Lindenberger, J., Rutzinger, M., Georges, C. & Petrini-Monteferri, F. (2015) Evaluating the potential of multispectral airborne lidar for topographic mapping and land cover classification. *ISPRS Annals of Photogrammetry, Remote Sensing and Spatial Information Sciences*, II-3/W5, 113–119. https://doi.org/10.5194/isprsannals-II-3-W5-113-2015.

Woodhouse, I.H., Nichol, C., Sinclair, P., Jack, J., Morsdorf, F., Malthus, T.J. & Patenaude, G. (2011) A multispectral canopy LiDAR demonstrator project. *IEEE Geoscience and Remote Sensing Letters*, 8(5), 839–843. https://doi.org/10.1109/LGRS.2011.2113312.

Laser Scanning – Riveiro and Lindenbergh
© 2020 ISPRS, ISBN 978-1-138-49604-0

Chapter 3

Multiplatform mobile laser scanning

Antero Kukko and Harri Kaartinen

ABSTRACT: Kinematic mapping solutions implemented by means of mobile laser scanning (MLS) provide the best accurate and timely structural information of our environment. This chapter discusses a multiplatform approach to the application of MLS in various surveying tasks to fill the requests for 3D data for terrain modeling, industrial scenes, construction sites, and infrastructure as well as urban mapping. Many of the applications involve civil engineering in urban areas and urban planning, which serve to make 3D city modeling, BIM creation (building information model), and related modeling probably the fastest-growing market in this field.

This presentation outlines the idea of multiplatform mobile laser scanning and solutions such as vehicle-, trolley-, and ATV-operated data acquisition, and boat-mounted equipment. With state-of-the-art sensors, the composed point clouds capture object details with good coverage and precision. Aspects of backpack MLS for surveying applications, as-built data collection, and use for natural sciences where the requirements include mobility, accessibility, and precision in variable terrain and climate conditions are presented. In addition to presenting technical description of the systems, we discuss the performance of the solutions in applications in urban mapping, forestry, as-built data, public works, and terrain modeling. Performance of the MLS is reported based on results of evaluations stemming from multiple case studies on a permanent MLS test field and in situ.

1 INTRODUCTION

1.1 *A brief background of MLS*

A land-based mobile mapping technology is based on integration of a collection of positioning, navigation and data sensors constituting a mobile mapping system (MMS) that is mounted on a kinematic platform. The most prominent sensors to collect these data are cameras and laser scanners (El-Sheimy, 2005; Petrie, 2010; Kukko, 2013). The global navigation satellite system (GNSS) receiver and inertial measurement unit (IMU) form usually the basis for the positioning sub-system (El-Sheimy, 2005) to observe the movements of the platform and produce sensor orientation data. System position and orientation at any given time, to a discrete sampling time interval according to the specifications of a given system, is used for direct georeferencing of the collected data (Puente *et al.*, 2011). In addition to a car or a van, the platform to give desired mobility for the mapping can be, e.g. an all-terrain vehicle (ATV), a boat, and a backpack (El-Sheimy, 2005; Kukko *et al.*, 2012) to name but a few examples.

The applications of MLS to environmental remote sensing have thus far mainly focused on vegetation and erosion studies and hydrology (Barber and Mills, 2007; Alho *et al.*, 2009; Alho *et al.*, 2011; Vaaja *et al.*, 2011), while a number of applications have been presented for urban road environments (Jaakkola *et al.*, 2008; Kukko *et al.*, 2009; Lehtomäki *et al.*, 2010; Lehtomäki *et al.*, 2011; Cabo *et al.*, 2014).

Most of the mapping applications in various fields stand to benefit from the accuracy and efficiency of MLS technology. Compared to traditional mapping methods utilizing digital aerial images and airborne laser scanning, the precision of the data collected can be greatly improved with MLS. Furthermore, the time and expense of geodetic measurements to survey a certain area with total stations and terrestrial lasers can be reduced. Beyond that, numerous advantages arise when using MLS data to produce high-resolution 3D models. This is demonstrated by the application examples later on.

1.2 *Comparison to the other terrestrial and airborne laser scanning*

MLS is capable of faster and more efficient 3D data acquisition than stationary terrestrial laser scanning (TLS). That makes MLS suitable for validation purposes such as airborne experiments or when dealing with areas covered by satellite data (c.f., Connor *et al.*, 2009; Kaasalainen *et al.*, 2008). Conducting a survey with TLS requires targets for accurate mutual co-registration of the individual scans, which takes time and care to place so that sufficient visibility and network geometry to them is guaranteed. In cases where data are needed to be delivered in a certain coordinate system, some targets, if not all, need to be surveyed separately. One of the major setbacks of TLS is the need to collect a large amount of redundant points due to mainly two reasons: (1) to guarantee mutual overlap with adequate horizontal span for reliable geometry for co-registering and (2) scan geometry from a single perspective with fixed angular sampling. For that reason, a large number of scan positions are often required to cover the objects of interest – usually in increasing numbers the more complex the scene becomes.

Considering the data acquisition compared to that of TLS, MLS provides a more efficient way for generating dense point clouds. This is due to a couple of characteristic differences. Mobility of a kinematic positioning makes MLS suitable for surveying and modeling of large areas. Mobility also reduces the time spent to cover complex environment, as the viewpoint can be freely selected. The kinematic laser scanner can in principle be virtually used as a paintbrush to illuminate the object surfaces with lidar to collect the geometry.

Direct georeferencing based on GNSS-IMU results in the data to have known coordinates per se. Self-sustained positioning reduces the need for placing targets on the scene for other purposes than for quality analysis, which requires only a reduced number according to task demands. For monitoring purposes over time, a set of permanent targets may however be required for traceability. This helps in preparations on the site, speeding up the conduct of the survey, in effect minimizing stoppages and delays in time critical applications, usually on industrial sites.

As the kinematic platform gives more freedom in selecting viewpoints for the survey, the resulting point cloud is more optimal in point distribution and has fewer data gaps in reduced time. Further, shorter ranges to the objects yield to smaller laser spot size and diminished effect of angular errors on the 3D point uncertainty, thus improving the precision of point cloud data.

3D models processed from the data collected by MLS offer usually high-resolution visualization with features typically lacking from the models produced using airborne laser scanning (ALS) data and/or aerial image. These data provide only coarse rendition with considerably lower point density, accuracy and precision. However, the view from ALS, or unmanned aerial vehicles (UAV), could capture objects that MLS cannot perceive from the ground perspective. For a visually pleasing and realistic urban model, for example, data is needed from both perspectives to render the objects completely.

Because of panoramic sensing geometry, MLS is able to see certain objects better than downwards-looking airborne applications. For example, vertical pole-like objects, such as pylons, may be difficult to detect with airborne systems (Ahokas *et al.*, 2002; Mills *et al.*, 2010) but can be extracted with higher accuracy using MLS (Lehtomäki *et al.*, 2010; Cabo *et al.*, 2014; Li *et al.*, 2018). On the other hand, recent UAV laser scanning applications with wide-field-of-view scanners at low flying altitudes can achieve similar data properties from low altitudes. Details inside the forest canopy, e.g. subdominant trees and saplings, branch size distribution, and stem curve, can be retrieved using MLS, which is not typically possible with sparse ALS data.

2 MULTIPLATFORM CONCEPT

As the sensors involved in MLS are usually costly, it is attractive to consider multipurpose use for them. For example, as a standalone laser scanner for TLS, or together with a navigation system composing an MLS system. To that end one may be interested in using the sensors on different configurations and platforms depending on the data needs, but also on the type of environments in which the systems need to operate. The aim for our studies at the FGI was from the beginning, since 2005, to develop an MLS system for operating with multiple platforms and to be versatile in various tasks using a multipurpose laser scanner. In the early years, this was contradictory to most other systems, with bulky operating and recording systems more or less permanently fixed on the selected platform.

Different objects of interest and complexity of the scenery, visibility and traversability set certain limitations to the standard approaches to be feasible. For example, applying MLS in geomorphology with obstacles and complexity of natural forms and shapes possesses a different environment than that of a street scene with well-defined trajectories and manmade features. In such case, the area of interest may only be reachable using a boat, an ATV or by foot, whereas in paved streets practically any vehicle can be operated. In the following pages, we will discuss a variety of different applications where multiplatform MLS becomes handy and helps in cost management alongside the improved data for the particular needs.

2.1 *Mobile laser scanning*

Compared to ALS, a technique widely used for terrain modeling and building extraction (Kaartinen *et al.*, 2005; Vosselman, 2002; Dorninger and Pfeifer, 2008) with typical point density up to a few tens of points per m^2, MLS collects point clouds with up to thousands of points per m^2 at 10 m distance from the scanner. This is due to high-pulse-rate scanning thanks to significantly shorter ranges ($\leq$ 100 m), the narrow beam divergence angle, and millimeter-level ranging precision, as seen in Figure 3.1 illustrating single scanner data from an urban scene. A 15° scan angle was used for the data with 1017 kHz pulse rate and 250 lines per second scanning (ca. 1.5 mrad angular resolution, 33 mm line spacing).

The accuracy of MLS point clouds is determined by the GNSS-IMU in a dynamic way: the errors vary over time. When GNSS satellites are available, 1–3 cm accuracy can be achieved with high-end MLS systems supported with post-processing of the trajectory, or using real-time kinematic (RTK) positioning in urban or semi-urban environments (Haala *et al.*, 2008; Kaartinen *et al.*, 2012). Similar behavior has been found for backpack MLS in an open ephemeral river valley and impact crater survey (Kukko *et al.*, 2015). Even when the MLS works perfectly, that 1–3 cm uncertainty could introduce multiple copies of objects in multipass data that could differ 6 cm from each other.

According to recent studies in boreal forest canopy cover, postprocessed positioning with a tactical-grade GNSS-IMU seems to provide 0.2–0.7 m absolute accuracy (Kukko *et al.*, 2017; Liang *et al.*, 2014a; Kaartinen *et al.*, 2015) and backpack MLS (Liang *et al.*, 2014b). In forest and urban scenes, the accuracy can at worst degrade to several meters (Gu *et al.*, 2015). The relative

Figure 3.1. Single-scanner MLS can capture building facades and street objects in great detail, here Empire-style buildings, street furniture and cobblestone-paved streets in Helsinki, Finland.

accuracy is dependent on the quality grading of the IMU, and subcentimeter data cannot be typically achieved with low-cost IMUs due to high rate of drift.

In a recent study by Kukko *et al.* (2017), the GNSS-IMU trajectory drift in forestry MLS data was solved in postmission processing by adapting graph SLAM (simultaneous localization and mapping). Features extracted from the point cloud computed using the initial trajectory were used to find correspondences between overlapping data measured at different times during the mapping. Similarly, UAV or ALS data could possibly be used in an optimization scheme to improve the MLS data suffering from poor satellite visibility.

2.2 *Aspects to MLS system development*

High-resolution laser scanning in combination with the advanced GNSS-IMU technology has revolutionized the field of mapping. Multiple commercial MLS systems have been introduced, e.g. Optech Lynx and Maverick, Mitsubishi MMS-X, StreetMapper and Riegl VMX and VMQ variants, Leica Pegasus and Siteco Road-Scanner systems, and numerous research platforms. Most of these were designed for fixed mounting on a single configuration.

Rapid development in sensor technology with enhancing performance and decreasing size and weight, e.g. Riegl VQ-250 and VUX-1HA, Velodyne HDL-64E and VLP-16, Faro Focus3D, Optech Lynx, Z+F Profiler 9012, is expected to help in automatic object detection and reconstruction, as demonstrated, e.g. by Li *et al.* (2018) in a case study on traffic sign labeling.

For current MLS applications, use of 2D scanners is the most common approach. The third dimension in to the data is obtained from tracking the movements of the platform. 2D scanners with 360° field of view are ideal for mobile laser scanning, as they give good data coverage. Such sensors are widely deployed for kinematic mapping, although sub-360° FOV scanners are also used considerably. To improve object capture, multisensor layouts are often put to use.

The latest trend in MLS sensors is multilayer scanning (e.g. Velodyne HDL-64E and VLP-16 and other such) producing multiple profiles per one scan following the trait of Ibeo LUX 4-layer bumper scanner from a decade back. Alternatively, but not common, oscillating-beam axis implementation with Risley prisms gives also along-track field of view, e.g. Neptec Opal with 360° × 45° scanning. This improves the spatial coverage of data and helps to overcome some of the occlusion effects present in single-profile data.

New MLS systems with increasingly varying component combinations are being introduced at breathtaking rates at the moment. StreetMapper was the first commercially available multiscanner mobile mapping system. It was initially aimed at highway asset inventory but has been deployed also for other applications since its first appearance (Hunter *et al.*, 2006; Kremer and Hunter, 2007). The StreetMapper system has been implemented with Riegl VQ-250 and later VUX-1HA scanner units for high-speed data collection with a similar functional construct adopted to the Optech Lynx, launched 2007, and later inherited by the Trimble MX-8 system. On phase-shift ranging instruments, FARO Focus3D seem to be dominant and is used in systems such as Siteco Road-Scanner3 and IGI SAM. Z+F profiler 9012 was the first phase-shift ranging device to have 360° FOV, and implemented into an MLS, e.g. by Leica.

Common to most of the MLS systems is that the instrumentation is mounted on a compact platform that can be removed and replaced. The compact structure helps in maintaining the relative calibration of different sensors involved, and system modularity is emphasized. Dual GNSS-antenna systems are preferred for faster and more accurate heading alignment, especially at slow-motion applications. The trend shows also an increase in the number of scanners mounted on single units to overcome occlusion detriment. Low-cost systems are also entering the market en masse, providing moderate performance with affordable pricing.

2.3 *Multiplatform MLS*

Mobile laser scanning has its past on the roads, so vehicle-mounted systems are the most prominent instance of the technology. With a wheeled vehicle, the data could be captured at road speeds day and night for the crew safety and job efficiency. The suitability of a particular system on a specific task depends on the selected sensors and the sensor layout itself. How well one could fit their system on the needs depends on the platform design and implementation.

One unique feature of the Roamer system design (Figure 3.2) was its adjustable scanner orientation. This was in part motivated due to the fact of partially obstructed field of view by the scanner base, but also to investigate the effects of scanner orientation on the data and modeling in different applications. For example, 30° or 45° off-nadir looking positions were mostly used for road surface and terrain modeling, while for urban modeling and forestry, the vertical scanner orientation was used. In Figure 3.2, these positions can be seen to the left and bottom of the figure.

The first Roamer system in 2006 was based on an iQsun/FARO LS 880HE80 scanner. The scanner was initially purchased for TLS purposes, but what then became so called 'helical mode' was requested as custom modification from the manufacturer. This modification was first implemented for the FGI Roamer. As the scanner was a TLS scanner, it was capable to provide wide field of view not seen before, and the phase shift ranging provided superior data rates at the time. Stop-and-go mode could be used for vehicle-mounted TLS, where the data was georeferenced using the GPS-IMU system (see e.g. Kukko *et al.*, 2007; Kaartinen *et al.*, 2012). This feature could be implemented using current scanners from, e.g. FARO, Riegl and Leica.

Figure 3.2. Roamer system was developed in 2005–2006 for enabling research on road environment and urban mapping and could be fitted with multiple cameras. The system was also deployed for snow surface remote sensing and mounted on a Ski-Doo sleigh (2nd top left). R2–R4 versions use significantly smaller and higher-performance FARO Focus3D and Riegl VUX-1HA scanners and positioning.

For the later versions of the Roamer, the use of a multipurpose scanner was pertained. Miniaturization of sensors resulted in smaller scanners to emerge, and FARO Focus3D series was a natural continuum. However, the multiorientation option was limited to only two prime scanner positions: scanner up and scanner down. This was achieved, making the sensor assembly into one solid piece, i.e. all the positioning sensors and scanner were mounted on a compact common frame. Nonetheless, the scan angle orientation could be altered relatively easily using angled GlobalTruss pieces, as can be seen in the top right of Figure 3.2. The system could easily be mounted basically on any platform without a need of recalibration. Also, as the Focus3D stored data on an SSD card, only a tablet computer was needed to record the positioning data. That helped greatly the integration and use in various applications.

One feature that currently prominently differentiates the Roamer from the commercial systems is the elevated scanner; depending on the setup, the scanner can be mounted up to 3.8 meters from the road surface. The elevated sensor aims at providing less occlusion on building walls behind parked cars and fences. High elevation also provides advantage of longer ground visibility in comparison to a standard setup. These implementation features are further discussed in Kukko (2013).

2.4 *Backpack systems*

Early Akhka prototypes (Figure 3.3) from 2010 show extremely bulky layout with 15 kg FARO Photon 120 scanner and were powered by a 6 kg motorcycle battery! Extensive projects were

Figure 3.3. Backpack platforms of Akhka evolution versions. Top left: Akhka, Top right: Akhka-R2, Bottom, Akhka-R4 with SLAM.

conducted using the platform nonetheless, for example scanning an 8-km-long section of Rambla de la Viuda river bed (c.f. Calle *et al.*, 2015; Kukko *et al.*, 2015) in Spain, representing a feat of 28 km of mapping in a single day! The extensive field works and data products obtained provided the proof of concept. As seen in Figure 3.3, the scanner position was fixed in the first version, or actually we never thought of swapping the scanner.

Timely, the sensor development reduced the sensor size and improved the scan and data output to result in the R2 version in 2013. R2 implemented around FARO Focus3D (120S and X330) with helical adapter kit introduced up and down orientation options for the scanner, so that forestry became an obvious application as we now could scan the tree tops completely (Figure 3.3 in top right shows R2 with X330 scanner). R3 again was framed in 2016 using a Riegl VUX-1HA scanner to provide time of flight ranging data and multiple echo detection to see past vegetation. Akhka-R4, seen in bottom of Figure 3.3 in duty of cultural heritage mapping at the Matsumoto castle in Japan, was launched 2017 and deploys a near navigation grade IMU and fittings for Riegl MiniVUX-UAV also. Experiments in varying configurations based on different combinations of Riegl VUX-1HA, Riegl MiniVUX-UAV and Velodyne VLP-16 have been conducted, including experiments with continuously rotating Focus3D to provide panoramic scanning.

3 APPLICATIONS

3.1 *Public infrastructure*

Public infrastructure and works provide the basis for everyday life and functions. Management of public property and utilities, streets, roads and cadastre are functions for reliable, safe and sustainable communities. In the following chapters, use of multiplatform MLS in various public works is discussed.

3.1.1 Powerline mapping

MLS has been applied in the mapping of power lines in and urban environment (Kim and Medioni, 2011; Cheng *et al.*, 2014; Guan *et al.*, 2016). However, very few reports on the application of MLS in rural environments including forest have been published. Matikainen *et al.* (2016) have published a review on various remote sensing methods, including laser scanning, for power line corridor surveys.

In an ongoing study based on data from ATV and backpack MLS of a forest and rural power lines, the developed automated power line detection (Figure 3.4) could reach detection comparable to those of the previous studies in urban environments. The results suggest that MLS is sufficient for inspection in such environments. The data was captured with the scan angle slightly tilted from the vertical, and the data reconstruct the whole corridor with poles and trees along with the power cables. Mobility challenges may occur even for an ATV in the form of large boulders or the terrain cannot bear the weight of the vehicle. Nonetheless, both platforms investigated could resolve for similar results.

In contrast to airborne or UAV operations, the MLS approach provides means for inspection also in challenging weather conditions and avoids need for experienced personnel and paperwork. It is also easier and cheaper to multiply ground-based systems than airborne units for improved data output. The Akhka-R2 setup used for the power line corridor can be seen in Figure 3.3, and the data acquired with the system is illustrated in Figure 3.4 along with the automatically extracted power line cables. The ATV system setup used for this study is shown in Figure 3.11.

3.1.2 Municipal infra-engineering: district heating pipeline documentation

Figure 3.5 shows data collected for a case study to investigate the feasibility of MLS and data properties for pipeline documentation. Documentation of underground installations becomes a necessity when more and more infrastructure and construction has a possibility for collision and

Figure 3.4. Power lines automatically detected from backpack MLS data.

Figure 3.5. A district heating pipeline junction documented with backpack MLS. High-density data captures the depth of the pipes, locations of valves, branching and welds, power cables and other structures and installations to be buried.

excavating possess a risk for the critical breakdowns. In this study, the district heating installation was monitored over a period of time to document different phases and progress of the work. Especially toward the end of such a project, the time constraints may become critical for rapidly progressing work, e.g. filling of the pipeline trench. For such purposes, MLS provides the required capability for quick data collection.

3.1.3 Dam structures

Dam structures are monitored for risk management and flood protection aims, often using permanent GNSS-stations and fixed targets for total station measurements. These give precise information of selected locations, but especially in earthfill embankments, more comprehensive monitoring of the full dam structure is required with high precision to be able to detect possible changes and react before possible structure failures.

For such end backpack laser scanning was tested on a section of 2.8-km-long embankment. The main structure could be scanned for example by utilizing an ATV-MLS, but some structures, like the concrete flood canal for controlling the flow out of the dam basin, could be only covered by foot (Figure 3.6).

3.1.4 Industrial structures

In an industrial complex, such as a harbor, the daily business deals with loading and unloading a mass of cargo and goods from trucks, trains and ships, constituting a busy environment with heavy machinery and cranes. Like any other large business plant, the area is also often large yet complex with structures that are hard to cover efficiently with TLS. Usual survey tasks are to map the existing installations and storage to support the daily operations and management of material flows. Rapid 3D mapping provides means for estimating the volumes of bulk materials, such as piles of iron pellets. Due to the large area to map, a vehicle-mounted MLS is an attractive approach for constructing the base mapping of such a site. However, in some areas, data completeness is hard to achieve due to the complexity of the site installations, or one can only pass on by heavy machines or on foot, so there is also a need for complementary platforms to use. In Figure 3.7, a backpack laser scanning is used to collect complementary data of certain installations.

Figure 3.6. Snapshot of a backpack MLS data from a 10-m-high embankment and a flood canal beneath.

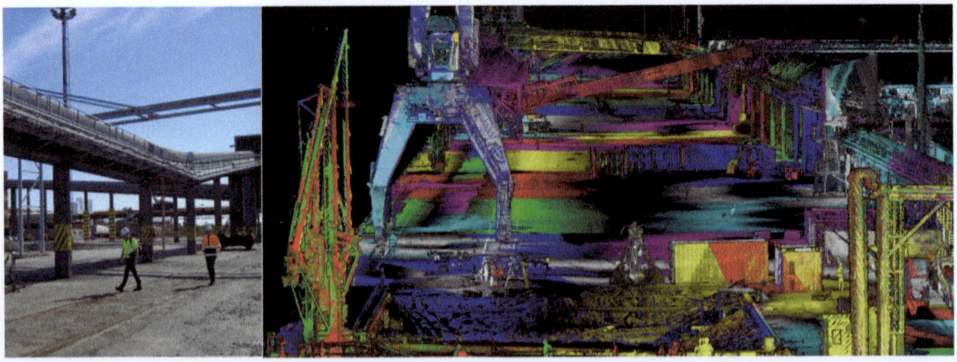

Figure 3.7. Left: Mapping a conveyor line at a harbor with Akhka-R2 backpack laser scanning system. Right: 3D data captures the details of the installations accurately, and a kinematic approach helps greatly in collecting complete data for modeling and plant design purposes.

3.2 *Urban mapping*

Mobile laser scanning is a prominent tool for urban mapping. Such application could provide means for cadastral surveys, street, bridge and building management, land use planning, urban silviculture and addressing, e.g. accessibility issues on public services and buildings. These tasks can be conducted at different scales in conjunction with other lidar and image data.

3.2.1 Regional mapping

Regional-scale urban mapping is usually carried out as part of an ALS mapping. Despite the ability of such a method to detect the buildings and topography, ALS typically lacks building wall and street details. Such data are more and more requested for regional-scale planning and land use mapping. Due to its speed and suitability for capturing such data from streets and pedestrian alleys, MLS is often deployed for such tasks. Figure 3.8 shows a section of MLS data collected over an urban region in Southern Finland. Despite the large area, the data could be captured in a couple of hours. To cover an urban scene in full including, e.g. walking passages and courtyards, a multiplatform concept could be realized by using for example ATV, trolley or backpack MLS.

3.2.2 Large-scale mapping

For certain purposes, one often requires more detailed and complete data. Such needs arise e.g. from city plans, architectural planning, areal development, road works and construction project documentation, cultural heritage even. For detailed 3D data and especially complete data coverage over the region of interest, comprising a single building or a block of houses, may not be achieved by a full-sized MLS system with limited accessibility. With a backpack system, the required details of buildings can be obtained, as seen in Figure 3.9. Each corner, protrusion and doorway of a particular building can be captured. However, what can be also seen is that MLS data needs to be complemented with airborne data for roof structures.

For a building modeling, one would prefer a high-speed and precise scanner to capture the facade details, railings, gutter, piping etc. The data shown in Figure 3.9 was captured using 95 Hz scan frequency and 488 kHz point measurement rate to gain 1.2 mrad angular resolution within the near-vertical scan profiles. Beyond modeling purposes, dense and precise point cloud of a building provides a good interpretation of the objects per se, and possible new plans and models, for example, can be fused with the point cloud data for augmented visualization without a need for completely modeling the scene, thus saving time and costs in the plan sketching phase, and allows participatory communication within the stakeholders involved.

Figure 3.8. Regional 3D data can be captured with vehicle MLS to complement ALS data driving the streets and alleys. Coloring illustrates the progress of the data collection, as each distinct color shaded with point intensity indicates a 30 s section of data.

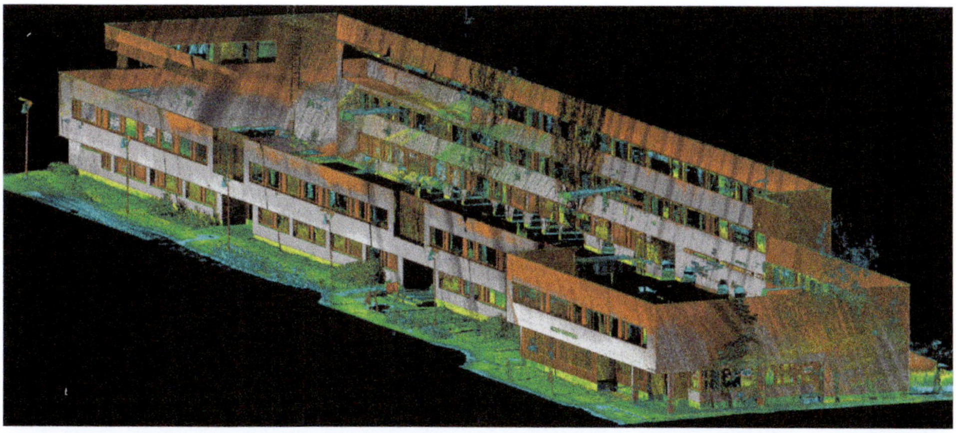

Figure 3.9. Detailed building data collected with a backpack MLS. Point colors reflect the point intensity.

3.3 *Forest economy*

Economy based on forest products is a significant source for jobs and income in the countries situated at the boreal forest zone. Up-to-date and accurate information for management of the forest resource is ever important for sustainable use and renewal. In combination with wide-area airborne data, ground-based techniques provide individual-tree-level stem volume and biomass estimates, and stem quality and assortment in terms of stem curve shape and branch distribution. While TLS provides timber information at plot-wise analysis, it lacks generality with only local data with an effort. Ground level 3D data, however, shows great potential in improving the accuracy of data and even new data products, and a kinematic platform can be used for the purpose. All-terrain vehicles make a good opportunity to be used in forest mapping, as they provide good fuel economy and maneuverability on varying forest terrain, as seen in Figure 3.11, where the scanner is mounted for

Figure 3.10. Laser pulse echo type captured with Riegl VUX-1HA scanner mounted on a backpack MLS. Echo-type information can be utilized along with the echo deviation information to classify the data (Red: Only, Blue: First of many, Green: Intermediate, Yellow: Last of many).

Figure 3.11. Single pine tree captured with MLS (left), a typical MLS installation on an ATV (middle), trajectory-optimized forest data (right).

vertical cross-track scanning to capture the tree stems and canopies, but also terrain data to be used for planning the harvesting operations.

When the forest is too dense for an ATV to pass, or the terrain is too rocky or sloped for safe driving, and steering off the obstacles becomes too much of a burden, mapping with a backpack gives flexibility in selecting the route among the trees and ability to cross ditches and small creeks en route. Figure 3.10 illustrates a cross-section of backpack MLS data from a Japanese cedar forest captured with Riegl VUX-1HA of Akhka-R4. The forest density at these latitudes differs greatly from boreal forests, and terrain is more mountainous. Laser echo information can be used for structural and ecological analysis and classification of the data. The scanning was carried out with 250 lines per second, 1017 kHz pulse rate and 125 m maximum range. The scan plane is slightly forward tilted on the operator's back, which helps in capturing even thin features. Too much tilting, however, is problematic at confined sample plots (typically 20 m in radius): a 45° scan angle misses half of the treetops in the plot when standing at the plot border facing the plot center. When mapping, one needs to pass the plot limit according to the maximum tree height and tilt angle ($tan(a_{tilt})*H_{max\_tree}$).

In the presence of drift when the data capture passes the trees multiple times for the purpose to capture the canopy and stem completely, the instances of each tree becomes blurred. That hampers the automated stem curve modeling but also the tree mapping and timber estimations (Liang *et al.*, 2014a, b), as single trees show multiple copies in the resulting maps if the position drift remains uncorrected. To improve the situation and enable automated data analysis, one can use e.g. graph optimization to improve the localization and thus the tree map accuracy (Kukko *et al.*, 2017). By graph optimization, the tree map (see Figure 3.11) quality could be improved to a millimeter level, which greatly helps in modeling the stem curve and volume but also to build an accurate forest map. However, it is worth noting that swaying of the trees in the wind during the mapping limits the top stem and canopy estimation accuracy.

3.4 *Industrial engineering*

As-built models are often needed from industrial facilities, whether it deals with planning new assemblies or maintaining the existing ones. Collecting data for as-built models can be very challenging in an industrial environment; objects to be mapped can be complex and labyrinthine, scales vary from small parts to several hundreds of meters of processing lines and halls, there may be restricted accessibility to certain areas, etc. Collecting occlusion-free data from such sites with static systems can be extremely time consuming and cumbersome.

As MLS systems cannot rely on satellite positioning indoors, SLAM technology provides a solution for indoor data collection. Today, most of the commercial SLAM-based MLS systems utilize rather low-grade laser scanners, which are not capable of capturing smaller details with good precision and accuracy. There are ongoing studies where data collected with a high-grade scanner is postprocessed using SLAM-algorithms (Kukko *et al.*, 2017), and this should also provide a solution for producing high-quality point clouds from complex indoor environments.

3.5 *Terrain modeling*

3.5.1 Fluvial geomorphology

For river environment mapping, the multiplatform concept has proven to be functional for collecting detailed topography for geomorphology (Alho *et al.*, 2009; Alho *et al.*, 2011; Vaaja *et al.*, 2013; Kukko *et al.*, 2012; Calle *et al.*, 2015). In the aforementioned studies, cars, ATVs (Figure 3.12), boats, UAVs and backpacks were used to collect seamless data of the selected reaches of rivers, typically several kilometers long. A typical approach has been to use both vehicles and backpack MLS to scan the features at desired locations for high-resolution DEMs, such as seen in Figure 3.13. The level of detail in the data allows for estimating certain hydraulic parameters over extended areas, which had not been possible earlier. Also the river bed changes induced by seasonal discharges can be quantified for hydraulic modeling (Lotsari *et al.*, 2018).

3.5.2 Permafrost landforms

Daily and seasonal freezing and thawing of the ground in the High Arctic wreaks havoc on the bare rocks and soil. The volume expansion of freezing water in pores and cracks forces rocks to break apart and the ground to fracture. These effects are causing potholes in roads and roof damage from backed-up gutters. In the High Arctic, freezing, thawing, cracking and heaving causes

Figure 3.12. An ATV platform for mapping fluvial geomorphology in Valencia, Spain.

Figure 3.13. Left: About 4-km-long section of meandering Pulmanki river in the Finnish Arctic measured in combined effort of boat and backpack mobile laser scanning. Right: High-resolution data allow for monitoring annual discharge-induced erosion and deposition changes, and provide a reach scale basis for simulation and modeling of certain hydraulic processes.

much movement within the active layer as freeze–thaw processes work to sort stones and soil into patterns.

These cracks propagate into ice wedges, which then form polygons that can cover large areas (Figure 3.14). On hillslopes, thawing creates debris flows and rockslides, etching into and eroding the landscapes. On marshlands, permafrost and the frost–thaw cycles have created palsa landforms (low, often oval, frost heaves containing permanently frozen ice lenses occurring in polar and sub-polar climates, Figure 3.14) whose current melting and cracking development, e.g. in the Finnish Arctic Lapland and elsewhere, is seen as an indicator of the climate change (Zanetti *et al.*, 2018; Kukko *et al.*, 2012).

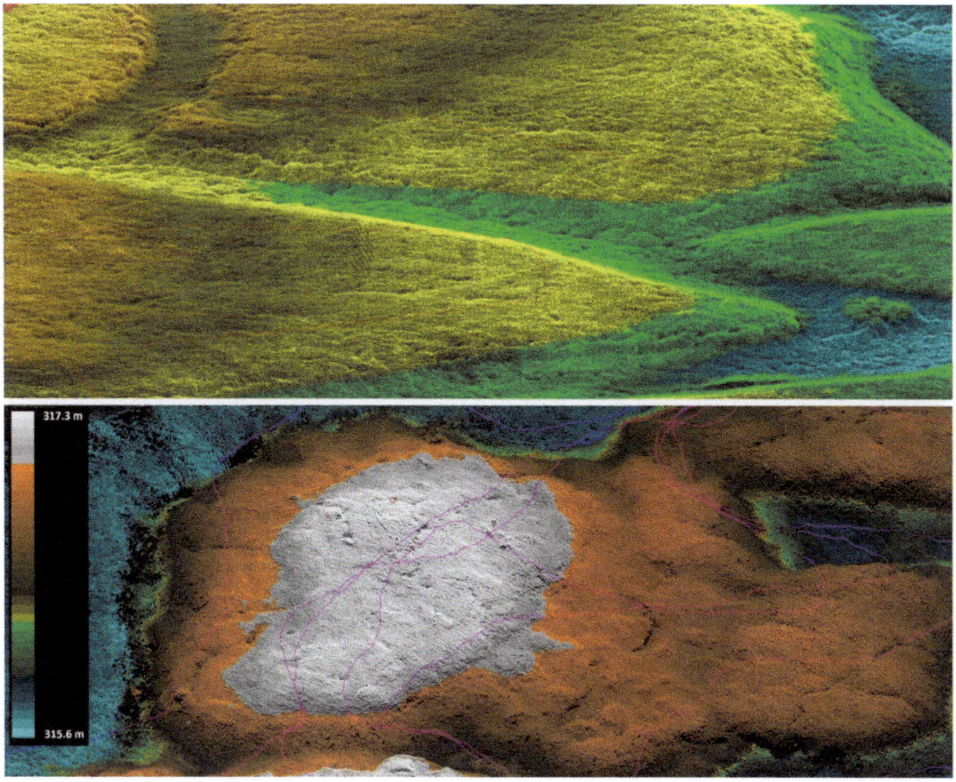

Figure 3.14. Top: Periglacial polygon ground at the Haughton impact structure in Devon Island, Nunavut, Canadian High Arctic, captured using backpack laser scanning with Akhka-R3. Bottom: Permafrost palsa landforms measured with Akhka backpack MLS in June 2011 at Vaisjeaggi mire in Utsjoki, Finland (from Kukko *et al.*, 2012).

The characterization of these phenomena (measuring diameter, spacing, trough depth and other morphometric parameters) was earlier done with painstaking and simple manual measurements. For example, year-on-year and decadal changes were measured using marked sticks shoved into the ground. The time-intensive nature of the measurements meant that only, small areas could be investigated, mostly in isolation from nearby features. Just recently multiplatform MLS has made it possible to acquire high-resolution topography on wide areas, and first results are presented in e.g. Zanetti *et al.* (2018) and Hawkswell *et al.* (2018).

4 SUMMARY

This chapter discusses aspects to multiplatform implementations of mobile laser scanning technology. MLS offers performance in diversified environments and data needs given that the system design allows for the versatility needed. There are currently multiple alternatives to choose from for laser sensors to capture the objects in 3D. However, the developer and system user alike need to recognize the different performance factors ultimately affecting the final data characteristics and quality. Furthermore, one must not be indifferent to the effect of sensor layout on the platform, as it greatly influences the point pattern, the distribution and the conduct of data capture.

As especially the higher grade performance sensors are expensive to acquire, it is oftentimes worth considering whether one could get sensors with multiple uses, or they provide multiple modes of operation. That helps in maintaining the investment costs by allowing a more varied project portfolio to be handled using the same instrumentation. MLS systems using TLS scanners in profiling, or helical, mode, can be used for static scanning also, e.g. for indoors to complete, e.g. BIM modeling tasks. Also, the versatility of the system can be increased when carefully considered in the system design, allowing multiple sensor orientations that could be easily altered so that different projects' ends are achieved.

At best MLS provides timely data within the given accuracy grade efficiently even over large extents of area. MLS data can be complemented with UAV/RPAS or ALS data, and as the GNSS-IMU based positioning brings the multisource data into a common frame per se, that helps mutual fine-tuning when needed.

It is evident that the proposed MLS approaches have the potential to speed up and improve the collection of 3D survey data and widen spatial coverage, with remarkable results as regards point density and quality. Future applications will have important parts to play in various modeling tasks in the vast fields of civil and transportation engineering, archaeology and geomatics, as well as in the monitoring and understanding of processes in different disciplines of natural sciences, e.g. cryosphere and glaciology, geography, hydrology, silviculture and agriculture.

REFERENCES

Ahokas, E., Kaartinen, H., Matikainen, L., Hyyppä, J. & Hyyppä, H. (2002) Accuracy of high-pulse-rate laser scanners for digital target models. *Proceedings of 21st Earsel Symposium, Observing Our Environment From Space: New Solutions for a New Millennium*. Balkema Publishers, Paris, France.

Alho, P., Kukko, A., Hyyppä, H., Kaartinen, H., Hyyppä, J. & Jaakkola, A. (2009) Application of boat-based laser scanning for river survey. *Earth Surface Processes and Landforms*, 34, 1831–1838.

Alho, P., Vaaja, M., Kukko, A., Kasvi, E., Kurkela, M., Hyyppä, J., Hyyppä, H. & Kaartinen, H. (2011) Mobile laser scanning in fluvial geomorphology: Mapping and change detection of point bars. *Zeitschift für Geomorphologie*, 55, 31–50.

Barber, D.M. & Mills, J.P. (2007) Vehicle based waveform laser scanning in a coastal environment. *Proceedings of the 5th International Symposium on Mobile Mapping Technology, 29–31 May 2007*. Pradua, Italy.

Cabo, C., Ordoñez, C., García-Cortés, S. & Martínez, J. (2014) An algorithm for automatic detection of pole-like street furniture objects from mobile laser scanner point clouds. *ISPRS Journal of Photogrammetry and Remote Sensing*, 87, 47–56.

Calle, M., Lotsari, E., Kukko, A., Alho, P., Kaartinen, H., Rodrigues-Lloveras, X. & Benito, G. (2015) Morphodynamics of an ephemeral gravel-bed stream combining mobile laser scanner, hydraulic simulations and geomorphological indicators. *Zeitschift für Geomorphologie*, 59, 3, 33–57.

Cheng, L., Tong, L., Wang, Y. & Li, M. (2014) Extraction of urban power lines from vehicle-borne lidar data. *Remote Sensing*, 6(4), 3302–3320.

Connor, L.N., Laxon, S.W., Ridout, A.L., Krabill, W.B. & McAdoo, D.C. (2009) Comparison of Envisat radar and airborne laser altimeter measurements over Arctic sea ice. *Remote Sensing of Environment*, 113, 563–570.

Dorninger, P. & Pfeifer, N.A. (2008) Comprehensive automated 3D approach for building extraction, reconstruction, and regularization from airborne laser scanning point clouds. *Sensors*, 8(11), 7323–7343. https://doi.org/10.3390/s8117323.

El-Sheimy, N. (2005) An overview of mobile mapping systems. *Proceedings of FIG Working Week 2005 and GSDI-8, 16–21 April 2005*. Cairo, Egypt.

Gu, Y., Hsu, L.-T. & Kamijo, S. (2015) Passive sensor integration for vehicle selflocalization in urban traffic environment. *Sensors*, 15, 30199–30220.

Guan, H., Yu, Y., Li, J., Ji, Z. & Zhang, Q. (2016) Extraction of power-transmission lines from vehicle-borne lidar data. *International Journal of Remote Sensing*, 37(1), 229–247.

Haala, N., Peter, M., Kremer, J. & Hunter, G. (2008) Mobile Lidar mapping for 3D point cloud collection in urban areas: A performance test. *International Archives of Photogrammetry, Remote Sensing and Spatial Information Sciences*, 37(B5), 1119–1124.

Hawkswell, J.E., Godin, E., Osinski, G.R., Zanetti, M. & Kukko, A. (2018) Comparative investigation of polygon morphology within the haughton impact structure, Devon Island with implications for Mars. *49th Lunar and Planetary Science Conference, 19–23 March 2018*. The Woodlands, TX, USA, #2899.

Hunter, G., Cox, C. & Kremer, J. (2006) Development of a commercial laser scanning mobile mapping system – StreetMapper. *Second International Workshop the Future of Remote Sensing, 17.18. October 2006*. Antwerp, Belgium.

Jaakkola, A., Hyyppä, J., Hyyppä, H. & Kukko, A. (2008) Retrieval algorithms for road surface modelling using laser-based mobile mapping. *Sensors*, 8, 5238–5249.

Kaartinen, H., Hyyppä, J., Gülch, E., Hyyppä, H., Matikainen, L., Vosselman, G., . . . Vester, K. (2005) EuroSDR building extraction comparison. *ISPRS Hannover Workshop 2005 on "High-Resolution Earth Imaging for Geospatial Information"*. Available from: https://www.isprs.org/proceedings/2005/hannover05/paper/papers.htm

Kaartinen, H., Hyyppä, J., Kukko, A., Jaakkola, A. & Hyyppä, H. (2012) Benchmarking the performance of mobile laser scanning systems using a permanent test field. *Sensors*, 12(9), 12814–12835.

Kaartinen, H., Hyyppä, J., Vastaranta, M., Kukko, A., Jaakkola, A., Yu, X., . . . Hyyppä, H. (2015) Accuracy of kinematic positioning using global satellite navigation systems under forest canopies. *Forests*, 6(9), 3218–3236.

Kaasalainen, S., Kaartinen, H. & Kukko, A. (2008) Snow cover change detection with laser scanning range and brightness measurements. *EARSeL eProceedings*, 7, 133–141.

Kim, E. & Medioni, G. (2011) Urban scene understanding from aerial and ground LIDAR data. *Machine Vision and Applications*, 22(4), 691–703.

Kremer, J. & Hunter, G. (2007) Performance of the StreetMapper Mobile LIDAR mapping system in "Real World" projects. In: Fritsch, D. (ed) *Photogrammetric Week, 2007*. University of Stuttgart, pp.215–225. Available from: https://phowo.ifp.uni-stuttgart.de/publications/phowo07/

Kukko, A. (2013) *Mobile Laser Scanning – System Development, Performance and Applications*. Kirkkonummi, 247 p. Available from: http://urn.fi/URN:ISBN:978-951-711-307-6.

Kukko, A., Andrei, C.O., Salminen, V.-M., Kaartinen, H., Chen, C., Rönnholm, P., . . . Čapek, K. (2007) Road environment mapping system of the Finnish Geodetic Institute – FGI Roamer. *International Archives of Photogrammetry, Remote Sensing and Spatial Information Sciences*, 36, 241–247.

Kukko, A., Jaakkola, A., Lehtomäki, M. & Kaartinen, H. (2009) Mobile mapping system and computing methods for modelling of road environment. *Proceedings of the 2009 Urban Remote Sensing Joint Event, 20–22 May 2009*. Shanghai, China.

Kukko, A., Kaartinen, H., Hyyppä, J. & Chen, Y. (2012) Multiplatform mobile laser scanning: Usability and performance. *Sensors*, 12(9), 11712–11733.

Kukko, A., Kaartinen, H., Hyyppä, J. & Zanetti, M. (2015) Backpack personal laser scanning system for grain-scale topographic mapping. *46th Lunar and Planetary Science Conference, 16–20 March 2015*. The Woodlands, TX, USA, #2407.

Kukko, A., Kaijaluoto, R., Kaartinen, H., Lehtola, V.V., Jaakkola, A. & Hyyppä, J. (2017) Graph SLAM correction for single scanner MLS forest data under boreal forest canopy. *ISPRS Journal of Photogrammetry and Remote Sensing*, 132, 199–209.

Lehtomäki, M., Jaakkola, A., Hyyppä, J., Kukko, A. & Kaartinen, H. (2010) Detection of vertical pole-like objects in a road environment using vehicle-based laser scanning data. *Remote Sensing*, 2, 641–664.

Lehtomäki, M., Jaakkola, A., Hyyppä, J., Kukko, A. & Kaartinen, H. (2011) Performance analysis of a pole and tree trunk detection method for mobile laser scanning data. *Proceedings of the ISPRS Workshop Laser Scanning 2011, 29–31 August 2011*. Calgary, AB, Canada.

Li, F., Oude Elberink, S. & Vosselman, G. (2018) Pole-Like road furniture detection and decomposition in mobile laser scanning data based on spatial relations. *Remote Sensing*, 10(4), 531. https://doi.org/10.3390/rs10040531.

Liang, X., Hyyppä, J., Kukko, A., Kaartinen, H., Jaakkola, A. & Yu, X. (2014a) The use of a mobile laser scanning system for mapping large forest plots. *IEEE Geoscience and Remote Sensing Letters*, 11(9), 504–1508.

Liang, X., Kukko, A., Kaartinen, H., Hyyppä, J., Yu, X., Jaakkola, A. & Wang, Y. (2014b) Possibilities of a personal laser scanning system for forest mapping and ecosystem services. *Sensors*, 14(1), 1228–1248.

Lotsari, E.S., Calle, M., Benito, G., Kukko, A., Kaartinen, H., Hyyppä, J., . . . Alho, P. (2018) Topographical change caused by moderate and small floods in a gravel bed ephemeral river - A depth-averaged morphodynamic simulation approach. *Earth Surface Dynamics*, 6, 163–185.

Matikainen, L., Lehtomäki, M., Ahokas, E., Hyyppä, J., Karjalainen, M., Jaakkola, A., . . . Heinonen, T. (2016) Remote sensing methods for power line corridor surveys. *ISPRS Journal of Photogrammetry and Remote Sensing*, 119, 10–31.

Mills, S.J., Castro, M.P.G., Li, Z., Cai, J., Hayward, R., Mejias, L. & Walker, R.A. (2010) Evaluation of aerial remote sensing techniques for vegetation management in power-line corridors. *IEEE Transactions on Geoscience and Remote Sensing*, 48(9), 3379–3390.

Petrie, G. (2010) An introduction to technology mobile mapping systems. *GeoInformatics*, 13, 32–43.

Puente, I., González-Jorge, H., Arias, P. & Armesto, J. (2011) Land-based mobile laser scanning systems: A review. In: Lichti, D.D. and Habib, A.F. (eds) ISPRS workshop laser scanning, Calgary, Canada, 29–31 August. *International Archives of Photogrammetry, Remote Sensing, and Spatial Information Sciences*, 38-5/W12, 163–168.

Vaaja, M., Hyyppä, J., Kukko, A., Kaartinen, H., Hyyppä, H. & Alho, P. (2011) Mapping topography changes and elevation accuracies using a mobile laser scanner. *Remote Sensing*, 3, 587–600.

Vaaja, M., Kukko, A., Kaartinen, H., Kurkela, M., Kasvi, E., Flener, C., . . . Alho, P. (2013) Data processing and quality evaluation of a boat-based mobile laser scanning system. *Sensors*, 13(9), 12497–12515.

Vosselman, G. (2002) Fusion of laser scanning data, maps, and aerial photographs for building reconstruction. *Proceedings of the International Geoscience and Remote Sensing Symposium (IGARSS), 24–28 June 2002*. Toronto, ON, Canada, pp.85–88.

Zanetti, M., Kukko, A., Neish, C. & Osinski, G. (2018) Comparative planetology Lidar unveils similarities of Earth and Mars. *Lidar Magazine*, 8(2), 10p. Available from: www.lidarmag.com/PDF/LIDARMagazine_Zanetti-ComparativePlanetology_Vol8No2.pdf.

Laser Scanning – Riveiro and Lindenbergh
© 2020 ISPRS, ISBN 978-1-138-49604-0

Chapter 4

Introduction to mobile mapping with portable systems

Erica Nocerino, Pablo Rodríguez-Gonzálvez and Fabio Menna

ABSTRACT: Portable mobile mapping systems (PMMSs) are an emerging technology, increasingly employed in different application fields. Their distinctive feature is the ability to acquire 3D information *on-the-move*, and, unlike vehicle-based MMSs, the possibility to efficiently document narrow, indoor and confined environments. The present chapter provides an overview of PMMSs, with a taxonomy of the available platforms and a description of hardware components and data processing algorithms. A review about 3D quality assessment and typical applications is also reported, with the aim of providing readers with the basic knowledge and understanding of potentialities and challenges of this promising technology.

1 INTRODUCTION

The ability of acquiring and recording precise, dense and geo-referenced 3D information is a constant request for a variety of applications, ranging from civil engineering and construction to cultural heritage, from environment to industry. The most effective approach to fulfill this need is represented by mobile mapping systems (MMSs), platforms able to acquire 3D information of large areas dynamically. An MMS is a 'compound' system consisting of three main elements (Puente *et al.*, 2013): a positioning and navigation unit for spatial referencing, mapping sensors for the acquisition of 3D/2D data (point coordinates and/or images), and a time referencing unit operating as central system for data synchronization and integration.

The term 'MMS' has been initially associated with systems which primary fit onto vehicles, such as cars, vans (Petrie, 2010), and boats (Vaaja *et al.*, 2013); however, also airborne light detection and ranging – lidar or vessel-based acoustic systems for seabed mapping – are considered MMSs.

Most of the MMSs rely on lidar sensors as 3D mapping units, but they are usually equipped also with cameras for retrieving color information of the scene. MMSs mounted onto vehicles represent the optimal choice for recording vast areas, but they are not always able to pass through narrow passages and are inapplicable to indoor scenarios. A more efficient alternative for 3D mapping in complex outdoor and indoor scenarios, characterized by rough terrain, small obstacles, and stairs, entails robotic platforms, whose application has however remained mainly limited to the research domain (Surmann *et al.*, 2003; Nüchter *et al.*, 2013; Ziparo *et al.*, 2013). Nowadays, very popular are MMSs designed and adapted to be easily carried or worn by persons walking through and, consequently, mapping the environment of interest. These systems are answering the growing need for mapping and monitoring complex environments, with harsh conditions, and in limited time, which requires agile systems to be carried around. In the last years, a plethora of systems has been developed in both the research and commercial domains, equipped with diverse sensors and implementing different acquisition strategies and software solutions.

This chapter focuses on those systems whose distinctive features can be summarized as follows: (i) they are conducted by an *operator*; (ii) they are *portable*; (iii) they are *mobile* and *dynamic*, i.e. they can be easily moved within the scene of interest in a *continuous* way (no need of *stop-and-go* measurements); (iv) the problem of navigation and mapping is automated and partly or even entirely based on the so called simultaneous localization and mapping (SLAM) algorithms; (iv) the provided data are *3D*. These systems are also defined as portable scanning solutions (LiDAR News, 2016) or indoor MMSs – IMMs (Tucci *et al.*, 2018). However, since their use is not confined to indoor scenarios, the term 'portable MMS' (PMMS) will be adopted here.

The chapter will first present a brief history of PMMSs and propose a classification according to their main features. Then, an overview of sensors and algorithmic solutions will be provided. Methods for the geometry quality assessment of data generated with PMMSs are reported in a dedicated section, followed by a review of typical applications and case studies.

1.1 *History*

Since their early development in the late 1980s, MMSs have been progressively improved to provide increasingly more precise and denser data in shorter time. Besides the progresses in optical sensors, one of the key advances in MMS is related to spatial referencing technology. While the very early applications were restricted to environments where the sensor positions were computed using ground control (Schwarz and El-Sheimy, 2004), thanks to advantages in satellite and inertial technology, today spatial referencing is commonly possible in previously unknown and undiscovered places. Also, the miniaturization and cost reduction of components have played a fundamental role in the spread of MMS, allowing for more and more flexible, portable, and low-cost systems. Ellum and El-Sheimy (2002) provide an overview of the developments of land-based MMSs, which first emerged in the academic research community, later followed by a number of systems developed and operated by private companies (Petrie, 2010).

To overcome the drawbacks of land-based, vehicle-mounted MMSs, especially high costs and complexity of use, one of the first prototypes of PMMS, in the form of a backpack system, was investigated at the beginning of the new millennium at the University of Calgary (Ellum and El-Sheimy, 2002). The backpack MMS was intended to compete in accuracy and costs with more traditional mapping methods while featuring the key advantages of data collection efficiency and flexibility typical of land-based, vehicle-mounted MMS. This system was equipped with a dual-frequency GPS receiver, digital magnetic compass (DMC) combining three MEMS accelerometers and three magnetic sensors, and a consumer-grade digital camera.

In the first years of 2010, to gather imagery inside museums and outdoor in extreme environment, Google sponsored the Trolley (Googleblog, 2011), the Trekker (The Verge, 2013) and Cartographer (TechCrunch, 2014) backpacks, which, however, did not gain much popularity.

The first high-performance mobile laser system was developed by the Finnish Geodetic Institute and was called Akhka-Backpack MLS (Kukko *et al.*, 2012). The authors adapted their original MMS to map a palsa landform in Finnish Lapland. While other commercial systems started to appear in the form of both backpacks and trolleys, it is worth mentioning the GNSS and IMU-free backpack proposed by Nüchter *et al.* (2015).

One of the most innovative design was represented by the Zebedee handheld scanning system, developed by a group of researchers at the Autonomous Systems Laboratory, CSIRO ICT Centre, Brisbane, Australia (Bosse *et al.*, 2012). The system consisted of a 2D laser scanner, also called scanning rangefinder or 2D lidar sensor, and an IMU sensor, i.e. the sensor head, mounted on a spring and either connected to a vehicle or carried by an operator.

1.2 *Vehicle-based versus portable MMSs*

Land-based, vehicle-mounted MMSs are a moving platform, where a direct georeferencing system (DGS) and remote sensors are integrated (Madeira *et al.*, 2012). Integration means that data acquired by the different sensors are synchronized, i.e. referred to a common time scale, which is provided by the DGS, and fused together to provide more reliable and accurate results. This way, the position and orientation of the platform can be determined along with the position in space of the captured scene. The basic difference between land-based vehicle-mounted MMS and PMMS is that the first relies on DGS, where GNSS receivers are key components and are mostly coupled with an inertial navigation system (INS), and other dead-reckoning devices (odometers,

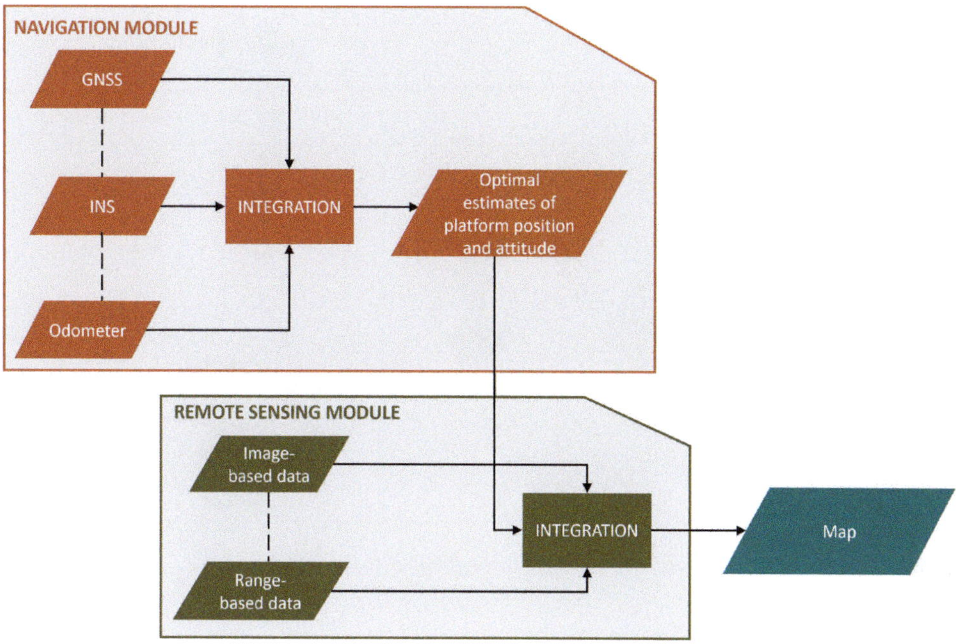

Figure 4.1. Flowchart of modules and data of a typical MMS (after Madeira *et al.*, 2012).

compasses, etc.). INS is made up of inertial measurement units (IMUs), typically containing three orthogonal rate-gyroscopes and three orthogonal accelerometers (Woodman, 2007). Remote (mapping) sensors are primarily video cameras and laser scanners. Figure 4.1 depicts a schematic flowchart of modules and data of a typical MMS.

Classic MMS provides data that are directly georeferenced, an important requisite for many geospatial applications but, while outdoor GNSS data can be effectively used with the other navigation sensors, indoor this is mostly impossible for obvious reasons. In a PMMS, the GNSS positioning system assumes a less important role, especially when it comes to mapping indoor spaces. Indeed, in these cases, system positioning, data integration, and 3D content generation are provided by the simultaneous localization and mapping (SLAM) algorithm.

1.3 *Taxonomy of PMMSs*

There are three common forms for PMMSs, available on the market and among research laboratories, namely *backpack* (Figure 4.2a), *handheld* (Figure 4.2b) and *trolley* (Figure 4.2c). Wearable *backpack* MMSs usually feature a conventional positioning and navigation unit integrating GNSS and inertial measurement unit (IMU) sensors; however, solutions with only IMU or even without positioning and navigation components exist (Nüchter *et al.*, 2015). Some systems are also equipped with cameras arranged in a spherical configuration providing panoramic imagery with reduced parallax effects. The main difference of *handheld* MMSs to backpack solutions is that the user holds the scanning device in hand. These systems are usually equipped with IMU-based navigation unit (no GNSS) and laser scanner and may also feature one or more ultra–wide-angle cameras. *Trolley* MMSs are mainly designed for indoor mapping applications, and, in this configuration, the different sensors (lidar, IMU, cameras, odometers, etc.) are fitted in a cart.

A list of the most popular portable MMSs developed in the research community and commercial sector is presented in Table 4.1. Considered in the table, as well as in this chapter, are systems which are operated by a person and rely on SLAM approaches to provide automatically the full 3D

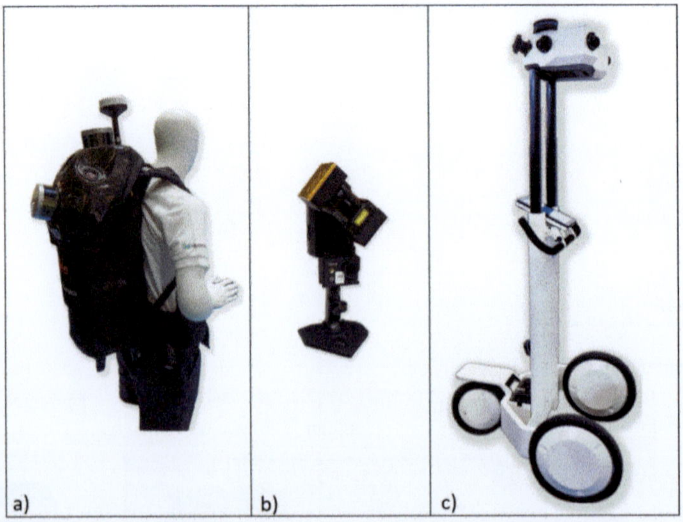

Figure 4.2. Example of backpack (a), handheld (b), and trolley (c) PMMSs.

Table 4.1 List of PMMSs

Name (website)	*Type*	*Domain*	*Navigation module*	*Remote sensing module*	*Related publications*
Akhka R2 (http://laserscanning.fi/laser-scanner-in-a-backpack/)	B	R	GNSS IMU	1 × laser scanner	Kukko *et al.* (2016)
Applanix TIMMS (www.applanix.com/products/timms-indoor-mapping.htm)	T	C	IMU Odometer	1 × laser scanner 1 × spherical camera	Babacan *et al.* (2017) Gong *et al.* (2014) Nakagawa *et al.* (2014)
FGI Slammer	T	R	GNSS IMU	2 × laser scanner	Lehtola *et al.* (2017)
HERON – 4 versions (www.gexcel.it/en/solutions/heron-mobile-mapping)	B	R/C	IMU	1 × laser scanner 1 × panoramic camera (not in all the versions)	
HERON	H	R/C	IMU	1 × laser scanner	
Kaarta STENCIL (www.kaarta.com/stencil/)	H	R/C	IMU	1 × laser scanner 1 × camera	Lehtola *et al.* (2017) Tucci *et al.* (2018)
Kaarta Contour (www.kaarta.com/contour/)	H	R/C	IMU	1 × laser scanner 1 × camera	
Leica Pegasus: Backpack (http://leica-geosystems.com/products/mobile-sensor-platforms/capture-platforms/leica-pegasus-backpack)	B	C	GNSS IMU	2 × laser scanner 5 × cameras	Lehtola *et al.* (2017) Masiero *et al.* (2018) Nocerino *et al.* (2017) Tucci *et al.* (2018)

Name (website)	*Type*	*Domain*	*Navigation module*	*Remote sensing module*	*Related publications*
NavVis 3D Mapping Trolley (www.navvis.com/m6)	T	C	IMU	3 × laser scanner 6 × cameras	Lehtola *et al.* (2017)
Robin (www.3dlasermapping.com/robin/)	B	C	GNSS IMU	1 × laser scanner 1 × camera	
Paracosm PX-80 (https://paracosm.io/)	H	C	IMU	1 × laser scanner 1 × wide-angle camera	
UVIGO Wearable Prototype	B	R	–	1 × laser scanner	Lagüela *et al.* (2018)
VIAMETRIS IMS3D (www.viametris.com/products/ims3d/)	T	C	IMU Odometer	3 × laser scanners 6 × cameras	Maboudi *et al.* (2017)
VIAMETRIS bMS3D (www.viametris.com/products/bms3d/)	B	C	GNSS IMU	2 × laser scanner 1 × spherical camera or 4 × cameras	
Würzburg Backpack	B	R	IMU	2 × laser scanner	Lehtola *et al.* (2017) Lauterbach *et al.* (2015)
Würzburg Backpack, IMU-free version	B	R	–	2 × laser scanner	Nüchter *et al.* (2015)
ZEB1 (http://geoslam.com/hardware-products/zeb1/)	H	R/C	IMU	1 × laser scanner	Chiabrando *et al.* (2017) Eyre *et al.* (2016) Farella (2016) Lehtola *et al.* (2017) Nikoohemat *et al.* (2017) Zlot and Bosse (2014) Zlot *et al.* (2014)
ZEB-REVO (http://geoslam.com/hardware-products/zeb-revo/)	H	C	IMU	1 × laser scanner 1 × wide-angle camera (optional, ZEB-CAM)	Caldwell (2017) Dewez *et al.* (2017) Eyre *et al.* (2016) Maboudi *et al.* (2017) Nocerino *et al.* (2017)

Note: B stands for backpack, T for trolley, H for handheld, R for research, C for commercial

of the environment. Conversely, robotic systems (e.g. Aalto VILMA, Lehtola *et al.*, 2016), those specifically designed to generate 2D maps (e.g. ViAmetris iMS2D), employing data integration solutions diverse from SLAM (e.g. the Vexcel Imaging UltraCam Panther) or requiring a '*stop-and-go*' acquisition approach (e.g. Matterport) are not considered. Thanks to the modularity of their design, some of the systems can be mounted on different platforms, such as drones, carts, etc.

2 NAVIGATION AND REMOTE SENSING MODULES IN PMMSS

Although systems purely based on photogrammetry are being developed by different manufacturers or research groups (Trimble v10; Imajing imajbox; Campos *et al.*, 2018; Masiero *et al.*, 2018; Pittore *et al.*, 2015), a typical portable MMS relies on laser scanning as a core technique for capturing 3D data. A minimum system configuration typically entails at least a remote sensor (mainly a laser scanner), one IMU device and a microcomputer for controlling the hardware, its synchronization, and raw sensor data storage. Indirect time of flight (TOF; Schulz, 2008) laser scanners are the most used due to the fast acquisition rate; IMU device of the solid-state microelectromechanical type (micro-electro-mechanical system – MEMS) are used due to the good trade-off between cost, size, and performances even if systems with more accurate INS (fiber optical gyroscope – FOG) are available (e.g. Akhka R2).

Systems specifically born for geospatial applications provide results directly in a cartographic coordinate system. For this reason, some systems feature a surveying-grade GNSS receiver and antenna (Leica Pegasus: Backpack, Akhka R2) that are typically part of an integrated GNSS + INS navigation module.

In the most basic configuration, the geometry of acquisition of a PMMS is similar to the one used by airborne laser scanning systems, consisting of a rotating head hosting a rangefinder (or similarly a 45-degree-inclined rotating mirror) whose laser beam propagates orthogonally to the path to be scanned. The combination of laser rotation and PMMS motion allows the laser beam to sense the area of interest. Made as modular systems, PMMSs from different manufacturers may be assembled using the very same modules used by other competitors, thus differentiating their own system for a peculiar geometric configuration of the sensors or for the integration of their hardware and software. The laser scanners used on MMSs can be of two types, (i) specifically manufactured for mobile mapping and robot applications (e.g. Velodyne, Hokuyo, RIEGL, SICK) or (ii) adapted from terrestrial laser scanning systems – TLSs (e.g. Faro Focus, RIEGL VZ-400). In the first case, these modules are much more compact and lightweight but typically not as accurate as the TLS. In the latter case, the rotation around the primary axis (the vertical one when placed upright on a tripod) is typically disabled and the instrument transformed into a 2D single-axis scanner (e.g. Akhka R2, Applanix TIMMS).

The current trend for developers is to minimize the overall weight and cost of the PMMSs by using single-axis rotating laser scanners (e.g. Velodyne PUCKTM VLP-16, Hokuyo UTM-30LX, RIEGL VUX-1HA, SICK MRS1000). A particular design has been introduced by Velodyne and SICK by arranging multiple laser diodes (*channels*) inside the rotating head in a fan-shaped configuration, which allows coverage of a larger field of view (up to 40 degrees) with respect to single-diode 2D laser scanners. Imaging sensors on MMSs consist of digital video cameras typically added in a panoramic configuration, sometimes through a spherical arrangement of the sensors. This configuration is mainly intended for colorizing the point cloud obtained with the laser scanner or providing a virtual immersive visualization of the acquired data. In few cases they are also used to perform visual SLAM or photogrammetric measurements.

3 THE DATA PROCESSING MODULE

As mentioned in section 2, available PMMSs are compound systems featuring the very same sensors and thus differentiating only for their integration and raw data processing. In this regard, software becomes very important in characterizing a particular MMS, and thus a distinguishing part. MMSs having reached a broader range of final users not necessarily familiar with surveying subjects, software applications are more and more required to be simple, with high automation and limited user interaction to minimize the user efforts in data production.

The processing and integration of data acquired by the different sensors constituting the navigation and remote sensing modules of a PMMS requires the implementation of a SLAM strategy. SLAM rises in the robotic community for autonomous navigation of robotic platforms (Cadena *et al.*, 2016), and has been adopted in several application fields, from indoor to outdoor, from underwater to airborne systems (Durrant-Whyte and Bailey, 2006). It is commonly defined as a 'chicken-or-egg problem', where a moving robot needs to simultaneously build a map of the environment and locate itself within the map, using the same generated map. The position estimate is possible thanks to the identification of landmarks, i.e. recognizable and re-observable features within the environment, which should be: (i) distinguishable from each other, (ii) in high number and redundancy in the environment, and (iii) stationary (Riisgaard and Blas, 2003).

Both the sensors' path and position of all the landmarks are estimated at the same time without any prior knowledge. The navigation module provides clues for estimating the system position and attitude (the state vector), while the remote sensing module generates the data to build the map from where the landmarks are extracted. The problem of solving the concurrent localization of the system in the unknown environment and the creation of the map of the same environment is, then, solved in a recursive and sequential two-step approach. A prediction of the time-updated system location and map is produced according to the navigation module input and the past trajectory and dynamics. The location and map predictions are confronted to and corrected with the updated measurements (observations) derived from the navigation and remote sensing modules. SLAM is formulated as a probabilistic problem, where both the vehicle position and attitude and landmark positions cannot be calculated following a deterministic approach but are estimated by defining probability distribution functions. The probability distributions are functions of sensors' characteristics, namely estimated noise and accuracy, which are then updated in the computational chain.

The SLAM problem can be formulated in two ways (Bresson *et al.*, 2017), which are most suitable for postprocessed or real-time computational approaches. In the full or off-line SLAM, the whole trajectory and the map is estimated providing all the sensor data and measurements. The complexity of the full SLAM problem grows with respect to the number of variables considered and is, therefore, not best suited for real-time computation. On the other hand, online SLAM updates pose and map with the last most recent estimates from the sensors. The incremental nature of the online SLAM problem makes it suitable for real-time applications. The computational solutions or estimation techniques can be separated into two main categories (Bresson *et al.*, 2017): (i) filter-based approaches, such as extended Kalman filter (EKF-SLAM), unscented KF (UKF), information filter, particle filter (e.g. Rao-Blackwellized particle filters or FastSLAM): (ii) optimization-based methods (bundle adjustment, graph-based SLAM). The filter-based approaches correspond to iterative processes and, as such, are mainly employed for online SLAM. On the contrary, optimization-based methods are usually applied to solve the full SLAM problem (Bresson *et al.*, 2017). The main steps involved in SLAM are graphically shown in Figure 4.3. Each of them can be implemented in a number of different algorithms (Riisgaard and Blas, 2003).

Crucial for SLAM approaches is the identification of the so-called *loop closure*, i.e. the detection of a previously mapped place and consequent relocalization of the system within the already-measured environment, also known as *kidnapped robot problem* (Bresson *et al.*, 2017). The *loop closure* allows one to reduce the drift accumulated in the SLAM solution over time (Newman and Ho, 2005) and is usually solved by a dedicated algorithm that runs independently from the estimation process. Unstructured environments, especially in GNSS-denied situations, such as urban canyons, under foliage, or underwater, represent challenging scenarios for SLAM based algorithms.

Three main approaches for data association and loop closure detection with a monocular SLAM system have been proposed in literature (Williams *et al.*, 2009): (i) map-to-map, (ii) image-to-image, and (iii) image-to-map. SLAM is applicable for both 2D and 3D motions, which corresponds to a further classification based on the number of degrees of freedom (DoF) of the sensor pose. A 3D pose estimate is defined by the position in the horizontal plane and a rotation and constraints

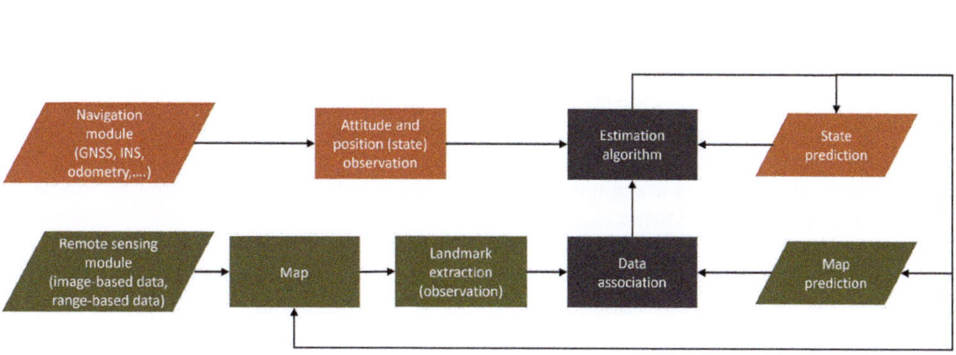

Figure 4.3. SLAM main steps.

of the system motion in a 2D map. On the contrary, a 6D pose considers all the six DoFs of a rigid body in a volumetric 3D space (Borrmann *et al.*, 2008). In a 6DSLAM, the (3D) maps measured at each step are registered using standard methods, such as the iterative closest point – ICP (Besl and McKay, 1992), least squares 3D surface (LS3D) matching method (Gruen and Akca, 2005), etc. (Bedkowski *et al.*, 2017).

When cameras are the only sensors used, the SLAM is based on visual information only and is therefore called visual SLAM (V-SLAM or vSLAM; Fuentes-Pacheco *et al.*, 2015; Taketomi *et al.*, 2017). According to Scaramuzza and Fraundorfer (2011) while SLAM and visual SLAM aim to obtain a global and consistent estimate of trajectory and map, visual odometry (VO) is mainly devoted to recover the path incrementally, potentially optimizing only over the last *n* poses (also called windowed bundle adjustment). VO can be implemented as a step for a complete SLAM algorithm, where also loop closure and possibly a global optimization step are performed.

As a final remark, it is worth mentioning that several SLAM frameworks exist, among which the best known in the scientific, research and education communities are the robotic operating system (ROS) and 3D Toolkit (3DTK). Moreover, to promote the accessibility and sharing of SLAM algorithms, the web-based project *OpenSLAM.org* provides interested researchers with a repository to publish and promote their work. Currently, more than 30 different implementations are available there.

4 3D DATA QUALITY ASSESSMENT

Following the proliferation of PMMSs, a number of technical and scientific works have attempted to ascertain the metric capabilities of PMMSs. Analyses are usually performed to assess the quality of the system as *a whole* and are therefore conducted on the main output provided by systems under evaluation, i.e. point clouds. In this way, the evaluation provides an indirect indication of sensors and system calibration and a direct assessment of data integration via the SLAM approach. Different procedures are generally adopted, which often entails the use of reference data against which the systems are tested.

According to the VIM (BIPM *et al.*, 2012), *verification* is the provision of objective evidence that a given item fulfils specified requirements; whereas *validation* refers to where the specified requirements are adequate for an intended use. Measurement *precision* is the closeness of agreement between measured quantity values obtained by replicate measurements on the same or similar objects under specified conditions. This definition is closely related to the repeatability. On the contrary, measurement accuracy indicates the closeness of agreement between a measured quantity value and a true quantity value of object being measured, called measurand. For practical purposes, true quantity, or ground truth, is provided by instruments and techniques with a higher accuracy than the tested one, also referred to as independent checks (Schofield and Breach, 2007). This ratio should be three or more times the a priori accuracy of the system under test; under these conditions, in the error propagation, the error from the ground truth can be dismissed (Rodríguez-Gonzálvez *et al.*, 2015).

Quality assessment goes through a multistep procedure, with different levels of completeness and complexity, aiming at verifying that the system follows the manufacturer specifications and is adequate for the application of interest. First, an investigation of random noise components should be conducted, which provides an indication about the relative accuracy of the system; then more extensive methods of accuracy evaluation are used to assess any systematic components not properly modelled and if outliers are properly tackled. Table 4.2 reports the most common approaches for quality assessment of PMMSs.

A simple method relies on fitting procedure of geometric primitives. Patches of planar surfaces are extracted from the original point cloud, either manually or automatically with several approaches (ordinary or robust least squares, PCA, etc.). Similarly, spherical targets or structural elements in the measured scene, such as columns, can be fitted through least square methods; their measured geometric parameters (diameter) can be compared to the nominal values. In both cases, the root mean square (RMS) resulting from the fitting computation can be assumed as a measure

Table 4.2 3D quality assessment methods for PMMSs

Method	*Sub-method*	*Evaluation*			*Reference*
		Precision	*Accuracy*	*Details*	
Fitting of geometric primitives	Single passage	X		Noise (mainly of the remote sensing module)	Nocerino *et al.* (2017)
	Multiple passages	X		Noise Data integration and processing	
Trajectory error			X	Data integration and processing	Lauterbach *et al.* (2015)
RMSE on checkpoints	Center of planar targets		X	Data integration and processing	Farella (2016)
	Center of spheres				Kukko *et al.* (2012)
LME	Reference measurements from laser distance meter		X	Data integration and processing	Farella (2016)
	Reference measurements from TLS				Nocerino *et al.* (2017)
Distances	From reference planes		X	Data integration and processing	Maboudi *et al.* (2017)
	Point cloud to point cloud		X	Data integration and processing	Lehtola *et al.* (2017) Masiero *et al.* (2018) Nocerino *et al.* (2017) Thomson *et al.* (2013) Tucci *et al.* (2018)

of system noise or precision, when the patches are extracted from a 'one-way' or single-passage acquisition approach. If the plane fitting procedure is performed on data acquired with more than one passage, then the RMS also provides a global indication about the system calibration and quality of data integration and processing approach (SLAM).

While fitting procedures are suitable to estimate random components of error, hence providing a quantitative measure of precision, to evaluate the accuracy of the system, the use of a comparative approach entailing reference data is necessary. Two important issues then arise, i.e. the definition of a proper ground truth and of a common coordinate reference frame. The ground truth definition implies the adoption of a sensor/instrument of a higher accuracy.

If the evaluation approach aims to specifically ascertain the positioning accuracy of the SLAM algorithm, a high-precision positioning and tracking system can be used to provide reference pose information in indoor environments (Lauterbach *et al.*, 2015).

RMS error (RMSE) on checkpoints and length measurement error (LME) have been also adopted as quantitative measures for accuracy assessment. Planar targets (Farella, 2016) and spheres (Kukko *et al.*, 2012) can be fitted by least squares to extract their centers, which are compared with coordinates measured with classic surveying (Farella, 2016) or GNSS (Kukko *et al.*, 2012). Distance measurements are extracted from PMMS point clouds and compared against reference values provided by a laser distance meter (Farella, 2016) or terrestrial laser scanner – TLS (Nocerino *et al.*, 2017) – to compute LMEs.

When the accuracy assessment attempts to fully exploit the ability of PMMSs to map vast areas, the most suited method is based on comparisons with point clouds obtained from TLSs (Lehtola *et al.*, 2017; Maboudi *et al.*, 2017; Masiero *et al.*, 2018; Tucci *et al.*, 2018); however, for outdoor verification, also land-based MMSs (Nocerino *et al.*, 2017) and unmanned aerial vehicles – UAVs (Rönnholm *et al.*, 2016) – have been employed. Often, a single scan from the reference TLS is not enough to cover the area mapped with the PMMS, and the point clouds from the different scanning positions are to be aligned together. This step is of crucial importance, since the uncertainty in the point clouds registration may negatively affect the quality of the reference data. Usually, the point clouds from the PMMS and the reference technique are not in the same coordinate reference system, and consequently, a registration procedure is required. This is commonly performed employing the well-known ICP algorithm. The RMS of the alignment is reported as an indication of the geometric quality of the PMMS (Thomson *et al.*, 2013). Once the data are aligned, several comparisons can be performed: signed distances of points to the reference planes extracted from the TLS data (Maboudi *et al.*, 2017); cloud-to-cloud absolute distances (Maboudi *et al.*, 2017; Masiero *et al.*, 2018); cloud-to-cloud signed distances (Nocerino *et al.*, 2017; Tucci *et al.*, 2018). Mean and standard deviation of the distances are the often-reported metrics (Maboudi *et al.*, 2017; Tucci *et al.*, 2018). However, when it is verified that the errors in the data do not fit a Gaussian distribution, a robust statistical analysis should be performed and statistical values such as the median and median absolute deviation (MAD) would provide a more rigorous hint about the accuracy of the system under investigation (Höhle and Höhle, 2009; Nocerino *et al.*, 2017; Rodríguez-Gonzálvez *et al.*, 2014a). Lehtola *et al.* (2017) propose a full point cloud comparative approach, developing a metric that automatically works over all the length scales.

A further step in the quality assessment of point clouds derived from PMMSs is the evaluation of the obtainable level of detail (LOD) or resolution, which directly influences the suitability of the system for modelling applications. For PMMSs, the point cloud resolution is hardly defined a priori, being a direct function of the acquisition distance and dynamics.

5 OVERVIEW OF APPLICATIONS

PMMSs were initially developed as low-cost and easy-to-use alternatives to land-based, vehicle-mounted MMSs and, thanks to these characteristics, have been employed in an increasing number of applications. The availability of so many PMMSs can make the selection of the proper system not trivial, a choice that should be dictated by the requirements of the applications in terms of accuracy, resolution, time and budget.

In this section, the key applications for PMMSs in engineering fields related to outdoor and indoor mapping are reported. Table 4.3 provides an overview and qualitative rating of PMMSs. Each system is scored with a rank from one to three stars, according to the applicability of the system to the specific case study. TLSs and land-based, vehicle-mounted MMSs are also listed for comparative reasons, being very common systems employed in engineering and also considered a benchmark to assess PMMSs' accuracy (section 4). An in-depth analysis excesses the scope of the present chapter; therefore, the following is intended as a general qualitative overview that may ease the final user choice.

Outdoor and urban mapping are dominated by land-based, vehicle mounted MMSs, which are largely employed to provide updated information about urban infrastructure and its maintenance. In areas with limited GNSS coverage, such as urban canyons or corridors and areas with restricted traffic circulation (city centers), portable MMSs can represent a valid alternative. An example of the use of a handheld system is shown in Nocerino *et al.* (2017). GNSS-free, SLAM-based systems can generate noticeable erroneous maps in vast and open areas with moving elements. So operator interaction is required to identify and correct the system trajectory and map. Zlot *et al.* (2014) employs a portable MMS to map dozens of buildings of various sizes spread across a rural area of approximately 400 × 250 m.

Indoor mapping and modelling are probably the main application scenario for PMMSs. Here, PMMSs represent the best compromise between resolution, accuracy, time, and costs. On the contrary, TLSs have a precision one order of magnitude higher than portable MMSs, but they are less

Table 4.3 Suitability of mapping techniques according to a qualitative scale: bad (*); acceptable (**); best suited (***)

APPLICATIONS		*TLSs*	*Land-based, vehicle-mounted MMSs*	*PMMSs*		
				Backpack with GNSS	*Handheld & Backpack w/o GNSS*	*Trolley*
3D city modelling	Open space	*	***	**	*	*
	Urban canyons	*	**	**	**	*
Building information model (BIM)		**	*	**	***	**
As-built mapping (construction sites)		**	*	***	***	*
Industrial facilities & factory mapping		**	*	***	***	*
Indoor navigation		**	*	*	***	**
Underground mapping		*	*	*	***	**
Post disaster assessment		*	*	**	***	*
Structural analysis & stability		***	*	**	**	**
Forest mapping		**	*	**	***	*
Forensic documentation		***	*	***	***	**

effective to map vast areas with narrow spaces and occlusions. 3D mapping and modelling of interior buildings is closely related to building information models (BIM) applications. The output models should comply with standard formats for 3D city models and BIM (Biljecki *et al.*, 2015). A comparison of eight indoor acquisition instruments to generate 3D point clouds is reported in Lehtola *et al.* (2017), which employed two handheld (Zebdebee; Kaarta Stencil), two backpack (Leica Pegasus: Backpack, Würzburg backpack), two trolley (NasVis, FGI slammer). Authors report that the point clouds with most precision are provided by trolley systems, even if these systems are less effective when multiple floors and steps are to be measured. Sepasgozar *et al.* (2014) suggests the use of handheld PMMSs for as-built modelling and BIM in contrast to TLS and other solutions that may require tripods, vehicles, and skilled operators. Babacan *et al.* (2017) applies convolutional neural networks to a 3D point cloud from a trolley system (Applanix TIMMS) to obtain a semantic segmentation of an indoor scenario with the aim of classifying constructive elements.

Industrial facilities and factories are characterized by elements such as tubes, cylindrical tanks, and prismatic elements, which can usually be reconducted to parametric geometries. This property can be successfully exploited to generate CAD-based models, which constitute the basis for BIM (Rodríguez-Gonzálvez *et al.*, 2014b) and virtual and augmented reality (VR/AR) applications (Bi *et al.*, 2010). So far, PMMSs application in mapping industrial facilities, like nuclear or electrical power plants or oil and gas factories, etc., is very limited. For remote inspection with nondestructive testing (NDT) techniques, as stated in Caldwell (2017), it is hard to identify an approach that can be optimal for all needs be encountered. However, the main advantage of PMMSs is the possibility to cope with occlusions and work in GNSS-denied environments.

Indoor navigation is closely related to the semantic labelling of indoor spaces. Indoor point clouds are the base for 3D models for navigation purposes, (e.g. maintenance or emergency/disaster management, multistory pathfinding). Zlatanova *et al.* (2013) reported the conceptual foundation for the description of interiors of buildings subdividing the indoor space into functional spaces to allow advanced navigation, identifying rooms and corridors and including semantical information. Nikoohemat *et al.* (2017) studied the issue of PMMS point clouds along with the trajectory path. The ZEB1 handheld system was employed in a three-story building for the detection of walls, floor, and ceiling. As result, a 3D model of walls and space partition is generated.

PMMSs, especially in the handheld form, have been successfully used for 3D mapping and modelling of underground passages, caves, and mines (Eyre *et al.*, 2016; Farella, 2016; Zlot and Bosse, 2014).

Structural health monitoring of infrastructures can provide important information to detect, in an early stage, anomalies that can threaten them. Due to accuracy and resolution limitations, the applicability of MMSs may be limited only to those cases where anomalies in the order of some centimeters on the structures can be observed, like for example in case of disaster analysis and postearthquake assessment, management and monitoring (Chiabrando *et al.*, 2017). In the case of bridges, and according to Ahlborn *et al.* (2010) the structural monitoring can be classified into in-situ, on-site, and remote monitoring. The first one is related to contact measurement methods as gauges, accelerometers, attached to the structure. PMMS are catalogued in the on-site monitoring, while remote monitoring is mainly referred to airborne sensors. The main drawback of PMMSs is the acquisition range: only short bridges can be documented with this approach. Another example of structural analysis based on PMMSs is reported in Dewez *et al.* (2017), where cavity-collapse risks of underground cavities are mapped by means of a hand-held GeoSLAM Zeb-Revo. It was tested in a 11 ha mapping of an underground quarry, being able to produce hazard maps at 1:5,000 scale. Authors reported that instrumental drift and distance inaccuracies arisen loops longer than 100 m.

6 FINAL REMARKS

The present chapter provided an overview on PMMSs, with the aim of highlighting the crucial aspects for their correct and efficient exploitation. PMMSs are compound systems featuring state-of-the-art hardware and software solutions, more and more utilized also by nonexperts thanks to their ease of use and versatility in many engineering and architectural applications.

Most systems share similar sensors (e.g. lidar units) but differ for the implemented data integration and processing approaches. The complex technology together with often undisclosed algorithms make the evaluation and validation of obtained 3D data not a simple task, especially for users who might not be familiar with surveying and mapping technology. The way the information from the navigation module is integrated, i.e. trusted more with respect to for example an indirect estimation of position and orientation from mapping sensors, is what characterizes the different PMMSs. Systems specifically designed for geospatial applications tend to rely more on navigation modules to determine the trajectory, with the important benefit that data are directly georeferenced. Systems relying only on SLAM cannot provide directly geo-referenced information but seem to be more reliable, if not the only feasible solution, in GNSS-denied scenarios. A final user should therefore consider that the delivered 3D data are always the result of two main errors sources: (i) errors related to the trajectory estimation either from SLAM or navigation module sensors or their integration and (ii) errors coming from the mapping sensors. These errors can be mitigated with acquisition protocols: (i) entailing *loop closure* or *round-trip* trajectories, (ii) decreasing the walking dynamics, and (iii) in case of scenes with repetitive geometries and smooth surfaces, introducing additional elements of dimension befitting the system resolution.

REFERENCES

3D Toolkit, 3DTK. Available from: http://slam6d.sourceforge.net/ [accessed 30th April 2018].

Ahlborn, T.M., Shuchman, R., Sutter, L.L., Brooks, C.N., Harris, D.K., Burns, J.W., . . . Oats, R.C. (2010) The state-of-the-practice of modern structural health monitoring for bridges: a comprehensive review. Available from: https://mtri.org/bridgecondition/doc/State-of-PracticeSHMforBridges(July2010).pdf

Babacan, K., Chen, L. & Sohn, G. (2017) Semantic segmentation of indoor point clouds using convolutional neural network. *ISPRS Annals of Photogrammetry, Remote Sensing and Spatial Information Sciences*, 101–108. Available from: https://doi.org/10.5194/isprs-annals-IV-4-W4-101-2017

Bedkowski, J., Röhling, T., Hoeller, F., Shulz, D. & Schneider, F.E. (2017) Benchmark of 6D SLAM (6D Simultaneous Localisation and Mapping) algorithms with robotic mobile mapping systems. *Foundations of Computing and Decision Sciences*, 42(3), 275–295.
Besl, P.J. & McKay, N.D. (1992) Method for registration of 3-D shapes. *Sensor Fusion IV: Control Paradigms and Data Structures*, Boston, MA. Society of Photo-Optical Instrumentation Engineers (SPIE Proceedings. Vol. 1611), Bellingham, WA, USA. pp.586–607.
Bi, Z.M. & Wang, L. (2010) Advances in 3D data acquisition and processing for industrial applications. *Robotics and Computer-Integrated Manufacturing*, 26(5), 403–413.
Biljecki, F., Stoter, J., Ledoux, H., Zlatanova, S. & Çöltekin, A. (2015) Applications of 3D city models: State of the art review. *ISPRS International Journal of Geo-Information*, 4(4), 2842–2889.
BIPM, I., IFCC, I., IUPAC, I. & ISO, O. (2012) *The International Vocabulary of Metrology – Basic and General Concepts and Associated Terms (VIM)*. 3rd ed. JCGM 200: 2012. JCGM (Joint Committee for Guides in Metrology). Available from: www.bipm.org/utils/common/documents/jcgm/JCGM_200_2012.pdf.
Borrmann, D., Elseberg, J., Lingemann, K., Nüchter, A. & Hertzberg, J. (2008) Globally consistent 3D mapping with scan matching. *Robotics and Autonomous Systems*, 56(2), 130–142.
Bosse, M., Zlot, R. & Flick, P. (2012) Zebedee: Design of a spring-mounted 3-d range sensor with application to mobile mapping. *IEEE Transactions on Robotics*, 28(5), 1104–1119.
Bresson, G., Alsayed, Z., Yu, L. & Glaser, S. (2017) Simultaneous localization and mapping: A survey of current trends in autonomous driving. *IEEE Transactions on Intelligent Vehicles*, 2(3), 194–220.
Cadena, C., Carlone, L., Carrillo, H., Latif, Y., Scaramuzza, D., Neira, J., . . . Leonard, J.J. (2016) Past, present, and future of simultaneous localization and mapping: Toward the robust-perception age. *IEEE Transactions on Robotics*, 32(6), 1309–1332.
Caldwell, R. (2017) Hull inspection techniques and strategy-remote inspection developments. *SPE Offshore Europe Conference & Exhibition*. Society of Petroleum Engineers.
Campos, M.B., Tommaselli, A.M.G., Honkavaara, E., Prol, F.D.S., Kaartinen, H., El Issaoui, A. & Hakala, T. (2018) A backpack-mounted omnidirectional camera with off-the-shelf navigation sensors for mobile terrestrial mapping: Development and forest application. *Sensors*, 18(3), 827.
Chiabrando, F., Sammartano, G. & Spanò, A. (2017) A comparison among different optimization levels in 3D multi-sensor models. A test case in emergency context: 2016 Italian earthquake. *The International Archives of Photogrammetry, Remote Sensing and Spatial Information Sciences*, 42, 155.
Dewez, T.J., Yart, S., Thuon, Y., Pannet, P. & Plat, E. (2017) Towards cavity-collapse hazard maps with Zeb-Revo handheld laser scanner point clouds. *The Photogrammetric Record*, 32(160), 354–376.
Durrant-Whyte, H. & Bailey, T. (2006) Simultaneous localization and mapping: Part I. *IEEE Robotics & Automation Magazine*, 13(2), 99–110.
Ellum, C.M. & El-Sheimy, N. (2002) Land-based integrated systems for mapping and GIS applications. *Survey Review*, 36(283), 323–339.
Eyre, M., Wetherelt, A. & Coggan, J. (2016) Evaluation of automated underground mapping solutions for mining and civil engineering applications. *Journal of Applied Remote Sensing*, 10(4), 046011.
Farella, E.M. (2016) 3D Mapping of underground environments with a hand-held laser scanner. *Proceedings of the SIFET Annual Conference*, Lecce, Italy, 8–10 June 2016.
Faro Focus. Available from: www.faro.com/en-gb/products/construction-bim-cim/faro-focus/ [accessed 30th April 2018].
Fuentes-Pacheco, J., Ruiz-Ascencio, J. & Rendón-Mancha, J.M. (2015) Visual simultaneous localization and mapping: A survey. *Artificial Intelligence Review*, 43(1), 55–81.
Gong, Y., Mao, W., Bi, J., Ji, W. & He, Z. (2014) Three-dimensional reconstruction of indoor whole elements based on mobile LiDAR point cloud data. *Lidar Remote Sensing for Environmental Monitoring XIV*. Vol. 9262. SPIE Asia-Pacific Remote Sensing, 2014, Beijing, China. p.92620O.
Googleblog (2011) *Street View Takes You Inside Museums*. Available from: https://maps.googleblog.com/2011/02/street-view-takes-you-inside-museums.html [accessed 30th April 2018].
Gruen, A. & Akca, D. (2005) Least squares 3D surface and curve matching. *ISPRS Journal of Photogrammetry and Remote Sensing*, 59(3), 151–174.
Höhle, J. & Höhle, M. (2009) Accuracy assessment of digital elevation models by means of robust statistical methods. *ISPRS Journal of Photogrammetry and Remote Sensing*, 64(4), 398–406.
Hokuyo. Available from: www.hokuyo-aut.jp/search/index.php?cate01 = 1 [accessed 30th April 2018].
Hokuyo UTM-30LX. Available from: www.hokuyo-aut.jp/search/single.php?serial = 169 [accessed 30th April 2018].

Imajing Imajbox. Available from: http://imajing.eu/mobile-mapping-solutions/mobile-mapping-system-imajbox/ [accessed 30th April 2018].

Kukko, A., Kaartinen, H., Hyyppä, J. & Chen, Y. (2012) Multiplatform mobile laser scanning: Usability and performance. *Sensors*, 12(9), 11712–11733.

Kukko, A., Kaartinen, H. & Virtanen, J.P. (2016) Laser scanning in a backpack: The evolution towards all-terrain personal laser scanners. *GIM International*. Available from: www.gim-international.com/content/article/laser-scanner-in-a-backpack.

Lagüela, S., Dorado, I., Gesto, M., Arias, P., González-Aguilera, D. & Lorenzo, H. (2018) Behavior analysis of novel wearable Indoor mapping system based on 3D-SLAM. *Sensors*, 18(3), 766.

Lauterbach, H.A., Borrmann, D., Heß, R., Eck, D., Schilling, K. & Nüchter, A. (2015) Evaluation of a back-pack-mounted 3D mobile scanning system. *Remote Sensing*, 7(10), 13753–13781.

Lehtola, V.V., Kaartinen, H., Nüchter, A., Kaijaluoto, R., Kukko, A., Litkey, P., . . . Kurkela, M. (2017) Comparison of the selected state-of-the-art 3D indoor scanning and point cloud generation methods. *Remote Sensing*, 9(8), 796.

Lehtola, V.V., Virtanen, J.P., Vaaja, M.T., Hyyppä, H. & Nüchter, A. (2016) Localization of a mobile laser scanner via dimensional reduction. *ISPRS Journal of Photogrammetry and Remote Sensing*, 121, 48–59.

LiDAR News. (2016) A portable scanning solution. Available from: http://lidarnews.com/articles/a-portable-scanning-solution/ [accessed 30th April 2018].

Maboudi, M., Bánhidi, D. & Gerke, M. (2017) Evaluation of indoor mobile mapping systems. *Workshop 3D-NorOst 2017* (20th Application-Oriented Workshop on Measuring, Modeling, Processing and Analysis of 3D-Data), Berlin, Germany. 7 and 8 December 2017.

Madeira, S., Gonçalves, J.A. & Bastos, L. (2012) Sensor integration in a low cost land mobile mapping system. *Sensors*, 12(3), 2935–2953.

Masiero, A., Fissore, F., Guarnieri, A., Pirotti, F., Visintini, D. & Vettore, A. (2018) Performance evaluation of two indoor mapping systems: Low-cost UWB-aided photogrammetry and backpack laser scanning. *Applied Sciences*, 8(3), 416.

Matterport. Available from: https://matterport.com/ [accessed 30th April 2018].

Nakagawa, M., Kataoka, K., Yamamoto, T., Shiozaki, M. & Ohhashi, T. (2014) Panoramic rendering-based polygon extraction from indoor mobile LiDAR data. *The International Archives of Photogrammetry, Remote Sensing and Spatial Information Sciences*, 40(4), 181.

Newman, P. & Ho, K. (April 2005) SLAM-loop closing with visually salient features. *Robotics and Automation, 2005: ICRA 2005. Proceedings of the 2005 IEEE International Conference on*. IEEE. pp.635–642.

Nikoohemat, S., Peter, M., Elberink, S.O. & Vosselman, G. (2017) Exploiting indoor mobile laser scanner trajectories for semantic interpretation of point clouds. *ISPRS Annals of the Photogrammetry, Remote Sensing and Spatial Information Sciences*, 4, 355.

Nocerino, E., Menna, F., Remondino, F., Toschi, I. & Rodríguez-Gonzálvez, P. (2017) Investigation of indoor and outdoor performance of two portable mobile mapping systems. *Videometrics, Range Imaging, and Applications XIV*. Vol. 10332. SPIE Optical Metrology, 2017, Munich, Germany. p.103320I.

Nüchter, A., Borrmann, D., Koch, P., Kühn, M. & May, S. (2015) A man-portable, IMU-free mobile mapping system. *ISPRS Annals of Photogrammetry, Remote Sensing & Spatial Information Sciences*, 2.

Nüchter, A., Elseberg, J. & Borrmann, D. (2013) Irma3D: An intelligent robot for mapping applications. *IFAC Proceedings Volumes*, 46(29), 119–124.

OpenSLAM.org. Available from: http://openslam.org/ [accessed 30th April 2018].

Petrie, G. (2010) An introduction to the technology: Mobile mapping systems. *Geoinformatics*, 13(1), 32.

Pittore, M., Wieland, M., Errize, M., Kariptas, C. & Güngör, I. (2015) Improving post-earthquake insurance claim management: A novel approach to prioritize geospatial data collection. *ISPRS International Journal of Geo-Information*, 4(4), 2401–2427.

Point Cloud Library, PCL. Available from: http://pointclouds.org/ [accessed 30th April 2018].

Puente, I., González-Jorge, H., Martínez-Sánchez, J. & Arias, P. (2013) Review of mobile mapping and surveying technologies. *Measurement*, 46(7), 2127–2145.

RIEGL. Available from: http://riegl.com/nc/products/mobile-scanning/produktdetail/product/scanner/50/ [accessed 30th April 2018].

RIEGL VZ-400. Available from: www.riegl.co.at/nc/products/mobile-scanning/produktdetail/product/scanner/50/ [accessed 30th April 2018].

Riisgaard, S. & Blas, M.R. (2003) SLAM for dummies. *A Tutorial Approach to Simultaneous Localization and Mapping*, 22(1–127), 126.

Robot Operating System, ROS. Available from: www.ros.org/ [accessed 30th April 2018].

Rodríguez-Gonzálvez, P., Garcia-Gago, J., Gomez-Lahoz, J. & González-Aguilera, D. (2014a). Confronting passive and active sensors with non-Gaussian statistics. *Sensors*, 14(8), 13759–13777.

Rodríguez-Gonzálvez, P., González-Aguilera, D., Hernández-López, D. & González-Jorge, H. (2015) Accuracy assessment of airborne laser scanner dataset by means of parametric and non-parametric statistical methods. *IET Science, Measurement & Technology*, 9(4), 505–513.

Rodríguez-Gonzálvez, P., Gonzalez-Aguilera, D., Lopez-Jimenez, G. & Picon-Cabrera, I. (2014b). Image-based modeling of built environment from an unmanned aerial system. *Automation in Construction*, 48, 44–52.

Rönnholm, P., Liang, X., Kukko, A., Jaakkola, A. & Hyyppä, J. (2016). Quality analysis and correction of mobile backpack laser scanning data. *ISPRS Annals of the Photogrammetry, Remote Sensing and Spatial Information Sciences*, 3, 41.

Scaramuzza, D. & Fraundorfer, F. (2011) Visual odometry. part i: The rst 30 years and fundamentals. *IEEE Robotics & Automation Magazine*, 18, 8092.

Schofield, W. & Breach, M. (2007) *Engineering Surveying*. CRC Press, New York, USA and Canada.

Schulz, T. (2008) *Calibration of a Terrestrial Laser Scanner for Engineering Geodesy*. Vol. 96. ETH Zurich, Zürich. Available from: https://doi.org/10.3929/ethz-a-005368245

Schwarz, K.P. & El-Sheimy, N. (2004) Mobile mapping systems–state of the art and future trends. *International Archives of Photogrammetry, Remote Sensing and Spatial Information Sciences*, 35(Part B), 10.

Sepasgozar, S.M., Lim, S. & Shirowzhan, S. (2014) Implementation of rapid as-built building information modeling using mobile LiDAR. *Construction Research Congress 2014, May 19–21, 2014, Atlanta, Georgia*. American Society of Civil Engineers, pp.209–218. Available from: https://doi.org/10.1061/9780784413517

SICK. Available from: www.sick.com/ag/en/detection-and-ranging-solutions/2d-lidar-sensors/c/g91900 [accessed 30th April 2018].

SICK MRS1000. Available from: www.sick.com/ag/en/detection-and-ranging-solutions/3d-lidar-sensors/mrs1000/c/g387152 [accessed 30th April 2018].

Surmann, H., Nchter, A., Lingemann, K. & Hertzberg, J. (2003) Kurt3D-an autonomous mobile robot for modelling the world in 3D. *ERCIM NEWS*, 55, 24–25.

Taketomi, T., Uchiyama, H. & Ikeda, S. (2017) Visual SLAM algorithms: A survey from 2010 to 2016. *IPSJ Transactions on Computer Vision and Applications*, 9(1), 16.

TechCrunch (2014) *Google Unveils the Cartographer, Its Indoor Mapping Backpack*. Available from: https://techcrunch.com/2014/09/04/google-unveils-the-cartographer-its-indoor-mapping-backpack/ [accessed 30th April 2018].

Thomson, C., Apostolopoulos, G., Backes, D. & Boehm, J. (2013) Mobile laser scanning for indoor modelling. *The ISPRS Annals of the Photogrammetry, Remote Sensing and Spatial Information Sciences*, 5, W2.

Trimble v10. Available from: www.trimble.com/survey/trimble-v10-imaging-rover.aspx [accessed 30th April 2018].

Tucci, G., Visintini, D., Bonora, V. & Parisi, E.I. (2018) Examination of indoor mobile mapping systems in a diversified internal/external test field. *Applied Sciences*, 8(3), 401.

Vaaja, M., Kukko, A., Kaartinen, H., Kurkela, M., Kasvi, E., Flener, C., . . . Alho, P. (2013) Data processing and quality evaluation of a boat-based mobile laser scanning system. *Sensors*, 13(9), 12497–12515.

Velodyne. Available from: http://velodynelidar.com/ [accessed 30th April 2018].

Velodyne PUCKTM VLP-16. Available from: http://velodynelidar.com/vlp-16.html [accessed 30th April 2018].

The Verge (2013) *Google Invites You to Borrow Its Trekker Street View Backpacks and Photograph the World*. Available from: www.theverge.com/2013/6/27/4471740/google-invites-you-to-borrow-its-trekker-street-view-backpacks-and [accessed 30th April 2018].

Vexcel Imaging. UltraCam Panther. Available from: www.vexcel-imaging.com/products/ultracam-panther/ [accessed 30th April 2018].

ViAmetris. iMS2D. Available from: www.viametris.com/products/ims2d [accessed 30th April 2018].

Williams, B., Cummins, M., Neira, J., Newman, P., Reid, I. & Tardós, J. (2009) A comparison of loop closing techniques in monocular SLAM. *Robotics and Autonomous Systems*, 57(12), 1188–1197.

Woodman, O.J. (2007) *An Introduction to Inertial Navigation* (No. UCAM-CL-TR-696). University of Cambridge, Computer Laboratory, Cambridge, UK.

Ziparo, V.A., Zaratti, M., Grisetti, G., Bonanni, T.M., Serafin, J., Di Cicco, M., . . . Stachniss, C. (2013) Exploration and mapping of catacombs with mobile robots. *Safety, Security, and Rescue Robotics (SSRR), 2013 IEEE International Symposium on*. IEEE, Linköping, Sweden. pp.1–2.

Zlatanova, S., Liu, L. & Sithole, G. (2013) A conceptual framework of space subdivision for indoor navigation. *Proceedings of the Fifth ACM SIGSPATIAL International Workshop on Indoor Spatial Awareness*. AC, Orlando, FL. pp.37–41.

Zlot, R. & Bosse, M. (2014) Efficient large-scale three-dimensional mobile mapping for underground mines. *Journal of Field Robotics*, 31(5), 758–779.

Zlot, R., Bosse, M., Greenop, K., Jarzab, Z., Juckes, E. & Roberts, J. (2014) Efficiently capturing large, complex cultural heritage sites with a handheld mobile 3D laser mapping system. *Journal of Cultural Heritage*, 15(6), 670–678.

Laser Scanning – Riveiro and Lindenbergh
© 2020 ISPRS, ISBN 978-1-138-49604-0

Chapter 5

Indoor modeling with laser scanning

Cheng Wang and Chenglu Wen

ABSTRACT: Indoor environment is the main area of people's work and life. Indoor modeling is to establish the vectorized description of the walls, ceiling, floor, windows, doors, and other building structures or furniture features. This chapter presents the recent advances in indoor modeling with laser scanning techniques. This chapter focuses on two essential parts: calibration for multibeam laser scanners and indoor building modeling using a backpacked laser scanning point cloud. We first propose a target-free automatic self-calibration approach for multibeam laser scanners. The proposed method uses the isomorphism constraint among laser data to optimize the calibration parameters and then uses the ambiguity judgment algorithm to solve the mismatch problem. We then present a semantic line framework-based modeling building method for laser scanning point cloud data. The proposed method first semantically labels the raw point clouds into the walls, ceiling, floor, and other objects. Then line structures are extracted from the labeled points to achieve an initial description of the building line framework. To optimize the detected line structures caused by occlusion, a conditional generative adversarial nets (cGAN) deep learning model is constructed. The line framework optimization model includes structure completion, extrusion removal, and regularization. The result of the optimization is also derived from a quality evaluation of the point cloud. Thus, the data collection and building model representation become a united task-driven loop. The proposed method eventually outputs a semantic line framework model and provides a layout for the interior of the building.

1 INTRODUCTION

With the growth of urban populations and the prevalence of large buildings, there is an increasing demand for up-to-date spatial information of indoor environments. Traditionally, 2D floor maps have been used as the main source of indoor spatial information. In recent years, 3D modeling and reconstruction of the interior of buildings provide essential 3D models for applications, such as location-based services, building maintenance, disaster rescue, and building renovation planning. The primary requirement for these applications is 3D indoor building models, which are composed of the primitives of the building interiors, such as the ceilings, floors, walls, windows, and doors, but not the additional objects in the indoor spaces, such as furniture. In this chapter, we focus on the primitive building structures and ignore the additional indoor objects.

Indoor 3D measurement can be collected with different sensors. Popular 3D point cloud measurement systems include stereo cameras, terrestrial laser scanning (TLS), hand-held laser scanning devices, and low-cost RGB-D cameras. Perez-Yus *et al.* (2016) used RGB-D and fisheye cameras to obtain a scaled 3D model with wide scene reconstruction. Liu *et al.* (2015) proposed a small set of monocular images of different rooms to form a 3D indoor model using a Markov Random Field model. To measure 3D data for indoor environments, movable or backpacked systems have been developed based on RGB-D cameras or laser scanners (Wen *et al.*, 2014; Wen *et al.*, 2016).

Several methods have been developed for the automated generation of 3D indoor models from point clouds (Jung *et al.*, 2015; Oesau *et al.*, 2014; Ochmann *et al.*, 2014; Xiong *et al.*, 2013; Mura *et al.*, 2014; Wang *et al.*, 2016). Babacan *et al.* (2016) demonstrated a method to automatically extract floor plans from raw point clouds without using the 3D structure of the indoor environment. Michailidis *et al.* (2016) presented a method to extract the wall openings (windows and doors) of interior scenes from indoor 3D point clouds. Their method directly extracts windows and doors from a single wall surface. These two methods can be applied only to a single-plane surface, such as a floor or wall. Ochmann *et al.* (2016) developed an automated approach for the reconstruction of parametric 3D building models from indoor point clouds. Armeni *et al.* (2016) proposed a semantic parsing method for an entire building based on a point cloud acquired by a 3D camera.

However, the following two challenges remain for the task of indoor 3D building modeling:

1. Data quality challenge. Indoor environments have no GNSS signals, so another positioning and orientation method is needed. On the other hand, indoor environments are composed of many independent spaces. The walls between these spaces obstruct vision from one space to another. Thus, the 3D data collection must move through different spaces and be measured from different locations. Compared with static TLS and near-ranged RGBD sensors, simultaneous localization and mapping (SLAM)-based mobile laser scanning solutions exhibit better efficiency, range coverage, and geometric consistency (Bosse *et al.*, 2012; Wen *et al.*, 2014; Wen *et al.*, 2016). However, due to the failure of SLAM processing and noises, low-quality sections in SLAM-based laser scanning data are still inevitable. At the same time, heavy occlusions are often on the floor, walls, doors, and windows due to obstacles from furniture.
2. The challenge of how to effectively represent the indoor model. The indoor building model requires each component to be labeled with its semantic meanings, which will be used in further applications. As primitives of the representation, planes and lines are essential elements for building a 3D model of an indoor scene (Jung *et al.*, 2015; Oesau *et al.*, 2014; Ochmann *et al.*, 2014; Xiong *et al.*, 2013; Mura *et al.*, 2014). However, due to a high level of incompleteness of the point clouds caused by occlusions from furniture, the state-of-the-art methods are ineffective for extracting correct lines and planes. Also, most existing indoor modeling methods use rule-based prior knowledge or assumptions, such as the ceiling and floor planes should be horizontal, the wall planes should be vertical, etc. However, in complex cases, these rules may be invalid.

This chapter first introduces a target-free automatic self-calibration approach (Gong *et al.*, 2018) for multibeam laser scanners using the isomorphism constraint and ambiguity judgment algorithm. The proposed calibration process is fast and fully automatic. The chapter then presents a novel line framework-based semantic indoor building modeling method using 3D indoor backpacked mapping point clouds in cluttered and occluded indoor environments (Wang *et al.*, 2018). The processing method includes three stages: patch-based semantic labeling, 3D line structure feature extraction, and line framework optimization. At the patch-based semantic labeling stage, a trained conditional random fields (CRFs)-based method automatically classifies the raw laser scanning point cloud into four categories: floors, walls, ceilings, and other objects. At the line structure extraction stage, 3D line structure features are extracted from labeled point clouds. A semantic level of details 3 (LOD3; Biljecki *et al.*, 2014) building model is obtained by providing a 3D line framework of the walls, ceiling, floor, windows, and open doors. At the line framework optimization stage, a conditional generative adversarial nets (cGAN)-based deep learning model is constructed and applied to the line framework to deal with structure completion, extrusion removal, and line regularization. Also, in this stage, to detect the failure of the SLAM mapping process, the result of line optimization is derived as the quality evaluation of point cloud data.

The rest of this chapter is structured as follows. Section 2 describes our laser scanning sensors and systems for indoor environments. Section 3 introduces SLAM-based mobile laser scanning. Section 4 and section 5 show the experimental results of effective calibration of multiple sensors and semantic building model extraction, respectively.

2 LASER SCANNING SENSORS AND SYSTEMS FOR INDOOR ENVIRONMENT

2.1 *Indoor backpacked laser scanning system*

Based on the indoor mobile laser scanning systems (Wen *et al.*, 2014; Wen *et al.*, 2016), an upgraded backpacked 3D laser scanning system is shown in Figure 5.1. Detailed hardware information for the upgraded backpack system is shown in Table 5.1.

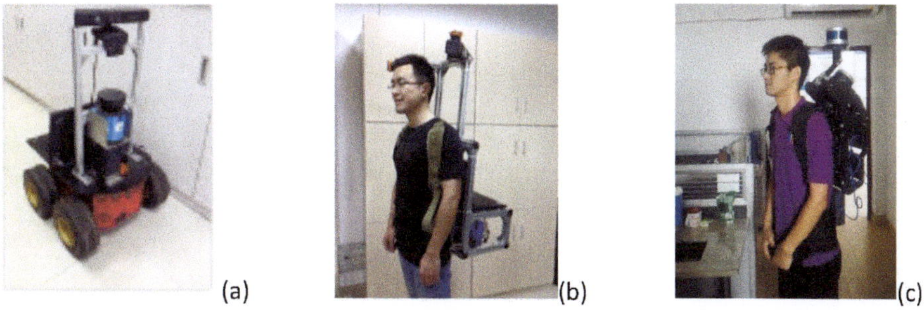

Figure 5.1. Indoor mobile laser scanning system. (a) Robot-based mapping system (Wen *et al.*, 2014). (b) Single-beam backpacked laser scanning system (Wen *et al.*, 2016). (c) Multibeam backpacked laser scanning system.

Table 5.1 Hardware information

Equipment	*Specifications*
Laser scanner	Two 16-beam laser scanners (Velodyne VLP-16)
Battery	Lithium battery, 12V 20AH, 127 × 72 × 52 mm, 0.973 kg
Processing unit	HP ENVY 15-ae125TX, 384 × 255 × 23 mm, 1.9 kg
Trestle	Carbon fiber, 3.5 kg

This system contains two 16-beam 3D laser scanners. Each laser scanner consists of 16 individual laser-detector pairs over the 30° (-15° to +15°) field of view. One laser scanner is placed horizontally to acquire the point cloud P_{Horiz} the other laser scanner is mounted at 45° below the horizontal one to acquire the point cloud P_{Titl}. Using Eq. (1), we merge the scanners to acquire the global point cloud P_{Global}, where T_{cali} is a transform matrix calculated between the two laser scanners (Gong *et al.*, 2018).

$$P_{Global} = P_{Horiz} + T_{cali} * P_{Titl} \tag{1}$$

A human operator, carrying the backpack mapping system, performs the survey at a normal walking speed. To achieve mapping results, the consecutive frames taken by the two laser scanners, from time to time, are registered by the LOAM algorithm (Zhang and Singh, 2014). Compared with our previous systems, the upgraded laser scanning system, by using multibeam laser scanners and an improved mapping algorithm, acquires indoor 3D point cloud data with higher density and efficiency.

2.2 *Characteristics of the indoor point cloud data*

Examples of the acquired 3D indoor laser scanning point cloud by the above backpack system are shown in Figure 5.2 (data available online: www.mi3dmap.net). Point clouds acquired in two complex underground parking areas are given in Figure 5.2 (a) and 5.2 (b). Point clouds acquired in a rectangular corridor with a closed-loop are shown in Figure 5.2 (c). Specifically, Scene 3 consists of two parts with different building heights. The left part of the building is higher than the right part of the building. Point clouds acquired by the backpack system for a multiroom scene are shown in Figure 5.2 (d). The mapping results indicate that the backpacked laser scanning system provides robust point cloud mapping results for different indoor scenes.

The nature of indoor and outdoor environments is very different. Compared with an open outdoor environment, an indoor environment is usually complex, narrow, and GNSS-denied. Especially, severe occlusion exists due to the presence of a large amount of furniture. Regarding the nature of the indoor environment, the characteristics of the indoor 3D point clouds acquired by our proposed backpacked laser scanning system are as follows: (1) data incompleteness due to the occlusion and contaminated areas caused by narrow space and (2) data uncertainty due to the cluttered background. Figure 5.3 shows the indoor laser scanning point clouds and the corresponding

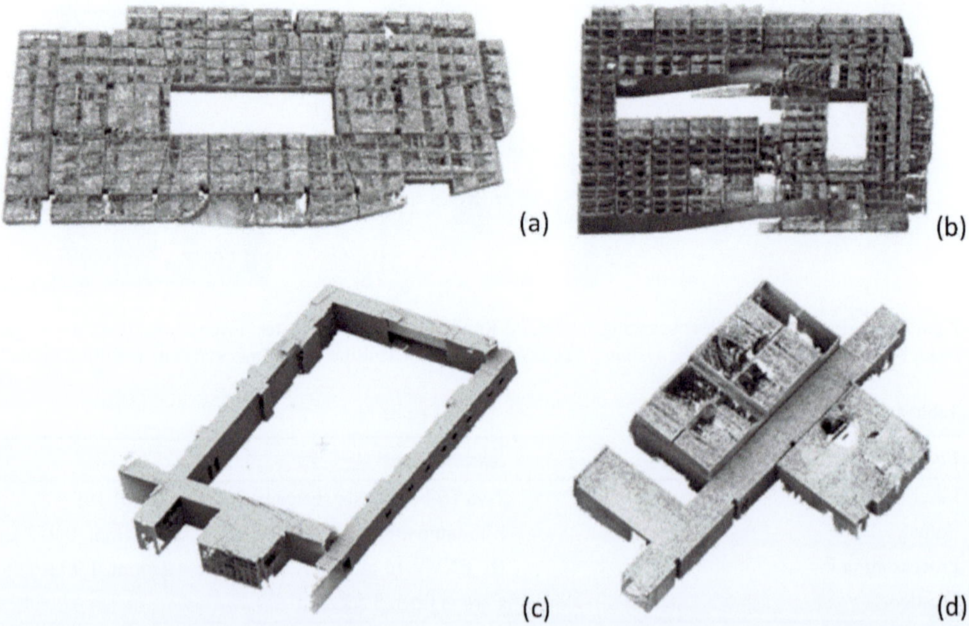

Figure 5.2. Examples of 3D point clouds built by our system. (a) Scene 1 – Underground parking area. (b) Scene 2 – Underground parking area. (c) Scene 3 – A closed-loop corridor. (d) Scene 4 – Connected rooms.

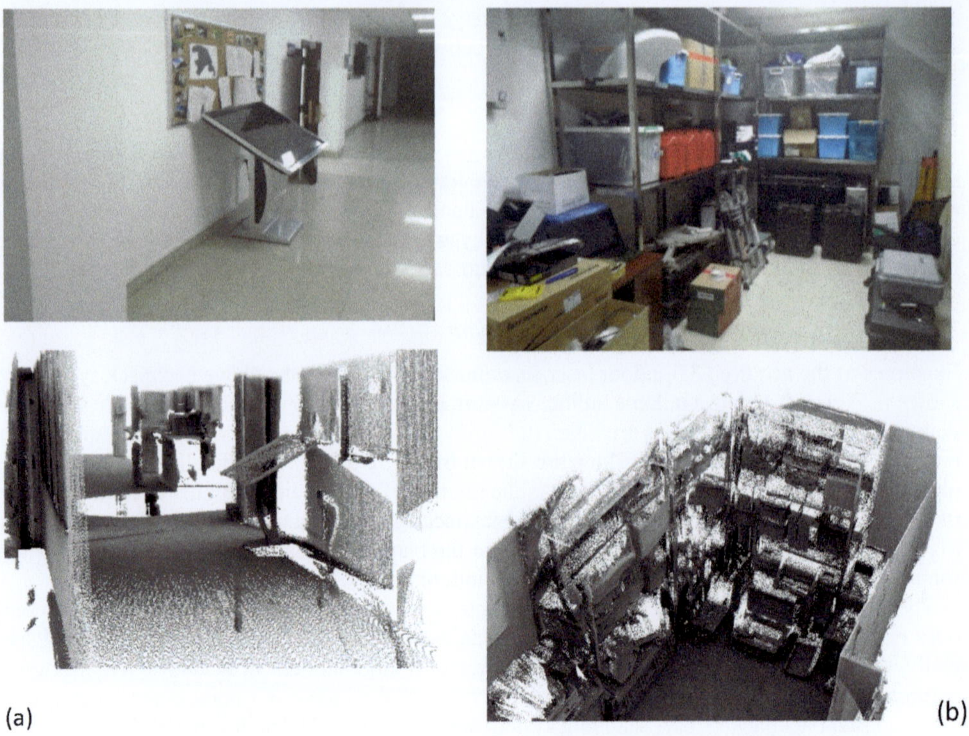

Figure 5.3. Typical examples of indoor laser scanning point clouds. (a) Corridor. (b) Room.

images of two typical indoor scenes. Figure 5.3 (a) shows simple occlusion by a monitor in a corridor. Figure 5.3 (b) shows an indoor scene with a high level of occlusion and a cluttered background. As shown in Figure 5.3 (b), the incompleteness and uncertainty of the data significantly increase when there are more occlusion and cluttered background.

3 SLAM-BASED MOBILE LASER SCANNING

In the application of laser-based SLAM (simultaneous localization and mapping) and 3D reconstruction (Zhuang *et al.*, 2013), to further improve the mapping process, multisensor data fusion brings more robustness and higher precision to the algorithm and captures the 3D environment in greater detail. Due to the limited view of a single laser scanner sensor, multiple laser scanner sensors are usually cross-mounted to more completely acquire 3D data. In multisensor systems, sensors are usually fixed in a rigid platform; each sensor has its own local coordinate. To unify the coordinate systems of different laser scanners, the 3-D coordinate transformation relation between the laser scanners must be accurately calibrated.

Previously, a standard approach to deal with the multisensor calibration is to achieve a common view of the sensors by introducing special calibration targets (Zhuang *et al.*, 2014). As a classical tool for laser scanner calibration, a pattern plane plays a significant role in the extrinsic calibration of laser scanners. Atanacio *et al.* (2011) and Muhammad and Lacroix (2010) proposed the feature constraint–based methods for multi-beam. He *et al.* used the multitype geometric feature (corner points, lines, and planes) based algorithm to handle the pairwise lidar calibration (He *et al.*, 2013). These methods provide accurate solutions for the calibration. However, they all require special design targets with specific geometric features. In multibeam laser scanner calibration, to determine the corresponding target points or corner features in the common view of the sensor from the sparse point cloud is very difficult. An automatic intrinsic calibration approach for a multibeam laser scanner is suggested by energy function but cannot handle extrinsic calibration (Nouira *et al.*, 2016).

In section 4, a target-free automatic self-calibration (TFASCal) approach is proposed for multibeam laser scanners. The approach uses one lidar mapping in the first step and then uses the isomorphism constraint among the data to optimize the calibration parameters with no target required. During calibration, generally, a mismatch in target registration may be introduced, which brings errors into the statistical model. To improve the accuracy of calibration, a mismatch elimination rule based on the statistical model is also developed. In a comparison of the target-based approach, the TFASCal approach eliminates the need of an external target and thus is fully automatic. Since the approach does not need the manually fitting to find the corresponding target point in sparse point cloud frame data, it also avoided the fitting error and achieved efficient and accurate self-calibration.

4 EFFECTIVE CALIBRATION OF MULTIPLE SENSORS

This section provides numerical results of calibration by using our target-free automatic self-calibration approach for multibeam laser scanners. Before giving the numerical results in section 4.4, we first introduce some preliminary backgrounds in sections 4.1 to 4.3.

4.1 *Data acquisition and multisensor coordinate system*

The calibration was performed on a backpack 3D scanning system (Figure 5.4 (a)). The system uses two 16-line 3D laser scanners (Velodyne VLP-16; Velodyne LiDAR Datasheet),[1] each composed of 16 laser-detector pairs individually aimed in 2° increments over the 30° (−15° to +15°) field of view

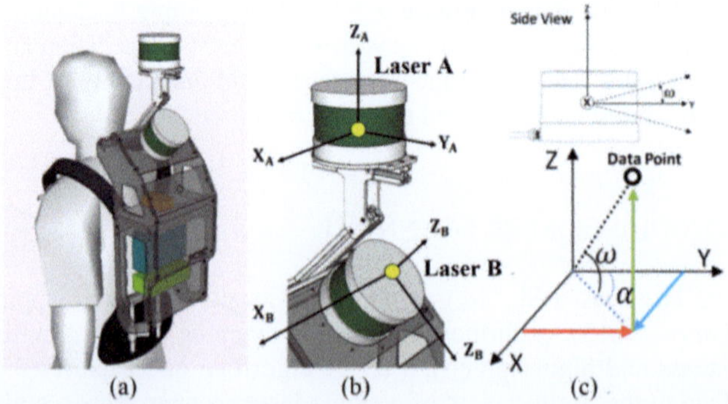

Figure 5.4. (a) The backpack 3D scanning system. (b) Laser scanner A and laser scanner B's coordinates. (c) Coordinate of the laser scanner.

of the laser scanner (Figure 5.4(b)). The point cloud $P(x, y, z)$ in the scanner's sensor coordinate system are identical to those calculated for VLP-16 as given in Equation (1) and shown in Figure 5.4(c).

Laser scanner A (Laser A) is mounted horizontally, and laser scanner B (Laser B) are mounted at 45° below Laser scanner A (Figure 5.4(b)). The point cloud coordination in the Cartesian coordinate system (X, Y, Z) is calculated for 3D laser scanners in Equation (2). Our goal is to determine a transform T_{cali} that places $P_{Laser\ A}$ and $P_{Laser\ A}$ in the same coordinate and merge them in P_{global} (Equation (3)).

$$\begin{bmatrix} X \\ Y \\ Z \end{bmatrix} = \begin{bmatrix} R * \cos(\omega) * sin(\alpha) \\ R * \cos(\omega) * cos(\alpha) \\ R * sin(\alpha) \end{bmatrix} \tag{2}$$

$$P_{global} = P_{Laser\ A} + T_{cali} * P_{Laser\ B} \tag{3}$$

4.2 *Automatic self-calibration algorithm*

Instead of using the target-based approach, the sensors coordinate transformation in a large quantity of data that is calculated in our method; that is, the calibration matrix is recursively computed in the continuous construction of the submap and its isomorphism constraint.

Based on the Generalized-ICP algorithm (Segal *et al.*, 2009) and lidar odometry and mapping (LOAM; Zhang and Singh, 2014) method, a precise local submap, M was built. Assuming T_A^n is laser scanner A's trajectory at the time $(0 \sim n)$ in the mapping algorithm, P_B^n is the laser scanner B's point cloud at the time, n. T_{guess} is the initial value of the coordinate transformation relation between the two sensors. T_{guess} is estimated by a coarse manual measurement. Among them, T_A^n and P_B^n are synchronized, with one-to-one correspondence. The output of the approach is to determine the exact calibration matrix T_{cali}.

The multilaser scanner system provides the following data format:

$$\begin{aligned} &Given \quad (M, T_A^n, P_B^n, T_guess)\ (n = 0 \sim t) \\ &And \quad T_A^n \sim P_B^n\ (synchronization) \\ &Find \quad T_{cali} \end{aligned}$$

With the isomorphism and rigid coordinate constraints, the mathematical model for automatic calibration is introduced by:

$$P_A^n = NN\left(M, T_A^n, P_B^n, T_{guess}\right) \quad (4)$$

$$T_{\text{cali}} = arg\min_{T_{cali}} \sum_n \| P_B^n * T_{cali} - P_A^n \|^2 \quad (5)$$

In Equation (4), NN is the nearest neighbor point search algorithm in FLANN (Fast Library for Approximate Nearest Neighbors).[2] Here we assume that data and trajectories are synchronized, laser scanner B's point cloud, P_B^n, is transformed to its location at time n in the submap M using T_A^n and T_{guess}. Then the nearest-neighbor search algorithm is applied to find the nearest neighbor point, P_A^n, on the sub map. Finally, the environmental consistency constraint is introduced to deduce Equation (5) to obtain T_{cali}^n.

For the corresponding point clouds, ICP method is used to obtain the minimum value with the initial transform T_{guess}:

$$T_{cali}^1 = ICP\left(P_B^1, P_A^1, T_{guess}\right), T_{guess} = T_{cali}^1 \quad (6)$$

$$T_{cali}^2 = ICP\left(P_B^2, P_A^2, T_{guess}\right), T_{guess} = T_{cali}^2 \quad (7)$$

$$T_{cali}^n = ICP\left(P_B^n, P_A^n, T_{guess}\right), Output: T_{cali}^n \quad (8)$$

4.3 *How to avoid mismatching*

The ambiguity of the calibration environment, for example, a long corridor, might fail the calibration algorithms. To deal with it, we proposed an ambiguity evaluation algorithm to filter the outliers and use multiple registration statistics to estimate the calibration matrix (Equations (9) and (10)). The random sample consensus (RANSAC; Fischler and Bolles, 1981) method is used to deal with the large error outliers. Then we calculate the mean value to determine the final calibration matrix T_{cali}^{Final}.

$$T_{cali}^{'} = RANSAC\left(T_{cali}^n\right) \quad (9)$$

$$T_{cali}^{Final} = Mean\left(T_{cali}^n\right) \quad (10)$$

4.4 *Experimental results*

For target-based calibration approach in (Muhammad and Lacroix, 2010), the size of targets "A" and "B" at 400 × 400 mm, with hole size 200 × 200 mm were designed and placed one meter away on both sides of the system, respectively (Figure 5.5(a)). As shown in Figure 5.5(b), the point cloud of multibeam laser scanners is sparse and unorganized, so it is impossible to locate the exact correspondence point in the common view area. Thus, the RANSAC fitting method was used to fit two squares (yellow) in the sparse point cloud to corresponding points on the square corner.

The target corner points $P_1 \sim P_8$ and $P_1^{'} \sim P_8^{'}$ (Figure 5.5(b), Figure 5.5(c)) were used to analyze the error of the calibration matrix. After calibration, the average distance between the corresponding points of the target and the total least mean square error were calculated. Two groups of experimental errors (X, Y, Z, roll, yaw, pitch error, and mean square error) are listed in TABLE I (scene b1 and scene b2).

Carrying the backpack system and walking along two corridor scenes, two sets of data (scene f1 and scene f2) were collected. Each set of data includes 550 scanning frames. The LOAM [8] method was used for short-term (550 frames) mapping of point cloud data collected by laser scanner A (red point cloud in Figure 5.5(d)). The data of laser scanner B (green point cloud in Figure 5.5(d)) were rotated using the initial value T_{guess} and synchronized transformed to the approximate nearest point location of the laser scanner A's local submap.

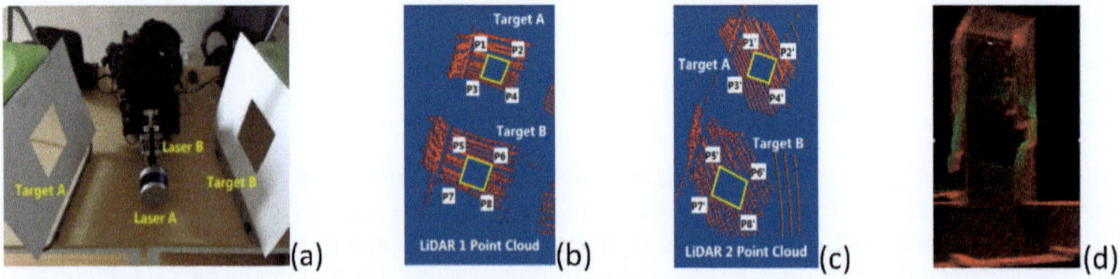

Figure 5.5. (a) Target-based calibration experimental scene. (b) Laser scanner A's point cloud. (c) Laser scanner B's point cloud. (d) Registering the point cloud (green) from laser scanner B to the local submap (red) from laser scanner A.

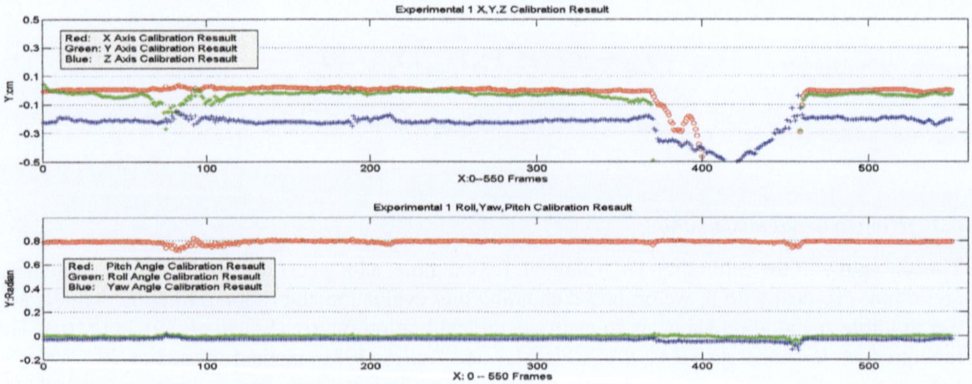

Figure 5.6. Automatic self-calibration results.

Figure 5.6 shows the automatic calibration result from 0~550 frames for scene f1 and scene f2. Our approach performs well between 100 and 360 frames and becomes unstable after 360 frames. A generalized-ICP (Segal *et al.*, 2009) was used in our approach to register the point cloud of laser scanner B to the local submap of laser scanner A. When the mapping system is moving in an environment where obvious structural features are absent (e.g. long corridors), the generalized-ICP becomes unstable, and causing mismatch and large errors. The ambiguity judgment algorithm was performed in this situation to remove outliers and achieve stable calibration results. Target corner points were used to analyze the error of the calibration matrix. The average distances between the corresponding points of the target and the total least mean square error were calculated. Two groups of experimental errors (X, Y, Z, roll, yaw, pitch error, and mean square error) are listed in Table 5.2.

In Table 5.2, RMS_T_ error is the x, y, z mean square root error, and RMS_R_ error is the roll, yaw, pitch root mean square error. Results indicate that in both rotation and drift situations, the TFASCal calibration approach achieves better accuracy and robustness than the target-based calibration approach.

TFASCal approach can work in both indoor and outdoor scenes that have rich structural features (lines or corners). Figure 5.7 displays the final multilaser scanner mapping results using the transformation matrix achieved by the proposed approach. The testing scene is a 30-meter-long, 2-meter-wide two-floor-corridor (Figure 5.7(a)). Orange and yellow point clouds were acquired by laser scanner A and laser scanner B, respectively. Our calibration approach was also tested on outdoor (Figure 5.7(b)) and basement scene (Figure 5.7(c)). The results indicate that the point clouds from the two calibrated sensors are well fused.

Table 5.2 Calibration errors of two methods

	Target-based Scene b1	*Target-based Scene b2*	*Target-free Scene f1*	*Target-free Scene f2*
x_error(mm)	15.896	21.960	6.372	7.877
y_error(mm)	20.271	23.389	5.943	5.276
z_error(mm)	7.377	10.257	3.129	6.134
Roll_error(deg)	0.053	0.066	0.018	0.029
Yaw_error(deg)	0.060	0.013	0.022	0.019
Pitch_error(deg)	0.082	0.096	0.037	0.024
RMS_T_error(mm)	26.795	33.682	9.258	11.291
RMS_R_error(mm)	0.114	0.116	0.046	0.045

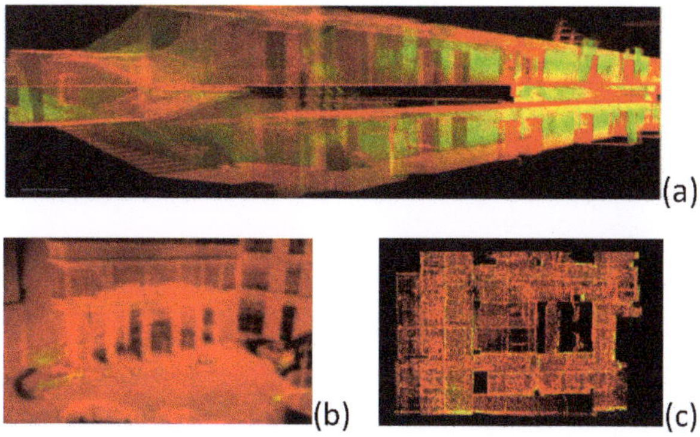

Figure 5.7. Multilaser scanners mapping results for different scenes. (a) Two-floor-corridor. (b) Outdoor. (c) Basement.

5 SEMANTIC BUILDING MODEL EXTRACTION

This section describes a semantic line framework–based indoor building modeling method. The pipeline is shown in Figure 5.8.

5.1 *Patch-based point cloud semantic labeling*

First, our proposed method provides prior knowledge of the building data by automatically and semantically labeling the 3D point cloud scenes via learning a labeling model from a certain number of training samples.

In the training stage, to yield improved classification results over locally independent classifiers, the learning framework of the AMNs for point cloud labeling is applied to exploit contextual information (Munoz *et al.*, 2009b). A category label $l_k \in \{l_1,\ldots,l_K\}$ is given to the 3D patch, *i*. To describe the 3D patches, fast point feature histogram (FPFH) descriptors (Rusu *et al.*, 2009) and the height information of the centroid of the points in a patch are computed to form feature vectors $\mathbf{x} = \{x_i, x_{ij}, x_i^c\}$, where $i = 1 \ldots N$, N is the number of 3D patches contained in a point cloud scene; x_i is the feature vector that describes the 3D patch, *i*, by giving statistics of the local distribution of the points in the 3D patch, *i*. x_{ij} is the feature vector describing the two spatially adjacent 3D patches *i* and *j*; x_i^c is the feature vector describing the clique, *c*, to which the 3D patch, *i*, belongs.

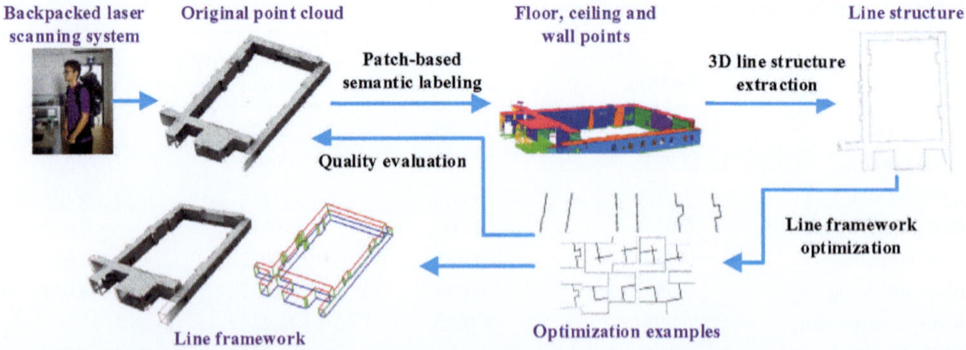

Figure 5.8. Pipeline of the semantic line framework–based indoor building modeling.

The assigned labels for the patches are defined as $y = \{y_1, \ldots, y_N\}$. The potential function used in the AMNs model is:

$$\Phi(x,y,W) = \Phi_n(x,y,W_n) + \Phi_e(x,y,W_e) + \Phi_c(x,y,W_c) \tag{11}$$

where Φ_n, Φ_e and Φ_c represents a node, a edge, and a clique potentials, respectively. $\boldsymbol{W} = [W_n, W_e, W_c]$ are the parameters in the AMNs model. Then log-linear potentials is used to represent the dependence of the node potentials on the extracted features as follows:

$$\log\left(\Phi_n(x,y,W_n)\right) = \sum_{i=1}^{N} \log\left(\phi_i\left(y_i^k\right)\right) = \sum_{i=1}^{N} W_n^k \cdot x_i \tag{12}$$

where $y_i^k = l_k$ (the label value assigned to node i), and $W_n^k \in A^{d_n}$ are the weights used when a node is assigned to l_k. Similarly, the potential over an edge models an associative/Pott's behavior that favors the two linked nodes taking on the same labels and penalizes as indicated by Equation (13).

$$\log\left(\Phi_e(x,y,W_e)\right) = \sum_{(i,j)\in E}^{N} \log\left(\phi_{ij}\left(y_i^k, y_j^o\right)\right)$$

$$\log\phi_{ij}\left(y_i^k, y_j^o\right) = \begin{cases} W_e^k \cdot x_{ij}, & l_k \neq l_o \\ 0, & l_k = l_o \end{cases} \tag{13}$$

where l_k and l_o are the labels of neighboring nodes i and j, and $y_j^o = l_o$. E is the edge set, where each edge is defined by two neighboring nodes. A P^n Pott model (Boykov *et al.*, 2001), which can be efficiently minimized, was used as an energy function. Following this model, the clique potentials $\forall c \in S$, where S is the clique set. $E_c(\mathbf{y}_c) = -\log\phi_c(\mathbf{y}_c)$ are defined by the high-order energy terms in the AMNs log-liner model, and:

$$\log\left(\Phi_c(x,y,W_c)\right) = \sum_{c\in C} \log\phi_c(y_c)$$

$$\log\phi_c(y_c) = \begin{cases} W_c^k \cdot x^c & , \text{if } \forall i \in c, y_i = l_k \\ 0 & , \text{otherwise} \end{cases} \tag{14}$$

For the AMN learning process, the objective function Equation (15) is optimized by the subgradient method and graph-cut inference method (Munoz *et al.*, 2009b).

$$\min_{w} \frac{\lambda}{2} \| W \|^2 + \max_{y}(\Phi(x,y,W) + \zeta(y,\hat{y})) - \Phi(x,\hat{y},W) \tag{15}$$

where $\zeta(y,\hat{y})$ is a loss function that computes the Hamming distance between the inferred labeling (y) and the true labeling ($\hat{y}$). λ is a regularization term.

For the AMN inference process, inferring category labels from unlabeled scenes is carried out in the labeling stage. The category labels, $\mathbf{y}^*$, for 3D patches are estimated effectively by maximizing Equation (7) via the α-expansion graph-cut method (Boykov *et al.*, 2001).

$$y^* = \arg\max_{y} P_w(y \mid x) = \arg\max_{y} (\Phi(x,y,W)) \tag{16}$$

For the four categories of point clouds obtained, only the points of the floor, ceiling, and wall are used in the further indoor modeling.

5.2 *Line structure extraction on 3D point cloud*

A line structure extraction method is developed directly on the 3D point cloud. We first extract a flat (but not specifically horizontal) boundary line using floor and ceiling points obtained from the above labeling results. However, due to the complexity of the indoor environment, the labeled floor and ceiling planes can be highly incomplete. In this situation, we merge the floor and ceiling points to obtain a completed plane only if the floor and ceiling planes are relatively parallel. If the floor and ceiling planes are not parallel, line extraction is applied directly to each plane.

First, we calculate the normal vector, $\vec{n_i}$, of every single point p_i. The dimensional coordinate system origin O(0,0,0) is the starting viewpoint. To calculate the normal vector $\vec{n_i}$ of point p_i, the k points nearest to the point p_i were selected as one plane, and the normal vector of the plane is taken as the normal vector of the point. Too many numbers of k may result in a lower accuracy of the calculated normal vector. Too few numbers of k result in a large computational cost and may result in a greater number of errors included during the calculation. For a normal laser scanning point cloud with the density of 1000 to 2000 point per square meter, set the value of k at 35.

Then the tangent plane of each point is calculated. For each input point, p_i, of the original point cloud P, the tangent plane T_{p_i} is expressed by its center point o_i, and the normal vector, $\vec{n_i}$, is as follows:

$$T_{p_i} = (o_i, \vec{n_i}) \tag{17}$$

The Euclidean distance from any point, p_i, to T_{p_i} in 3D space is calculated as follows:

$$dist(p_i, T_{p_i}) = |(p_i - o_i) * \vec{n_i}| \tag{18}$$

For the set of K neighborhoods of p_i: B_k (p_i), the best-fitting plane is obtained by solving the following equation:

$$\underset{T_{pi}}{\text{argmin}} \sum_{p_i \in B_k(p_i)} \text{dist}(p_i, T_{p_i})^2 \tag{19}$$

Because of all tangent planes, T_{p_i}, for all p_i have been computed, randomly select one point from the original point cloud as a seed point. Next, select one point, p_i, and create a new facet, $f_i = \left(\left\{K_i^m\right\}, x_i, \vec{T_i}\right)$ for the seed point, p_i, where $\vec{T_i}$ is the unit normal vector of T_{pi}. Then an improved region-growing procedure commencing with f_i is carried out. Then add each adjacent point, p_j, to the facet, f_i, if p_j is not used until now and satisfies the following three criteria: (i) the angle between $\vec{T_i}$ and $\vec{T_j}$ does not exceed the tolerance θ, (ii) the distance from p_j to p_i does not exceed R_{seed}, and (iii) the orthogonal distance from p_j to f_i is smaller than $\sigma/2$. Here, R_{seed} is used to constrain the radius of the facets to ensure that the large facets can be segmented into smaller pieces.

When the facet f_i is determined, choose another point p_k from among the last points of the original points and repeat the steps to find another facet f_k. This iterative process is performed until most of the points have been divided into different facets as shown in Algorithm I:

Algorithm I: Generating Facets

Input: Original point set P, point to facet distance threshold σ, the tolerance angle θ between tangent planes, *the number of neighboring points (k)*

Output: facets set F

1. For each p_i In P Do
2. Calculate T_{pi} using Equation (10)
3. Used(P) ← false
4. For each p_i In P Do
5. If Used(p_i) ≠ false Continue
6. $f_i \leftarrow \left(\left\{K_i^m\right\}, x_i, \vec{T_i}\right)$
7. For each p_j In P Do
8. If $\begin{cases} angle\left(T_i, T_j\right) < \theta \\ dist\left(p_i, f_i\right) < \sigma/2 \\ dist\left(p_i, p_j\right) < R_{seed} \end{cases}$ Then
9. $f_i \leftarrow f_i \cup p_j$
10. Used (p_j) ← *true*
11. End If
12. End For
13. End For

The next step is to extract line segments from these facets. Lin's work (Lin *et al.*, 2015; Lin *et al.*, 2017) is the state-of-the-art line segment extraction for large-scale point clouds. Lin's original method works specifically for extracting the lines of building exteriors from high-resolution and high-accuracy point clouds. To apply Lin's method to relatively low-resolution and low-accuracy indoor laser scanning point clouds, first increase the area of the facets (the parameter, R_{seed}, mentioned above) to reduce the number of lines in the internal plane and maintain as long a border line as possible. Second, decrease the angle, θ, between two tangent planes to separate different tangent planes better. Last, generate a continuous head-to-tail straight line to approximate the edge of the curve structure. In Lin's method, several discontinuous straight lines with gaps are generated in this stage.

To obtain line vectors, the first step is to extract the boundary points of each facet, f_i. The vertices of the α-shape of f_i are presented as the boundary points of the f_i. However, these boundary points can also contain the intersecting points of two adjacent coplanar facets. To overcome this problem, a facet F_i of f_i is defined, which contains adjacent coplanar facets and the current facet, f_i. Then we extract the α-shape points of F_i and compute the desired intersecting points, P_i.

The next step is to obtain a line segment based on the boundary points, P. Because of the relatively low-cost single/multibeam laser scanner used in data acquisition, indoor mobile laser scanning point clouds have the characteristics of relatively low density and limited precision, which results in messy extracted lines and multiple close parallel lines. Rather than directly group the boundary points into line segments, we group the boundary points into a cylinder to filter false detections by a cylinder-based alignment method (Lin *et al.*, 2017). To reduce line extraction false positives and ensure a good line segment result, the number of false alarms (NFA) algorithm (Desolneux *et al.*, 2000; Von Gioi *et al.*, 2010) is extended to 3D and kept only one line for each cylinder.

5.3 *Wall opening detection and line framework formation*

The wall line results are different from the floor and ceiling line results. We need to retain as many floor or ceiling lines as possible. Because the wall borderlines have been extracted already from the floor and ceiling point clouds, the wall line results require only the door and window lines. Therefore, it is better to drop the wall borderlines and retain only the internal lines from the windows and doors. For indoor scenes, the windows and doors are mostly rectangular. Only two intersecting edges are necessary to determine a rectangle. As for the original wall line extraction results, a k-means method is applied to capture potential door and window lines. For each wall plane, to find the best line extraction result, a different k value is set from 0 to 9. When the potential door or window lines are obtained, the longest line length is calculated to determine if it belongs to a door or a window. In this step, the detection results are refined using the hypothesis that doors and windows are rectangular. The last step is line framework formation.

5.4 *Wall opening detection and line framework formation*

The results obtained using the line extraction framework are usually imperfect because of the occlusion and cluttered background (Figure 5.9). The problems of line structure extraction are summarized as follows: (1) Irregular structures (a parallel or orthogonal relationship between some lines) are due to the uncertainty and noise level in the data (Figure 5.9(a)). (2). Incomplete structures and disconnected lines exist because of the occlusion (Figure 5.9(b)). (3) Extrusions remain because of the uncertainty and noise in the data (Figure 5.9(c)).

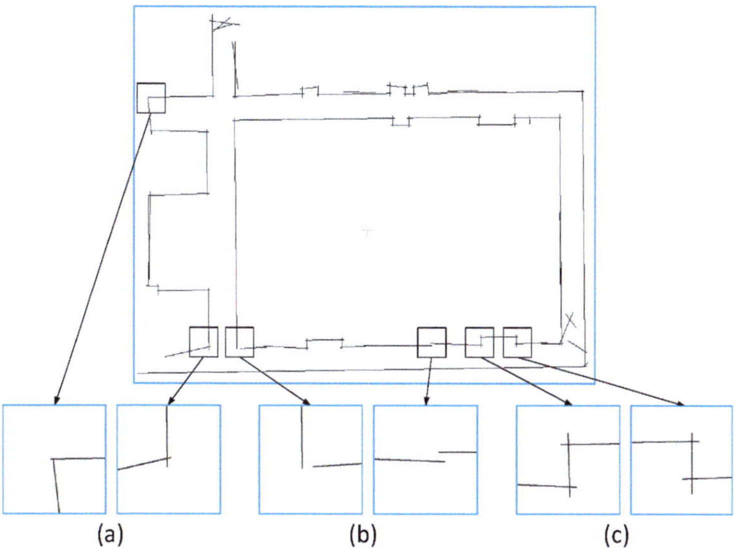

Figure 5.9. Examples of imperfect and incomplete line structure extraction.

Usually, a line regularization step, commonly using the rule-based method, is required after line extraction. The general line framework of a building meets some specific building rules, and lines can be further refined by the regularization based on these rules. However, the abovementioned rule-based line regularization method requires a large amount of human interaction and depends on pre-defined assumptions. Likewise, in complex indoor environments, fixed rules often become invalid.

5.5 *cGAN deep learning model*

To remove the extrusions, to complete and to regularize the line structure, we introduce a conditional generative adversarial nets (cGAN)-based deep learning model to optimize the imperfect line framework. The GAN (Goodfellow *et al.*, 2014) model trains a generator, *G,* to produce outputs that are indistinguishable from the "real" sample and trains a discriminator, *D*, to distinguish the outputs of the generator as much as possible. The cGAN model (Isola *et al.*, 2017), an improved network of the GAN model, originally completes the translation between semantic labels and photos, architectural labels and photos, edges and photos, etc. We applied the cGAN model to optimize the line structure.

A GAN learns the mapping from the random noise vector, *z*, to output, *y*; cGAN learns a mapping from the input, *x*, and random noise vector, *z*, to output, *y*. The same as GAN, the cGAN model learns a loss function automatically to satisfy different tasks without designing a new loss function. cGAN uses an objective function as follows:

$$G^* = arg\min_G \max_D \mathcal{L}_{cGAN}(G,D) + \lambda \mathcal{L}_{L1}(G) \tag{20}$$

$$\mathcal{L}_{L1}(G) = \mathbb{E}_{x,y\sim pdata(x,y),z\sim p_z(Z)}\left[\| y - G(x,z) \|_1\right] \tag{21}$$

By introducing a bound term, *L1* distance, the outputs of the generator, *D*, are not only similar to the real sample but also are more closely related to the conditional input sample. In generator architecture, a U-net encoder–decoder network is adapted because that U-net makes better use of the low-level information (Ronneberger *et al.*, 2015). A *Patch*GAN (Isola *et al.*, 2017) is designed as the new discriminator architecture to improve the efficiency of the discriminator. This new discriminator architecture attempts to classify whether each $N \times N$ patch in an image is real or fake then averages all responses to provide the ultimate output of *D*.

5.6 *Line framework optimization using cGAN model*

Since the cGAN model is working in a 2D plane, all the line structures extracted from each point cloud category (floor, wall, and ceiling) are first projected onto their own planes. To project each point, the coordinate system of every point is transformed from the previous *oxyz* coordinate system to a new *o'x'y'z'* coordinate system. In the new coordinate system, to achieve projection, the z' coordinate of each point is set to zero. The detailed steps to project are as follows: Firstly, a point $o'(x_0, y_0, z_0)$ is randomly chosen in the plane as the new origin. Then two orthogonal unit vectors $u'_x = (u'_{x1}, u'_{x2}, u'_{x3})$ and $u'_y = (u'_{y1}, u'_{y2}, u'_{y3})$ are chosen in the plane as the new x' axis and the new y' axis; the starting point of these vectors is o'. Next, a unit vector $u'_z = (u'_{z1}, u'_{z2}, u'_{z3})$ is chosen as the new z' axis from the normal vector of the plane; the starting point of this vector is also o'. Finally, a translation matrix $T = \begin{bmatrix} 1 & 0 & 0 & 0 \\ 0 & 1 & 0 & 0 \\ 0 & 0 & 1 & 0 \\ -x_0 & -y_0 & -z_0 & 1 \end{bmatrix}$ and a rotation matrix $R = \begin{bmatrix} u'_{x1} & u'_{x2} & u'_{x3} & 0 \\ u'_{y1} & u'_{y2} & u'_{y3} & 0 \\ u'_{z1} & u'_{z2} & u'_{z3} & 0 \\ 0 & 0 & 0 & 1 \end{bmatrix}$ are obtained, resulting in new coordinates for each point as follows:

$$(x', y', z', 1) = (x, y, z, 1) \cdot T \cdot R \tag{22}$$

After obtaining the projection, x' and y' are converted into rows and columns in a 2D image. The 2D image is divided into several 256 × 256 subimages. These subimages are classified by a VGG-16 convolutional neural network (Simonyan and Zisserman, 2014) to extract features and use three full connected layers to classify the features. The convolution layers use a 3 × 3 kernel and add batch normalization. Max-pooling is used to do down-sampling. During the training, 2000 training samples were used. The batch size and epoch were set to be 32 and 200, respectively. The results are the input to different cGAN models (see Figure 5.10). After obtaining the optimized 2D lines by cGAN models, the pixels on the optimized 2D lines are transformed back to 3D points by Eq. (22). Finally, the 3D points are fitted to the 3D lines by the linear least squares fitting algorithm.

During this process, the main precision loss comes from the process of projecting 3D lines onto 2D lines. The original 3D lines are vectors in 3D space, and the 2D lines on the image are several sets of discrete pixels. Different pixels per unit length of a 3D line results in a different precision loss. In general, the more pixels per unit length of a 3D line converted, the smaller the resulting precision loss. A precision loss comparison for Scene 1 and Scene 3 with 50, 100, and 200 pixels per meter of a 3D line is given in Table 5.3. The average distance from all projected 3D points onto the original 3D lines is used to measure the precision loss. In this chapter, taking into consideration both computational cost and precision loss, a meter of the 3D line is converted into 200 pixels. The average precision loss is about 1 mm. The results indicate that the impact of the precision loss during projection is relatively small.

To complete the structure optimization task, we construct three cGAN modules: structure completion, extrusion removal, and line regularization modules (see Figure 5.11). Each module only deals with the imperfections in each data. The generator is a symmetrical fully convolutional network containing 16 convolutional layers with 4 × 4 convolution kernels. The first eight layers of the generator form an encoder; the second eight layers form a decoder. The discriminator network consists of four convolutional layers; the final layer outputs the discrimination result by a sigmoid activation function.

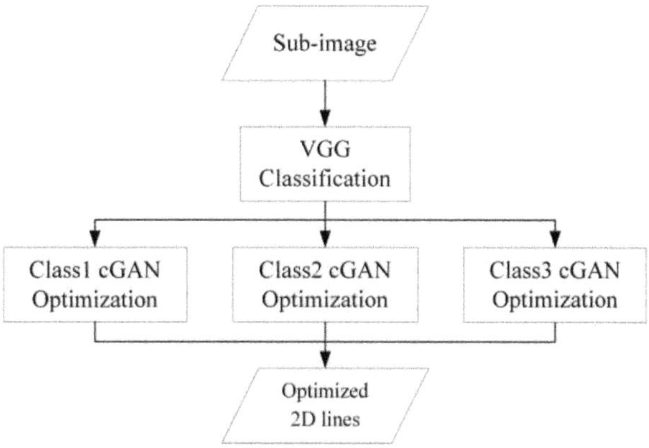

Figure 5.10. The flowchart of processing 2D data.

Table 5.3 Precision loss comparison

Pixels per meter of a 3D line	*Average distance (Scene1)*	*Average distance (Scene3)*
50	5.13 mm	4.81 mm
100	2.61 mm	2.56 mm
200	1.34 mm	1.17 mm

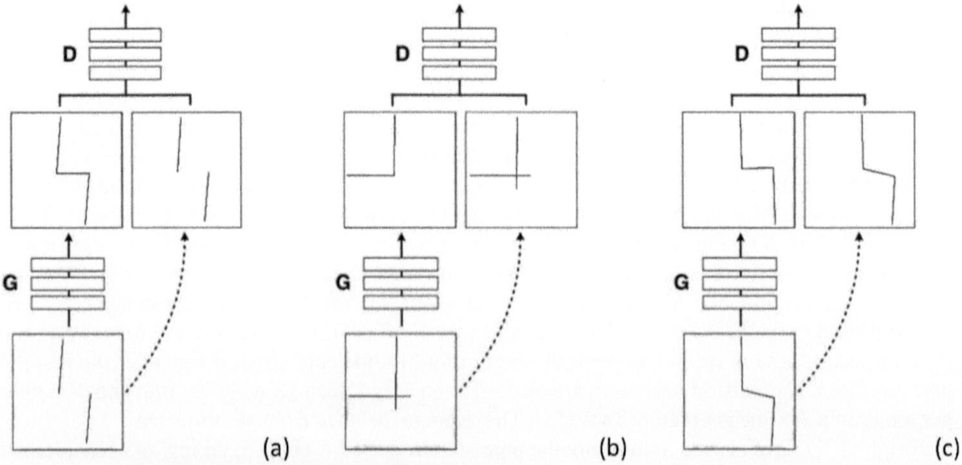

Figure 5.11. Three cGAN modules for framework structure optimization. (a) Structure completion module. (b) Extrusion removal module. (c) Line regularization module.

All training samples are divided into three categories to meet different optimization cases. Training data of the structure completion module consists of samples with two disconnected lines and the corresponding connected lines. The training data of the extrusion removal module consists of samples with extra parts and the corresponding samples without extra parts. The training data for line regularization consist of the samples with unparallel or nonorthogonal lines and the corresponding samples after regularization. To ensure sufficient training samples for the cGAN deep learning model, we prepared the training samples in two ways. A small number of training samples is manually created by cutting from the rule-based manual regularization results. A large set of training samples is generated by a computer program using manually developed rules-related building principles. Some examples of these rules are: an indoor structure should be a closed structure; the boundary of an indoor structure is unique; indoor framework lines should be continuous; building interior framework lines are either parallel or orthogonal.

Specifically, this framework optimization method can be further applied to other building structures by easily replacing the training samples generated from the new building structure. Meanwhile, since the imperfections of the data are detected automatically by the cGAN models, the data requires optimization at the same time. The data requiring optimization usually refers to the point cloud data with low quality. Meanwhile, the quality of the indoor laser scanning data is affected directly by the mapping process. For example, the varied walking speed of the human operator may result in inconsistent point cloud density. A too-fast turning angle of the system may fail the SLAM process and eventually lead to misregistration of point cloud frames. With these references, we can trace back to the original mapping process and assess the data quality, as well as filter the low-quality point cloud data. With the close-looped data quality evaluation and filtering, the data collection and model representation become a united task-driven loop.

5.7 *Experiments and results*

5.7.1 Patch-based point cloud semantic labeling

The labeling model was trained on labeled point clouds extracted from different indoor 3D scenarios. The learning parameters were determined by classifying validation point clouds. The training data consists of more than 300K points. The class label for each 3D point in the training samples was manually selected and labeled for learning parameters in AMNs. Some examples of training samples are shown in Figure 5.12.

Labeling results for Scene 1 and Scene 2 are given in Figure 5.13 and Figure 5.14, respectively. The original point cloud of the parking area (Figure 5.13(a)) is labeled by the proposed labeling method (Figure 5.13(b)). To better demonstrate the labeling results, we provide each category of point cloud data separately in Figure 5.13(c). The segmented point clouds are labeled into the ground (blue), the wall (green), the ceiling (red), and others (yellow). To quantitatively assess the accuracy and correctness of the semantic labeling results on a test dataset, we selected the following three measures: Precision, Recall, and F1-measure. Precision describes the percentage of true positives in the ground truth. Recall depicts the percentage of true positives in the semantic

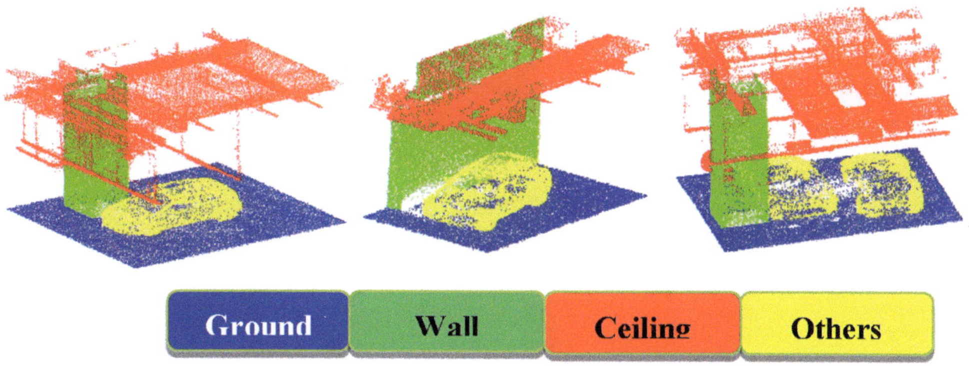

Figure 5.12. Examples of training samples. Different colors represent different categories. Blue points represent the floor, green points represent the wall, red points represent the ceiling, and yellow points represent other objects.

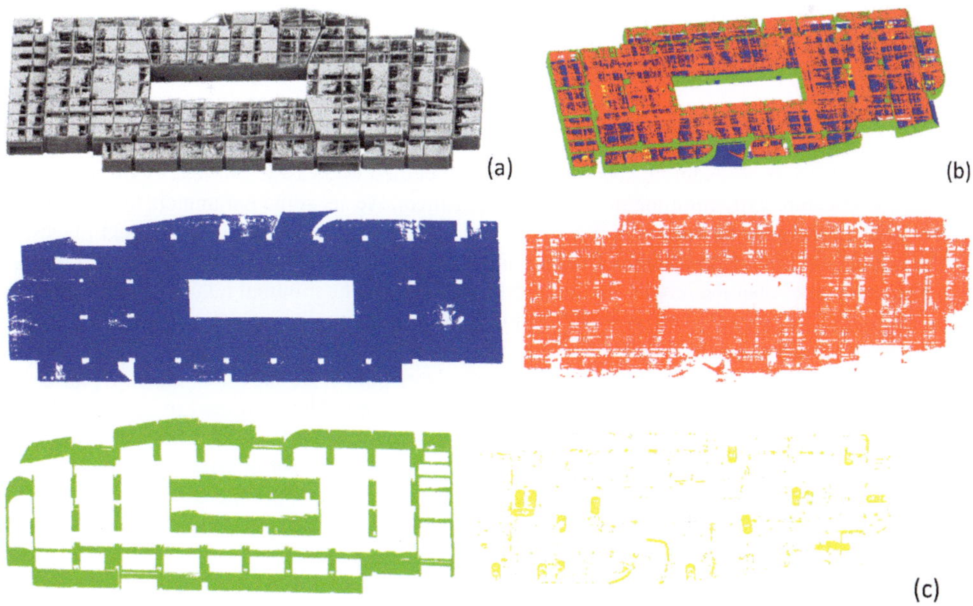

Figure 5.13. Labeling results of indoor Scene 1: (a) Original point cloud. (b) Semantic labeling results. (c) Labeling results for different categories: ground (blue), wall (green), ceiling (red), others (yellow).

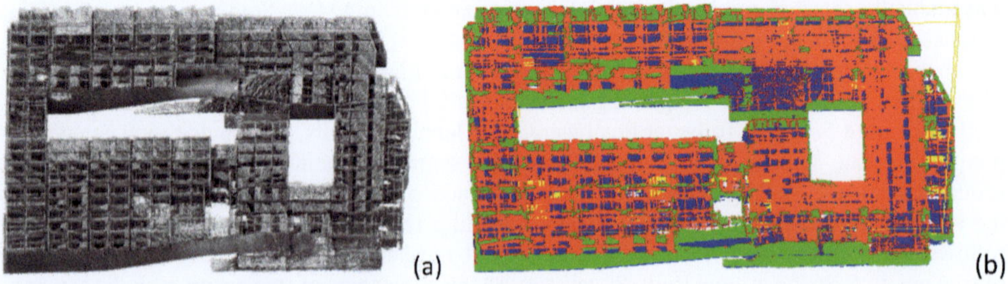

Figure 5.14. Labeling results of indoor Scene 2: (a) Original point cloud. (b) Semantic labeling results. Color code: ground (blue), wall (green), ceiling (red), others (yellow).

labeling results. F1-measure is an overall measure. The three measures are calculated on points as follows:

$$\text{precision} = \frac{TP}{TP + FN} \tag{23}$$

$$\text{recall} = \frac{TP}{TP + FP} \tag{24}$$

$$F_{1-measure} = \frac{2 * precision * recall}{precision + recall} \tag{25}$$

Where TP, FN, and FP represent the number of true positives, false negatives, and false positives, respectively. The quantitative evaluation results using these three measures of indoor Scene 1 and indoor Scene 2 are shown in Table 5.4 and Table 5.5. The proposed semantic labeling method achieved, in labeling Scene 1, an average precision, recall, and $F_{1\text{-}measure}$ values of 0.86, 0.89, and 0.87, respectively, and, in Scene 20.90, 0.84 and 0.87, respectively.

5.7.2 Line framework construction

Consider Scene 3 as an example scene to show the detailed line extraction results from the proposed line structure extraction method. In this experiment, we set some parameters to be constant (See Table 5.6). As we mentioned in Section 5.2, θ is the tolerance between the tangent planes, σ is the orthogonal Euclidean distance from a point to the tangent plane. Lines are discarded if their lengths are smaller than the length threshold. Min clusters are the minimal number of points in the tangent plane. The tangent plane is abandoned if it has a fewer number of points than min clusters.

The main parameters affecting the line extraction results are R_{seed} and K values. R_{seed} is used to determine the size of a facet; K is used to filter the isolated points. As shown in Table 5.7, different R_{seed} and K values lead to a different number of lines extracted. Also indicated in Table 5.7 is the larger the $[R_{seed}, K]$ value, the fewer number of lines extracted from floor, ceiling, and wall point clouds. However, $[R_{seed}, K]$ should not be so large that some important missing lines are avoided. Considering the above situations, we set $[R_{seed}, K] = [3.5, 35]$ in the experimental data to obtain the final results. The line extraction results on the floor, ceiling, and wall point clouds of Scene 3 are shown in Figure 5.15, Figure 5.16, and Figure 5.17, respectively.

In addition, Table 5.8 provides the runtime of line extraction for each point cloud category in Scene 3. The workstation is with a Windows 10, CPU Intel® Core™ i5-4460@3.20GHz, 12G RAM. The number of lines and the running time is also provided for Scenes 1, 2, 3, and 4 in Table 5.9. The greater the number of points, the greater the number of lines extracted, and the total running time is longer.

Table 5.4 Labeling confusion matrix for Scene 1

		Inferred Label				*Recall*
		Floor	*Wall*	*Ceiling*	*Others*	
True Label	Floor	152025	556	3459	362	0.97
	Wall	2361	124452	8801	3311	0.91
	Ceiling	9885	11274	222147	3242	0.90
	Others	766	145	3444	14170	0.77
Precision		0.92	0.91	0.93	0.67	

Table 5.5 Labeling confusion matrix for Scene 2

		Inferred Label				*Recall*
		Floor	*Wall*	*Ceiling*	*Others*	
True Label	Floor	363190	1079	28176	401	0.93
	Wall	3421	322084	29145	1284	0.91
	Ceiling	11762	18899	401651	1750	0.93
	Others	455	5338	6675	16230	0.57
Precision		0.96	0.93	0.86	0.83	

Table 5.6 Important parameters used in the experiment

θ(°)	*σ(m)*	*Length-threshold(m)*	*Min Clusters(number)*
10	0.2	0.1	30

Table 5.7 Different R_{seed} and K values *versus* different line number extracted

Scene 3	R_{seed} *(m)*	*K (number)*	*Line Number*
Ground-plane	1.0	15	338
Ground-plane	1.0	30	235
Ground-plane	3.0	15	292
Ground-plane	3.0	30	243
Ceiling-plane1	1.0	15	245
Ceiling-plane1	1.0	30	168
Ceiling-plane1	3.0	15	232
Ceiling-plane1	3.0	30	158
Wall-plane	1.0	15	81
Wall-plane	1.0	30	79
Wall-plane	3.0	15	43
Wall-plane	3.0	30	40

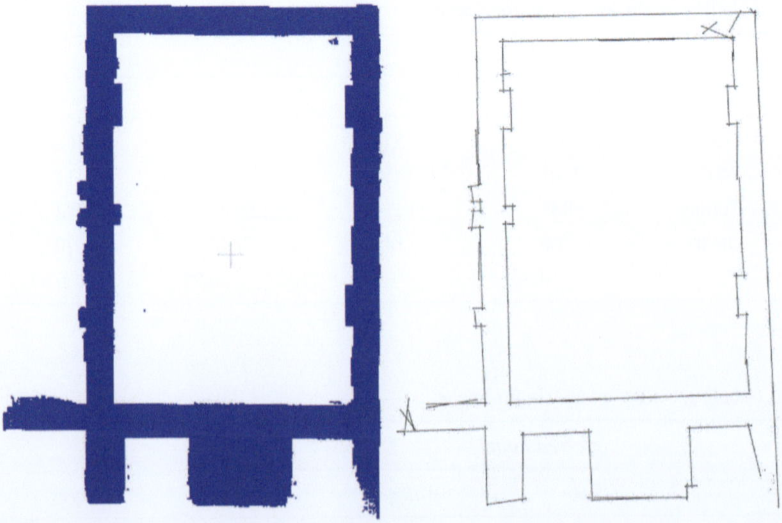

Figure 5.15. Line extraction results on floor point cloud.

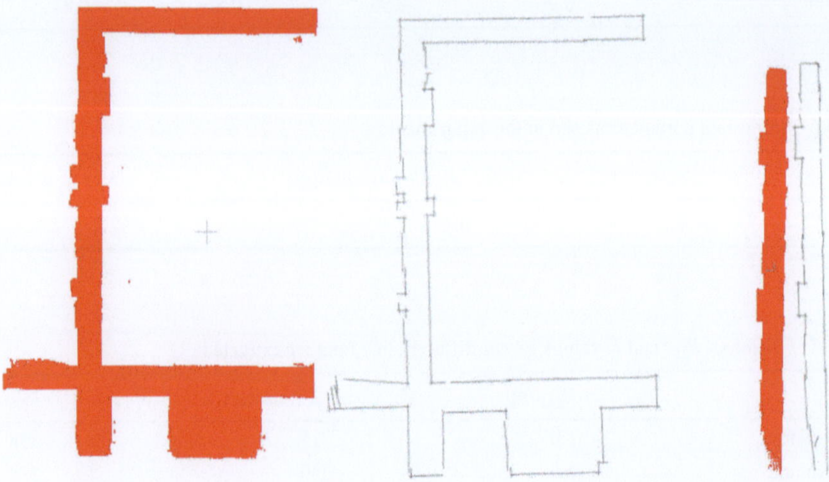

Figure 5.16. Line extraction results on ceiling point cloud 1 (first two) and ceiling point cloud 2 (latter two).

Figure 5.17. Line extraction results on wall point cloud.

Table 5.8 Number of lines extracted and running time for each category point cloud of Scene 3

Description	*Number of Points*	*Number of Lines*	*Line Segmentation(s)*	*Line Extraction(s)*	*Total Running Time(s)*
Ground plane	0.92 million	424	38.35	29.296	67.65
Floor plane 1	0.69 million	351	31.45	18.044	49.49
Floor plan 2	0.23 million	117	23.91	6.341	30.25
Wall plan	0.06 million	79	12.93	1.541	14.47

Table 5.9 Number of lines extracted and running time for different scenes

Description	*Number of Points*	*Number of lines*	*Line Segmentation(s)*	*Line Extraction(s)*	*Total Running Time(s)*
Scene 1	7.90 million	2652	292.19	84.88	377.07
Scene 2	3.85 million	1819	166.44	62.73	229.17
Scene 3	2.10 million	1652	114.66	69.99	184.66
Scene 4	8.62 million	1289	456.74	87.64	544.38

5.7.3 Line framework optimization

The line extracted are always contaminated by occlusions or noises, we introduce the cGAN (Isola and *et al.*, 2017) to optimize the line framework. The training data are automatically generated by a computer program. The training data falls into three categories:

1 Structure completion: We first generated two disconnected lines in a 256 × 256 pixel size image as input samples, then we connected them as target samples.
2 Extrusion removal: We first generated a corner as target samples (ground truth), then we lengthened two lines to obtain extrusion parts as input samples.
3 Line regularization: We first generated some lines, which are unparalleled or non-orthogonal as input samples, and then we adjusted them to be parallel or orthogonal as target samples.

For each category, there are 1,000 samples in a training set. During the training, the batch size and epoch of the cGAN model were set to be 4 and 500, respectively. The average training time for each category ranged from 6 to 8 hours on two Nvidia Titan X GPUs. Some examples of training samples are shown in Figure 5.18.

The testing sample size and time consumption are shown in Table 5.10. For each category, 500 samples are used for testing. The running time for each category is about 23 seconds. In addition, based on the principle of rule-base regularization, the target sample results were also provided manually. Then we compared the generated target results to the manual target results point by point and counted the number of points overlapping in the two results. Lastly, we used the proportion of nonoverlapping points to evaluate the performance of the proposed cGAN-based line framework optimization method.

Average percentages of nonoverlapping points of each category from 500 testing samples are also given in Table 5.10. Overall, the structure completion module has the lowest average non-overlapping percentage of 0.08%; the extrusion removal module has an average nonoverlapping percentage of 0.17%, and the line regularization module has the highest average nonoverlapping percentage of 0.52%. The results are understandable, because even for a human operator, it is more

Figure 5.18. Examples of training samples. On top are the input samples, and at the bottom are the target samples.

Table 5.10 Testing sample size, time and average nonoverlapping percentage

Module	*Testing Sample Size*	*Time(s)*	*Average Nonoverlapping Percentage (%)*
Structure Completion	500	23.34	0.08
Extrusion Removal	500	23.09	0.17
Line Regularization	500	23.38	0.52

difficult to regularize a line structure than to complete a line and remove the extrusions. In general, the average non-overlapping percentages of the three modules are all less than 1%, which indicates the cGAN model is effective for line framework optimization.

To further analyze the performance of the proposed framework optimization method, we give some challenging testing results for each category (see Figure 5.19). For example, for structure completion, even though the distance between two lines is relatively long, it can still be used to connect the lines (Figure 5.19(a)). Since line regularization is a more difficult task, we give a fail testing example in Figure 5.19(c). As shown in the figure, the line is partially regularized, but a part of the line remains missing.

After optimizing the line framework, the indoor framework structure becomes more completed and accurate; in addition, the corners are extracted. The before and after framework optimization of floor lines extracted from Scene 3 is shown in Figure 5.20(a). The before and after framework optimizations of ceiling lines extracted from Scene 3 is shown in Figure 5.20(b) and Figure 5.20(c). The before and after framework optimization of window and door lines extracted from Scene 3 is shown in Figure 5.20(d). Shown in Figure 5.21 are the combined line framework optimization results, which indicate that the proposed line framework optimization achieves good results.

6 CONCLUSION

Modeling the indoor environment is an essential task to support the information systems for indoor activities or applications. The indoor features include the building features such as walls, ceiling, floor, windows, doors, and the movable furniture features. This chapter focuses on the building features and ignores the movable furniture features. The backpacked indoor mapping system is the state-of-the-art equipment to measure large-scale indoor environments. This chapter starts with an introduction of backpacked indoor mapping system design then presents two major key issues: the calibration for multibeam laser scanners and indoor building modeling from point clouds.

This chapter first described a target-free automatic self-calibration (TFASCal) approach for multibeam laser scanners. Target-free is important for real-world applications. TFASCal is more convenient than the target-based calibration approach and provides enough accuracy for indoor modeling. The chapter then presented a semantic line framework modeling method for indoor

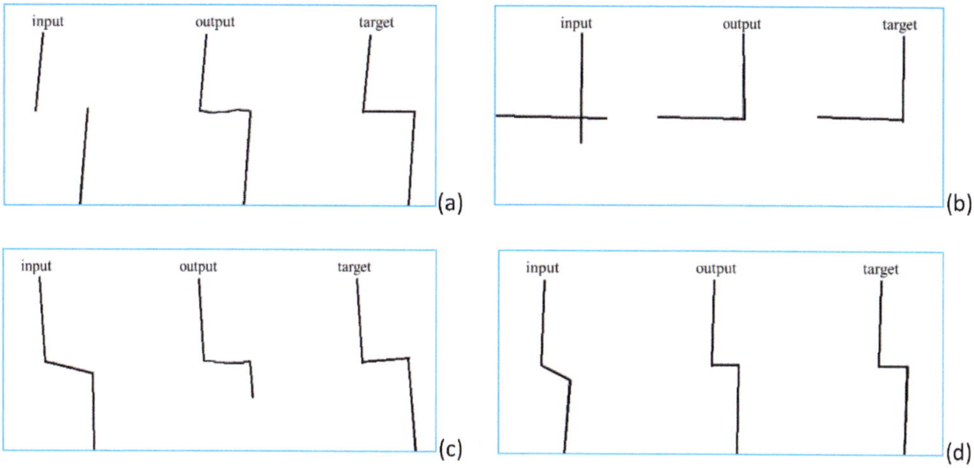

Figure 5.19. Examples of testing results. (a) Structure completion. (b) Extrusion removal. (c) and (d) Line regularization.

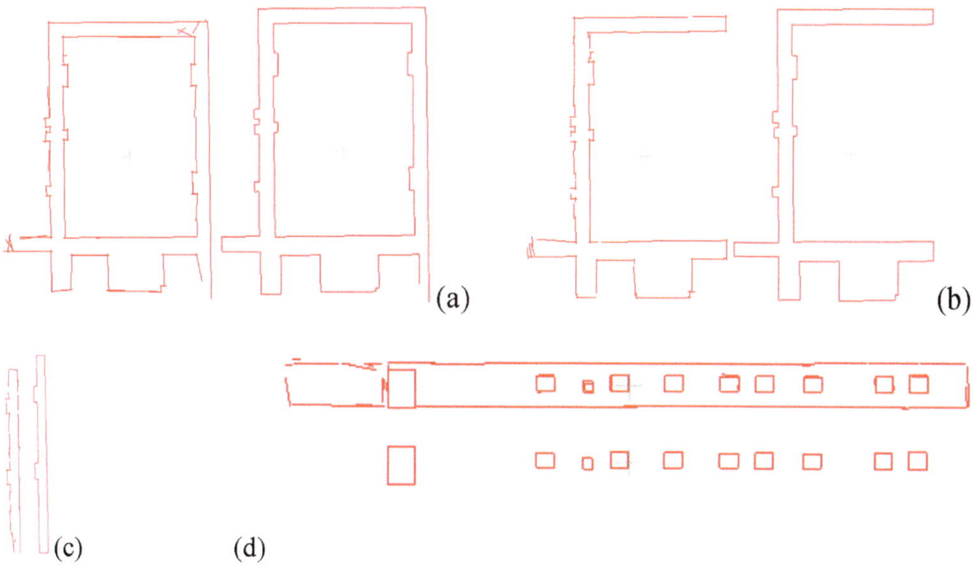

Figure 5.20. The before and after framework optimization of lines extracted from Scene 3. (a) Floor lines. (b) and (c) Ceiling lines. (d) Door and window lines.

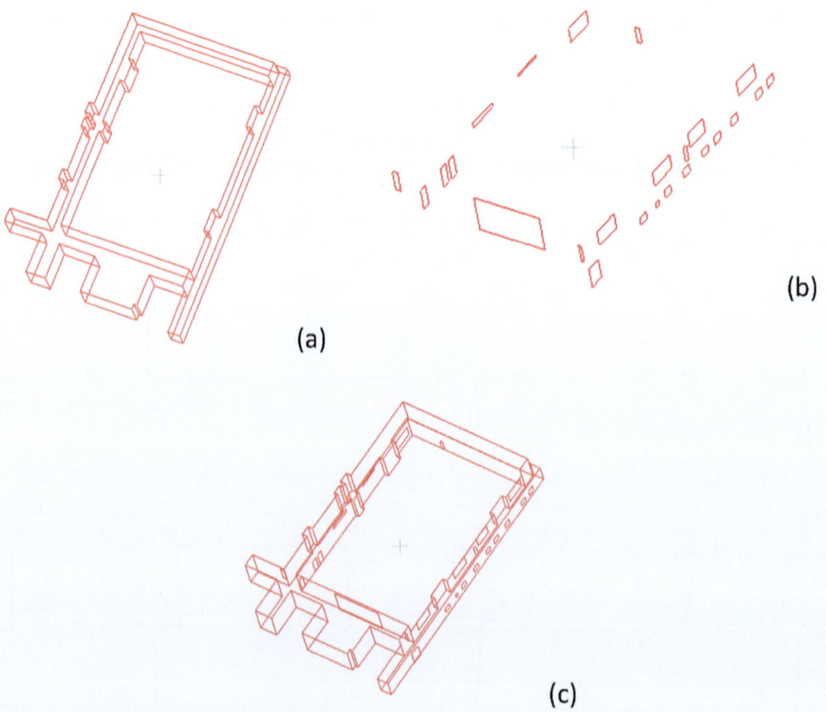

Figure 5.21. Line framework after optimization. (a) Floor, ceiling, and wall lines. (b) Windows and doors. (c) Combined line framework.

environments for point clouds. In the pipeline semantic line framework modeling, we first proposed a patch-based labeling result provided by the CRFs-based method. Then, to provide a rich structural representation of the cluttered building interior, we directly extracted a 3D line structure from the 3D point cloud. The extracted line structures are evitably contaminated by occlusion and disconnection. To cope with this degradation, we proposed a cGAN-based deep learning model to optimize the candidate line framework. By using different training data, the optimization method can deal with different types of environment. Line optimization can further instruct the point cloud quality evaluation to form a closed loop of the data collection and the representation of the indoor environment. Experimental results show that the presented modeling approach extracts a good line-framework semantic model for typical indoor point clouds acquired by the backpacked laser scanning systems.

The future work in indoor modeling might be in the following aspects: first, the laser scanning system can only measure the surface of the indoor environment, so registering or combining the surface model to the construction model or design model will be useful for BIM applications. Second, the current works focuses on the major frameworks such as walls, ceilings, doors, windows or floors, however, there are other fixed features, such as switches, lamps, plugins and wall signs that are also often wanted by applications. The detection, classification, and identification of these features is an interesting research direction.

NOTES

1 Velodyne LiDAR Datasheet. Available from: www.velodyne.com.
2 FLANN – Fast Library for Approximate Nearest Neighbors, www.cs.ubc.ca/research/flann/.

REFERENCES

Armeni, I., Sener, O., Zamir, A., Jiang, H., Brilakis, I., Fisher, M. & Savarese, S. (2016) 3D semantic parsing of large-Scale indoor spaces. *Proceedings of the IEEE Conference on Computer Vision and Pattern Recognition*. pp.1534–1543. Available from: https://ieeexplore.ieee.org/document/7780539

Atanacio, J., Nez, G., Gonza, B., Lez, J.J. & Hurtado, J.B.B. (2011) LiDAR Velodyne HDL-64e calibration using pattern planes. *International Journal of Advanced Robotic Systems*, 8(5), 70–82.

Babacan, K., Jung, J., Wichmann, A., Jahromi, B.A., Shahbazi, M., Sohn, G. & Kada, M. (2016) Towards object driven floor plan extraction from laser point cloud, International Archives of the Photogrammetry. *Remote Sensing and Spatial Information Sciences*, 41.

Bosse, M., Zlot, R. & Flick P. (2012) Zebedee: Design of a spring-mounter 3D range sensor with application to mobile mapping. *IEEE Transaction on Robotics*, 28(5), 1104–1119.

Boykov, Y., Veksler, O. & Zabih, R. (2001) Fast approximate energy minimization via graph cuts. *IEEE Transactions on Pattern Analysis and Machine Intelligence*, 23(11), 1222–1239.

Desolneux, A., Moisan, L. & Morel, J.M. (2000) Meaningful alignments. *International Journal of Computer Vision*, 40(1), 7–23.

Fischler, M.A. & Bolles, R.C. (1981) Random sample consensus: A paradigm for model fitting with applications to image analysis and automated cartography. *Communications of the ACM*, 24(6), 381–395.

Gong, Z., Wen, C., Wang, C. & Li, J., (2018) A target-free automatic self-calibration approach for multibeam laser scanners. *IEEE Transactions on Instrumentation and Measurement*, 67(1), 238–240.

Goodfellow, I., Pouget-Abadie, J., Mirza, M., Xu, B., Warde-Farley, D. & Ozair, S. (2014) Generative adversarial networks. *Proceedings of Advances in Neural Information Processing Systems*, December 2014, pp.2672–2680.

He, M.H., Zhao, H., Davoine, F. & Cui, J. (2013) Pairwise LIDAR calibration using multi-type 3D geometric features in natural scene. *Proceedings of International Conference on Intelligent Robots and Systems*, vol. 8215, November 2013, pp. 1828–1835.

Isola, P., Zhu, J.Y., Zhou, T. & Efros, A.A. (2017) Image-to-image translation with conditional adversarial networks. *Proceedings of the IEEE Conference on Computer Vision and Pattern Recognition*, July 2017, pp.5967–5976.

Jung, J., Hong, S., Yoon, S., Kim, J. & Heo, J. (2015) Automated 3D wireframe modeling of indoor structures from point clouds using constrained least-squares adjustment for as-built BIM. *Journal of Computing in Civil Engineering*, 30(4), 04015074.

Lin, Y., Wang, C., Chen, B., Zai, D. & Li, J. (2017) Facet segmentation-based line segment extraction for large-scale point clouds. *IEEE Transactions on Geoscience and Remote Sensing*, 55(9), 4839–4854.

Lin, Y., Wang, C., Cheng, J., Chen, B., Jia, F., Chen, Z. & Li, J. (2015) Line segment extraction for large scale unorganized point clouds. *ISPRS Journal of Photogrammetry and Remote Sensing*, 102, 172–183.

Liu, C., Schwing, A.G., Kundu, K., Urtasun, R. & Fidler, S. (2015) Rent3d: Floor-plan priors for monocular layout estimation. *Proceedings of the IEEE Conference on Computer Vision and Pattern Recognition*, June 2015, pp.3413–3421.

Michailidis, G.T. & Pajarola, R. (2016) Bayesian graph-cut optimization for wall surfaces reconstruction in indoor environments. *The Visual Computer*, 1347–1355.

Muhammad, N. & Lacroix, S. (2010) Calibration of a rotating multi-beam lidar. *Proceedings of the IEEE International Conference on Intelligent Robots and Systems*, October 2010, pp.5648–5653.

Munoz, D., Vandapel, N. & Hebert, M. (2009b) Onboard contextual classification of 3-D point clouds with learned high-order Markov random fields. *Proceedings of the IEEE Conference on Robotics and Automation*, May 2009, pp.2009–2016.

Mura, C., Mattausch, O., Villanueva, A.J., Gobbetti, E. & Pajarola, R. (2014) Automatic room detection and reconstruction in cluttered indoor environments with complex room layouts. *Computers and Graphics*, 44, 20–32.

Nouira, H., Deschaud, J.E. & Goulette, F. (2016) Point cloud refinement with a target-free intrinsic calibration of a mobile multi-beam LIDAR system. *XLI-B3*. pp.359–366. Available from: https://www.researchgate.net/publication/307522010_POINT_CLOUD_REFINEMENT_WITH_A_TARGET-FREE_INTRINSIC_CALIBRATION_OF_A_MOBILE_MULTI-BEAM_LIDAR_SYSTEM

Ochmann, S., Vock, R., Wessel, R. & Klein, R. (2016) Automatic reconstruction of parametric building models from indoor point clouds. *Computers and Graphics*, 54, 94–103.

Ochmann, S., Vock, R., Wessel, R., Tamke, M. & Klein, R. (2014) Automatic generation of structural building descriptions from 3D point cloud scans. *Proceedings of the IEEE Conference on Computer Graphics Theory and Applications*. pp.1–8.

Oesau, S., Lafarge, F. & Alliez, P. (2014) Indoor scene reconstruction using feature sensitive primitive extraction and graph-cut. *ISPRS Journal of Photogrammetry and Remote Sensing*, 90, 68–82.

Perez-Yus, A., Lopez-Nicolas, G. & Guerrero, J.J. (2016) Peripheral expansion of depth information via layout estimation with fisheye camera. *Proceedings of the European Conference on Computer Vision*, October 2016, pp.396–412.

Ronneberger, O., Fischer, P. & Brox, T. (2015) U-net: Convolutional networks for biomedical image segmentation. *Proceedings of the International Conference on Medical Image Computing and Computer-Assisted Intervention*, October 2015, pp.234–241.

Rusu, R.B., Blodow, N. & Beetz, M. (2009) Fast Point Feature Histograms (FPFH) for 3D registration. *Proceedings of the IEEE Conference on Robotics and Automation*, May 2009, pp.3212–3217.

Segal, A., Haehnel, D., & Thrun, S. (2009). Generalized-icp. In Robotics: science and systems, Vol. 2, No. 4, June 2009, p. 435.

Simonyan, K. & Zisserman, A. (2014) Very deep convolutional networks for large-scale image recognition. *Proceedings of the International Conference on Learning Representations*, 2014, arXiv preprint arXiv:1409.1556.

Von Gioi, R.G., Jakubowicz, J., Morel, J.M. & Randall, G. (2010) LSD: A fast line segment detector with a false detection control. *IEEE Transactions on Pattern Analysis and Machine Intelligence*, 32(4), 722–732.

Wang, C., Hou, S., Wen, C., Gong, Z., Li, Q., Sun, X. & Li, J. (2018) Semantic line framework-based indoor building modeling using backpacked laser scanning point cloud. *ISPRS Journal of Photogrammetry & Remote Sensing*, in press. https://doi.org/10.1016/j.isprsjprs.2018.03.025.

Wang, W., Sakurada, K. & Kawaguchi, N. (2016) Incremental and enhanced scanline-based segmentation method for surface reconstruction of sparse LiDAR data. *Remote Sensing*, 8(11), 967.

Wen, C., Pan, S., Wang, C. & Li, J. (2016) An indoor backpack system for 2-D and 3-D mapping of building interiors. *IEEE Geoscience and Remote Sensing Letters*, 13(7), 992–996.

Wen, C., Qin, L., Zhu, Q., Wang, C. & Li, J.J. (2014) Three-dimensional indoor mobile mapping with fusion of two-dimensional laser scanner and RGB-D camera data. *IEEE Geoscience and Remote Sensing Letters*, 11(4), 843–847.

Xiong, X., Adan, A., Akinci, B. & Huber, D. (2013) Automatic creation of semantically rich 3D building models from laser scanner data. *Automation in Construction*, 31, 325–337.

Zhang, J. & Singh, S. (2014) LOAM: Lidar odometry and mapping in real-time. *Robotics: Science and Systems Conference (RSS)*. Available from: https://www.researchgate.net/publication/282704722_LOAM_Lidar_Odometry_and_Mapping_in_Real-time

Zhuang, Y.N., Jiang, N., Hu, H. & Yan, F. (2013) 3-d-laser-based scene measurement and place recognition for mobile robots in dynamic indoor environments. *IEEE Transactions on Instrumentation and Measurement*, 62(2), 438–450.

Zhuang, Y.N., Yan, F. & Hu, H. (2014) Automatic extrinsic self-calibration for fusing data from monocular vision and 3-D laser scanner. *IEEE Transactions on Instrumentation and Measurement*, 63(7), 1874–1876.

Laser Scanning – Riveiro and Lindenbergh
© 2020 ISPRS, ISBN 978-1-138-49604-0

Chapter 6

Geometric point cloud quality

Derek D. Lichti

ABSTRACT: Terrestrial laser scanning is a versatile technology for rapidly collecting accurate and dense three-dimensional geometric measurements of structures. Coupled with powerful processing software to manipulate collected point clouds, laser scanning offers an attractive means to analyze structures for dimensional changes. However, a sound understanding of fundamental concepts including errors, geometric network design and error propagation is needed in order to make informed decisions concerning structural dimensions, or changes in dimensions, inferred from terrestrial laser scanner data. This chapter provides an overview of these concepts that aims to provide expert users with a concise reference and to inform nonexpert users of the many factors influencing geometric point cloud quality. Topics covered include fundamental definitions, geometric modelling of laser scanner observations, registration, random and systematic error sources and error propagation. A real-world example designed to illustrate the propagation of random errors completes the chapter.

1 INTRODUCTION

Laser scanning is a rapidly evolving technology that has been applied for a diverse range of applications, structural monitoring in particular due to the portability, high accuracy and high spatial and temporal resolutions of the instruments. Some notable examples include laboratory material testing (Gordon and Lichti, 2007; Olsen *et al.*, 2009; Park *et al.*, 2007), in situ measurement of dams (Alba *et al.*, 2006), pavement (Barbarella *et al.*, 2017), radio telescopes (Holst *et al.*, 2017) and large industrial sites (Hullo *et al.*, 2015) and others as described in detail in a recent review (Mukupa *et al.*, 2016).

Laser scanning systems comprise two salient components: the instrument for collecting data and the accompanying processing software for manipulation of the data. Many different instruments on different platforms (static and mobile) are available from a range of manufacturers. The focus in this chapter is terrestrial laser scanning (TLS). TLS is defined herein as a static instrument mounted atop a tripod that collects three-dimensional (3D) measurements – a point cloud – throughout a large angular field of view by deflecting the laser rangefinder in two orthogonal dimensions.

TLS instrument technology is now readily available and has advanced considerably in the last two decades. Scanners are now much more compact, lightweight (5–10 kg, but as low as 1 kg) and easy to use. The specifications and capabilities of different models of course vary considerably between manufacturers, but it can be generally stated that modern TLS instruments can acquire up to one million points per second with sub-centimetre precision. Some scanners are capable of making measurements up to a few hundreds of metres, while others can measure range up to a few kilometres. Integrated sensors (GNSS receiver, tilt sensors, laser plummet) provide additional information for instrument orientation as well as colour information collected with digital camera to aid data interpretation. Most modern instruments can collect data within a near-spherical field of view to provide millions of point measurements of the surfaces of a structure.

Powerful point cloud processing software allows operators to manipulate (e.g. scan registration, point cloud editing) and model (e.g. TIN surfaces, geometric primitives) laser scanner data in order to provide information about a structure. Such information may be used to determine the current state of the structure in terms of its dimensions and/or the presence of any surficial defects. The collection of 3D observations at multiple points in time (epochs) allows quantification of a structure's displacement or deformation. Laser scanning technology is thus a powerful tool to support informed decision-making about the health of a structure.

In addition to the aforementioned benefits, other factors make laser scanning technology attractive for structural monitoring. Potential users may be very impressed by the manufacturer-reported instrument specifications, not to mention the rich detail captured in dense 3D point clouds. Moreover, TLS instruments are very user friendly, so expert knowledge in geomatics engineering is not necessarily required for their operation to collect data of a complex scene. It can be easily argued, though, that structural monitoring is an application area in which the possession of such expertise is strongly recommended. Inaccurate measurements may result in a defective or faulty structure passing a health monitoring inspection.

The objective of geometric deformation analysis is to estimate deformation at the limit of the highest achievable precision at a specified confidence level (Caspary *et al.*, 1990). This can be realized with judicious instrument selection, strong observation network configuration design, careful observation procedure, application of systematic error corrections and rigorous analysis methodology. These are not trivial processes and require expert knowledge. Incorrect conclusions drawn from the deformation analysis can have disastrous consequences if rigorous procedures are not followed.

In the context of TLS, users should be aware of all contributing error sources in order to clearly understand the real capabilities of such measurement systems. Moreover, they should understand and apply fundamentals of network design to maximize point precision and, hence, maximize deformation detection sensitivity. The aim of this chapter is to provide a concise overview of contributing factors to TLS data quality in terms of fundamental methodology. The intended audience comprises two groups:

1 Geomatics engineering professionals who already understand the relevant methodologies and wish to apply TLS for structural engineering measurement tasks; and
2 Professionals with non-geomatics backgrounds (e.g. structural engineers) who lack knowledge of geomatics methods but desire a fundamental understanding of them to be able to properly utilize TLS.

This chapter is organized as follows. First, the geometric observations collected by TLS instruments and derived co-ordinates are presented. Some important fundamental concepts are discussed next, because these terms are often used interchangeably (incorrectly), misunderstood and/or abused, so it is important to provide clear definitions. Since a single scan is rarely sufficient for a monitoring project, the process of registering multiple scans into a common co-ordinate system is treated. This is followed by random and systematic error models for the TLS observations. Geometric network design, the location of registration targets and instrument stations, is a critical part of TLS structural monitoring, so it is treated next, with more emphasis on the datum and configuration problems. Least-squares estimation methodology brings together the observation equations, error models and datum definition to solve the registration problem. A short section follows on point radiation and the associated variance propagation. The chapter concludes with a small, real-world example to demonstrate the contributions of different sources of error to point cloud precision. The subject of the example is a portion of an aqueduct structure that has been scanned as part of its health assessment.

As is often the case with a book such as this, the number of pages is limited. It is therefore not possible to treat any of the topics with considerable depth. Relevant references comprising fundamental textbooks, landmark papers and recent publications reporting the latest developments are thus cited throughout the chapter so readers can find additional details.

2 TLS OBSERVATIONS

The raw geometric observations collected by a TLS instrument are range, ρ, horizontal direction, θ, and elevation angle, α (Figure 6.1). Alternatively, the angle with respect to the instrument's vertical axis, the zenith angle, may be used instead of the elevation angle. Pulse time-of-flight

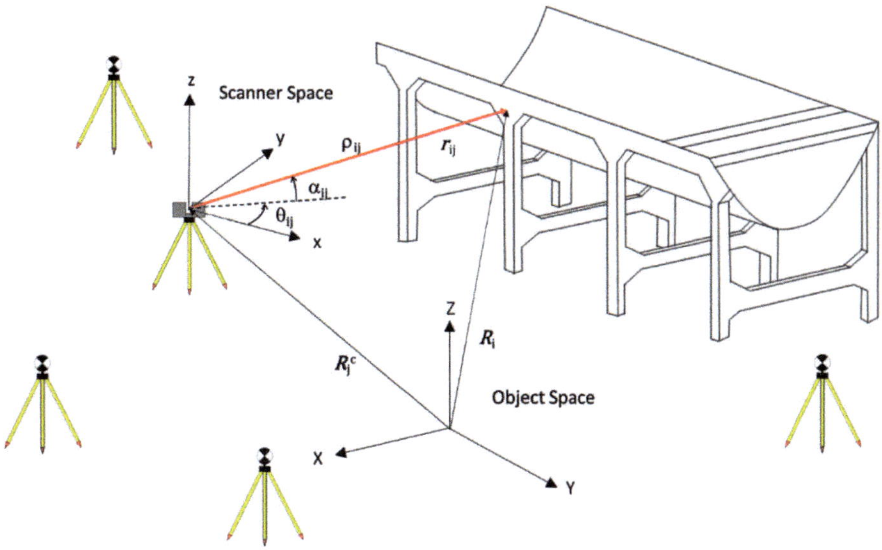

Figure 6.1. Terrestrial laser scanning for structural monitoring.

and phase difference are the range-finding methods used in TLS. Some instruments feature both. Optical encoders measure the angular position of the beam deflection mechanism, which may be a rotating mirror (monogonal or polygonal), an oscillating mirror or the entire instrument head. All measurements are made with respect to the instrument's internal co-ordinate system. The spherical co-ordinates are unseen by most users since scanner software typically outputs Cartesian co-ordinates, (x,y,z), which are easily derived as follows.

$$\boldsymbol{r}_{ij} = \begin{bmatrix} x \\ y \\ z \end{bmatrix}_{ij} = \begin{bmatrix} \rho cos\alpha cos\theta \\ \rho cos\alpha sin\theta \\ \rho sin\alpha \end{bmatrix}_{ij} \tag{1}$$

3 RELEVANT FUNDAMENTAL CONCEPTS

TLS measurements are affected by systematic, gross and random errors. Systematic errors are repeatable phenomena (biases) that arise due to imperfect instrument manufacture or assembly and environmental conditions. In principle, these errors can be modelled and corrections applied to remove or reduce their effect on the observations. For some error sources, such as non-orthogonality of instrumental axes, the physical cause is apparent and corresponding functional models are readily developed. For others, such as multi-path reflections, a universal correction model does not exist. Such errors are considered gross errors (or outliers or blunders) and must be identified and removed either manually or semi-automatically. For TLS scan registration, the ability to detect gross errors is maximized with strong geometric network design. For manual point cloud editing, expert knowledge and/or local statistical measures are used. The third category is random errors, which cannot be predicted. Their behaviour is described with a stochastic model in terms of moments of their probability density function. See Mikhail and Ackermann (1976), for example, for more details.

It is unfortunately a frequent occurrence that fundamental terminology is used incorrectly in the context of TLS point clouds. Accuracy, precision and resolution are often used interchangeably. Precision describes closeness to the sample mean. In the first instance this is the variance (or standard deviation) of the raw geometric TLS observations (range and angles). Variance propagation (Mikhail and Ackermann, 1976) is used to propagate observation precision through the TLS positioning models to determine point cloud co-ordinate precision. As will be seen, this is a function of many variables.

Accuracy is the closeness to the true value of a quantity. Of course, true values can never be known, so an independent source of information is required. For TLS, checkpoints surveyed by total station instruments are commonly used for this purpose. Three-dimensional co-ordinates of target points determined by TLS can be compared with those from the survey to quantify accuracy in terms of statistics computed from the co-ordinate differences. Inclusion of redundant observations is a critical aspect in the design of a deformation monitoring network as it not only improves precision but also allows the detection of biases in the observations. Reliability theory allows one to quantify the minimally detectable bias that may be identified for a given network design (Teunissen, 2006).

Resolution is related to information content. A TLS instrument can be considered as a device that collects range measurements in equal increments of arc and produces a range image. The finest level of detail that can be resolved from the range image is governed by the Nyquist interval, twice the sampling interval. The laser beam width acts analogously to an integrating detector of an imaging sensor in that it introduces uncertainty about the exact angular location of the range measurement and, thus, reduces resolution. Other factors such as the quantization of the measurements from the angle encoder device add additional uncertainty. All factors can be combined via transfer functions to obtain a final resolution measure. See Chen *et al.* (2017), Lichti and Jamtsho (2006) and Pesci *et al.* (2011) for further details.

4 REGISTRATION

Rarely is a single scan sufficient to completely cover a large and/or complex scene. In order to achieve complete coverage of an extended and/or complex structure with high resolution, multiple scans captured from different locations are required. Each scan must be transformed from its internal scanner space into a common, externally defined system, object space. Both are defined as right-handed co-ordinate systems in TLS methodology. A 3D rigid body transformation comprising three translational and three rotational degrees of freedom is the model used for scan registration. The rotation may be parameterized by Euler angles, angle and axis or quaternion. See Lichti and Skaloud (2010) for details.

The rigid body parameters can be directly observed, estimated from observations of independently measured targets or estimated from the point cloud data. The direct observation of position can be achieved by forced centring over a known control point or with observations collected with an integrated GNSS sensor. Angular orientation is handled by instrument levelling and making a back-sight to a reference target. Indirect observation, the focus hereafter, relies upon surveyed targets to estimate the six transformation parameters, which can be reduced to four if the instrument is precisely levelled. If sufficient overlap exists between two adjacent scans, the iterative closest point, ICP, method (Besl and McKay, 1992; Chen and Medioni, 1992) or one of its many variants can be used. The basis of the ICP is to register one scan to the other by minimizing the sum of squared distances between matching closest points or matching point and local surface.

The transformation of object point i observed in scanner space j can be parameterized at least two ways. The most common is in terms of derived scanner space Cartesian co-ordinates.

$$\begin{bmatrix} x \\ y \\ z \end{bmatrix}_{ij} = \boldsymbol{r}_{ij} = \boldsymbol{M}_j \left(\begin{bmatrix} X \\ Y \\ Z \end{bmatrix}_i - \begin{bmatrix} X \\ Y \\ Z \end{bmatrix}_j^c \right) = \boldsymbol{M}_j \left(\boldsymbol{R}_i - \boldsymbol{R}_j^c \right) \tag{2}$$

where $\boldsymbol{r}_{ij}$ is the observed scanner space vector of point i from scanner location j; $\boldsymbol{R}_i$ is the object space vector of point i; $\boldsymbol{R}_j^c$ is the object space vector of scanner location j; and $\boldsymbol{M}_j$ is the rotation matrix from object space to scanner space j.

The six registration parameters are typically estimated from observations of surveyed artificial targets placed throughout the TLS instrument's field of view. Since the TLS Cartesian co-ordinates are derived from the spherical co-ordinates, they are inherently correlated. Ignoring the correlation leads to optimistic precision estimates. Moreover, the precision of the Cartesian coordinates is usually assumed to be homogeneous, which is not true. The more fundamentally sound approach is to parameterize the transformation in terms of spherical coordinates, which more accurately models the observation process and allows construction of a proper stochastic model for each observation and, hence, more rigorous variance propagation.

$$\rho_{ij} + \varepsilon_{\rho_{ij}} = \sqrt{x_{ij}^2 + y_{ij}^2 + z_{ij}^2} + \Delta\rho_{ij} \tag{3}$$

$$\theta_{ij} + \varepsilon_{\theta_{ij}} = \arctan\left(\frac{y_{ij}}{x_{ij}}\right) + \Delta\theta_{ij} \tag{4}$$

$$\alpha_{ij} + \varepsilon_{\alpha_{ij}} = \arctan\left(\frac{z_{ij}}{\sqrt{x_{ij}^2 + y_{ij}^2}}\right) + \Delta\alpha_{ij} \tag{5}$$

Note different forms of the angle observation equations may exist depending on instrument architecture (Al-Manasir and Lichti, 2015; Lichti, 2010). The observation equations have been augmented with random error terms ($\varepsilon_{\rho_{ij}}$, $\varepsilon_{\theta_{ij}}$, $\varepsilon_{\alpha_{ij}}$) and systematic components ($\Delta\rho_{ij}$, $\Delta\theta_{ij}$, $\Delta\alpha_{ij}$).

5 ERROR MODELS

The random error behaviour is encapsulated by the stochastic model: zero-mean with variance that may be a function of several variables. Range observations are influenced by instrument jitter (Amann *et al.*, 2001), range, surface reflectivity, the atmosphere and incidence angle (Soudarissanane *et al.*, 2011). Wujanz proposed a return-signal intensity-based model for range errors that accounts for range, reflectivity and incidence angle (Wujanz *et al.*, 2017). Surface reflectivity is generally unknown, so a priori construction of a stochastic model on this basis for network design is challenging. Angular observations are affected by jitter, wobble (Montagu, 2004), the laser beam width and the atmosphere. Additional random error sources include instrument levelling and the registration target co-ordinates.

The systematic errors comprise both instrumental and environmental sources. As mentioned, some environmental errors are difficult to model and control and can only be removed by data editing. Range biases due to ambient atmospheric conditions can be corrected with standard models (Wang *et al.*, 2016) but are frequently small due to the short distances over which TLS data are typically collected. Instrumental errors can be categorized as individual component errors and assembly errors. Examples of errors due to individual components include scale errors and, for phase difference rangefinders, periodic range errors. Assembly errors include axis offsets, eccentricities, non-orthogonality and wobble. Several instrumental systematic error models comprising many additional parameters (APs) have been proposed (Lichti, 2007; Medić *et al.*, 2017; Muralikrishnan *et al.*, 2015; Reshetyuk, 2010), but a model comprising only a basic set of four APs (Lichti, 2010) is presented here:

$$\Delta\rho_{ij} = a_0 \tag{6}$$

$$\Delta\theta_{ij} = b_1 \sec\alpha_{ij} + b_2 \tan\alpha_{ij} \tag{7}$$

$$\Delta\alpha_{ij} = c_0 \tag{8}$$

where a_0 is the rangefinder offset; b_1 is the collimation axis error; b_2 is the trunnion axis error; and c_0 is the vertical circle index error.

The usual procedure in the closely related fields of photogrammetry and geodesy is to estimate instrumental errors beforehand via a calibration procedure then apply corrections to subsequently acquired data. See Lichti (2007 and Reshetyuk (2010) for details about the self-calibration approach. On-site self-calibration of the parameters is possible with appropriate network design. Uncertainty in the calibration parameters should be included in the variance propagation.

6 NETWORK DESIGN

A TLS network is a geometric configuration of inter-connected instrument stations and targets around the structure to be measured. It comprises two sets of parameters: the scan registration or orientation parameters and the 3D target coordinates. The design of a network can be decomposed into four processes (Fraser, 1984). Datum definition is the zero order design (ZOD) problem. The datum is the position, orientation and scale of a network within a chosen co-ordinate system. TLS networks inherently contain datum defects since the scanner-space observations provide relative rather than absolute positioning information. TLS networks suffer from three position and three orientation defects. Unlike photogrammetric networks, there is no scale defect since network scale is defined by the range measurements. Datum defects cause a singular system of equations in the registration problem.

The datum defects are removed either by eliminating unknowns or by adding additional constraints. Direct sensor orientation provides additional observations to define the datum elements. In target-based registration, the typical convention is to treat the target co-ordinates as constant (i.e. fixed) quantities, thereby removing them from the registration problem. As a general rule, the accuracy of the surveyed target co-ordinates should be an order of magnitude higher than what can be achieved by scanning. Achieving this condition can be challenging, though the target co-ordinates can be treated as observable quantities and their precision included in the error propagation.

Usually more than the minimum number of co-ordinates is constrained when fixed targets are used. Over-constraining the datum in this way can lead to biases in the network co-ordinates that could impact the estimation of deformation. A minimally constrained datum is therefore preferred, though the choice of which six co-ordinates to fix may be arbitrary and has an impact on the co-ordinates and their precision. Inner constraints (Teunissen, 2006) is an approach that allows all network parameters, or a subset thereof, to contribute to the datum without introducing biases and yields optimum precision for the constrained parameters (Fraser, 1982).

First-order design (FOD) is the process of determining the geometric configuration of network stations needed to meet project specifications. In the TLS context, this involves deciding upon the number, the location and the orientation of instrument stations to ensure complete object coverage and, ideally, to minimize high-incidence angles, and the placement of targets to accurately register the individual scans into object space. Obviously, the goal is to minimize the number of instrument setups and targets to reduce labour costs. Recent efforts from several researchers have focused on optimal TLS network FOD (Ahn and Wohn, 2016; Heidari Mozaffar and Varshosaz, 2016; Jia and Lichti, 2019; Soudarissanane and Lindenbergh, 2011; Wujanz and Neitzel, 2016).

Some basic rules should be followed in the first order design of TLS networks. Although six co-ordinates from three points is the minimum requirement for the solution to the registration problem, at least four targets should be observed from each scan in order to provide some redundancy. Targets should be distributed as far apart as possible. Linear target arrangements must be avoided since this condition leads to a singular system of equations.

The final two processes are second-order design (SOD) and third-order design (TOD). SOD is concerned with the stochastic model design and is governed by instrument choice, target measurement algorithm and observation procedure. Once an initial design has been identified, the aim of TOD is to identify how it can be enhanced. Additional targets and/or instrument locations may be considered.

7 METHODOLOGY

The framework for TLS network estimation problems such as registration (as well as instrument self-calibration) is the constrained Gauss-Markoff model, which comprises both functional and stochastic models. Complete details of the estimation procedure are beyond the scope; readers are referred to Mikhail and Ackermann (1976) and Teunissen (2003).

The functional model expresses the observation vector $\boldsymbol{\ell}$, plus random error vector ε, as some function of unknown parameters, $\boldsymbol{x}$.

$$\boldsymbol{\ell}+\varepsilon=f(x) \tag{9}$$

The observation vector comprises all TLS spherical co-ordinate observations, i.e.

$$\boldsymbol{\ell}=\begin{bmatrix}\rho_{11} & \theta_{11} & \alpha_{11} & \rho_{21} & \theta_{21} & \alpha_{21} & \ldots\end{bmatrix}^T \tag{10}$$

The parameter vector can be partitioned into three sets of variables

$$\boldsymbol{x}=\begin{bmatrix}\boldsymbol{x}_e^T & \boldsymbol{x}_a^T & \boldsymbol{x}_o^T\end{bmatrix} \tag{11}$$

where $\boldsymbol{x}_e$ is the set of registration parameters; $\boldsymbol{x}_a$ is the set of APs; and $\boldsymbol{x}_o$ comprises the object point co-ordinates. For registration from targets having fixed and known co-ordinates, only $\boldsymbol{x}_e$ is estimated, and the parameters for each scan can be determined independently. If the target co-ordinates are treated as unknowns, they must be observed or subjected to constraints to define the network datum. In this case, $\boldsymbol{x}_o$ is also estimated, and all variables are determined simultaneously. Though outside the scope of this chapter, all three parameter sets are simultaneously estimated in self-calibration. The functional relationship between the observations and the parameters is usually non-linear, which can be handled with a truncated (zero- and first-order terms only) Taylor series approximation, evaluated at a carefully chosen point of expansion, and an iterative solution procedure.

The accompanying datum constraint model expresses the functional relationship among (object space) parameters to define the datum.

$$\mathbf{0}=g(x) \tag{12}$$

The stochastic model is specified in terms of the first and second order the moments of the probability density function of the observation errors

$$E\{\varepsilon\}=\mathbf{0} \tag{13}$$

$$E\{\varepsilon\varepsilon^T\}=C_\ell \tag{14}$$

All observation errors are assumed to be uncorrelated. The full covariance matrix of observations, $\boldsymbol{C}_\ell$, is block diagonal comprising 3 × 3 blocks. Each 3 × 3 submatrix on the diagonal is also diagonal with the following form

$$\boldsymbol{C}_{\ell_{ij}} = \begin{bmatrix} \sigma_\rho^2 sec^2\beta_{ij} & 0 & 0 \\ 0 & \sigma_\theta^2 & 0 \\ 0 & 0 & \sigma_\alpha^2 \end{bmatrix} \tag{15}$$

where σ_ρ^2 is the range error variance at normal incidence; β_{ij} is the incidence angle of observation i collected from scan j, which can be estimated from the point position vector and a local estimate of the surface normal. The angular variances, σ_ρ^2 and σ_α^2, comprise the lumped effect of the aforementioned contributing sources.

The precision of the estimated parameters is given by Mikhail and Ackermann (1976).

$$\boldsymbol{C}_x = \left(\frac{\partial \boldsymbol{f}}{\partial \boldsymbol{x}}^T \boldsymbol{C}_\ell^{-1} \frac{\partial \boldsymbol{f}}{\partial \boldsymbol{x}} \right)^{-1} \tag{16}$$

where the partial derivative indicates the Jacobian of $\boldsymbol{f}$.

8 RADIATION

Radiation is the task of uniquely determining the 3D co-ordinates of an object space point from a scan once the registration parameters have been estimated. The transformation of an observed point in the cloud can be computed by rearranging Equation 2.

$$\boldsymbol{R}_i = \boldsymbol{M}_j^T \boldsymbol{r}_{ij} + \boldsymbol{R}_j^c \tag{17}$$

Note that although the registration process is redundant, radiation is not. The covariance matrix of a radiated object point i, $\boldsymbol{R}_i$, is a function of several variables:

$$\boldsymbol{C}_{R_i} = \frac{\partial \boldsymbol{R}_i}{\partial \boldsymbol{x}_e} \boldsymbol{C}_{x_j} \frac{\partial \boldsymbol{R}_i}{\partial \boldsymbol{x}_e}^T + \frac{\partial \boldsymbol{R}_i}{\partial \ell_{ij}} \boldsymbol{C}_{\ell_{ij}} \frac{\partial \boldsymbol{R}_i}{\partial \ell_{ij}}^T + \frac{\partial \boldsymbol{R}_i}{\partial \boldsymbol{x}_a} \boldsymbol{C}_{x_a} \frac{\partial \boldsymbol{R}_i}{\partial \boldsymbol{x}_a}^T \tag{18}$$

where $\boldsymbol{C}_{xj}$ is the covariance matrix of registration parameters for scan j; and $\boldsymbol{C}_{\ell_{ij}}$ the 3 × 3 observation covariance matrix for point i in scan j; and $\boldsymbol{C}_{x_a}$ is the covariance matrix of the APs estimated from self-calibration.

9 EXAMPLE

A numerical example with real data is presented to illustrate the impact of the many influences on point cloud precision. The Brooks Aqueduct is a National and Provincial Historic Site in southeastern Alberta, Canada. It was constructed in the early twentieth century as part of an irrigation network serving the agricultural industry of the region. The reinforced concrete structure is an elevated, 6.5 m wide flume that carried water across a 20-m-deep, 3.2-km-wide valley (Manz *et al.*, 1989). Abandoned late in the twentieth century, the aqueduct has fallen into a state of disrepair. In 2017, TLS data were collected of two sections of the aqueduct to assess the current geometric state of the structure.

A subset of these data has been used to demonstrate the propagation of random errors stemming from various sources in a TLS survey of a structure. It comprises an 18-m-long section of the

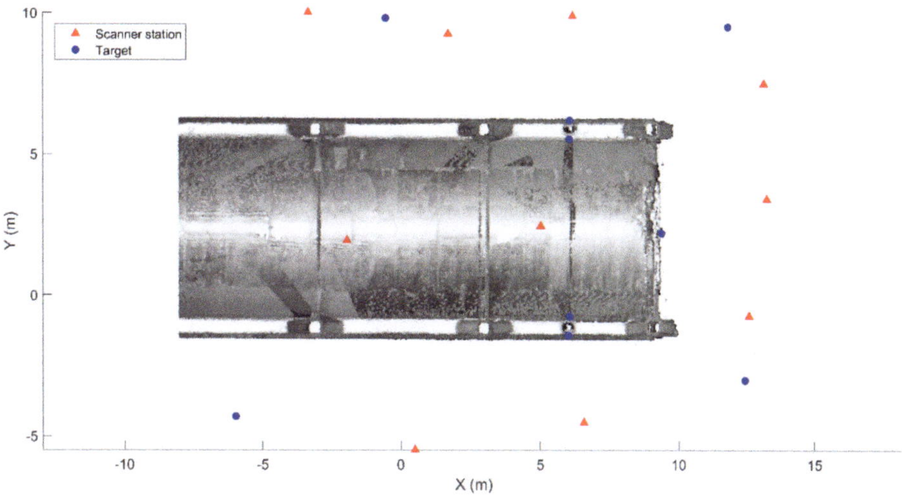

Figure 6.2. Plan view of the Brooks Aqueduct point cloud, scan locations and target locations.

aqueduct with maximum height of 8 m above ground. The section was scanned from 10 different locations around the aqueduct. Nine signalized targets were used for scanner orientation. Between four and nine targets were visible per scan. The point cloud, scan locations and targets are shown in Figure 6.2. A local, right-handed project coordinate system was adopted for this exercise. The scanner observations were parameterized in terms of spherical co-ordinates. Euler angles were used to model scanner rotation, and the transformation parameters were estimated from fixed control points without instrument levelling. The instrument is assumed to have been pre-calibrated. The registered point cloud comprises approximately 500 000 points after manual removal of the returns from the ground and vegetation and extraneous outliers. Estimates of the surface normal used in the incidence angle computation were computed from local best fit plane using principal components analysis (Shakarji, 1998).

TLS networks usually feature overlap of coverage from adjacent scans. As a result, co-ordinate quality and data density can vary considerably throughout a point cloud. Even though these data were acquired such that there was considerable overlap, it has been ignored in this example for the sake of presentation clarity. The points observed by each scan were artificially defined with non-overlapping polygons (Figure 6.3). As will be seen, this simplification leads to discontinuities in the graphical representation of scan data quality.

This dataset also features artificially introduced weak scanning geometry. In reality the western end of the aqueduct was scanned from another station (not shown). However, it is considered to have been scanned from fewer scans under sub-optimal conditions to demonstrate the impact of the incidence angle on point precision.

Table 6.1 gives the realistic precision estimates for the spherical observations and APs. Range observation precision modelled only as a function of incidence angle since a priori surface reflectivity is difficult to obtain and the distances are short (maximum 11.3 m), so the range-dependent contribution can be assumed to be insignificant. The angular observation variances include the contributions of all cited influences. All observation standard deviations are realistic but do not correspond to any particular instrument. The AP standard deviations (Table 6.2) are also realistic. All errors (in the observations and the APs) are assumed to be uncorrelated, which is also a very realistic assumption. Registration covariance matrices rigorously were computed from the solution with fixed target coordinates. The same observation standard deviations were used for the registration.

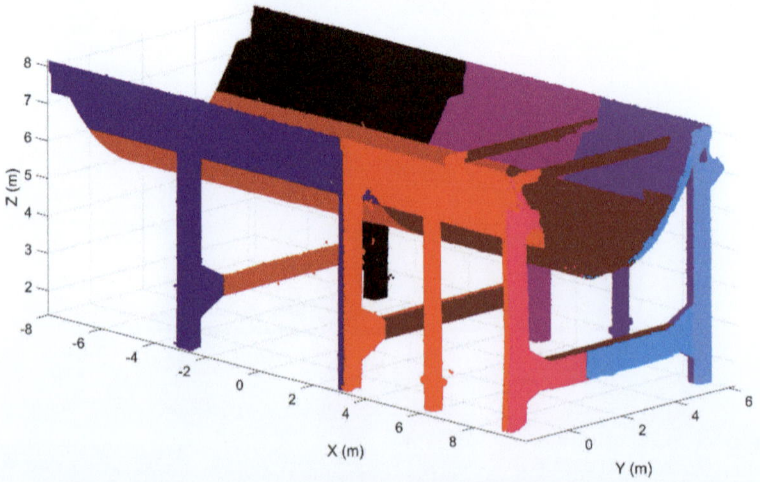

Figure 6.3. Brooks Aqueduct point cloud. Colour denotes points acquired from the same scan.

Table 6.1 Simulation parameters – observation precision

	σ_ρ *(mm)*	σ_θ *('')*	σ_α *('')*
Observations	±1	±30	±30

Table 6.2 Simulation parameters – AP precision.

	σ_{a0} *(mm)*	σ_{b1} *('')*	σ_{b2} *('')*	σ_{c0} *('')*
APs	±0.1	±10	±10	±10

For each point in the cloud, the 3 × 3 point covariance matrix was computed according to Equation 18. The square root of the largest eigenvalue was extracted and scaled by the root of the critical value of the Chi-squared distribution with three degrees of freedom, i.e. $\sqrt{\chi^2_{0.95,3}} = 2.796$ to obtain the semi-major axis of the 95% confidence ellipsoid as a single point precision measure. In addition, the eigenvalues were also computed from the covariance matrices of three contributing components: the observations, the registration parameters and the additional parameters. These are shown in Figures 6.4 through 6.7.

Overall point precision in the Brooks Aqueduct example ranges from ±3 mm to ±32 mm, with a root mean squared (RMS) value of ±12 mm. Precision varies smoothly within segments of the structure and degrades significantly with increasing incidence angle, notably at the west end. Discontinuities in precision exist at two locations. The discontinuities at the boundary between surfaces are expected due to significant change in incidence angle or change in the scan from which the data were acquired. The precision discontinuities that are visible within surfaces are due to the pre-defined scan extent polygons and are thus artificial.

The largest contribution to point cloud uncertainty is from the observations themselves, most notably the range measurements due to the incidence angle effect. The range of values is virtually the same as that of the overall precision. The registration component is the next largest, but the values range from ±2 mm to ±13 mm. Larger values occur where the aqueduct was farther from

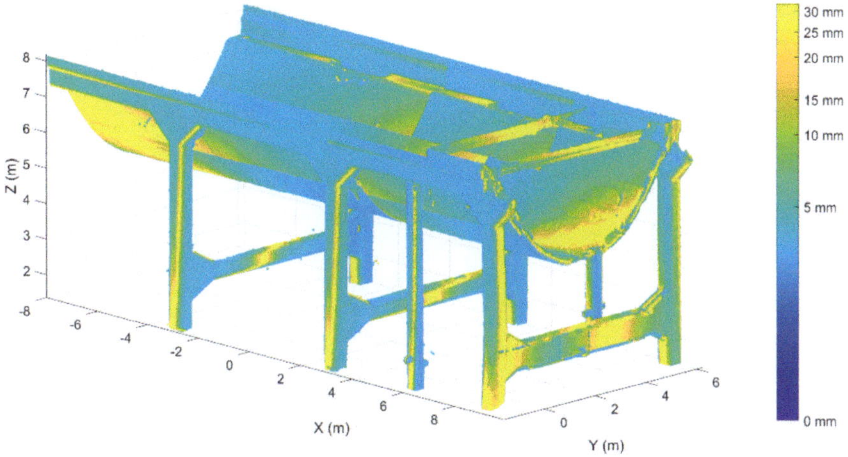

Figure 6.4. Point precision, observation component.

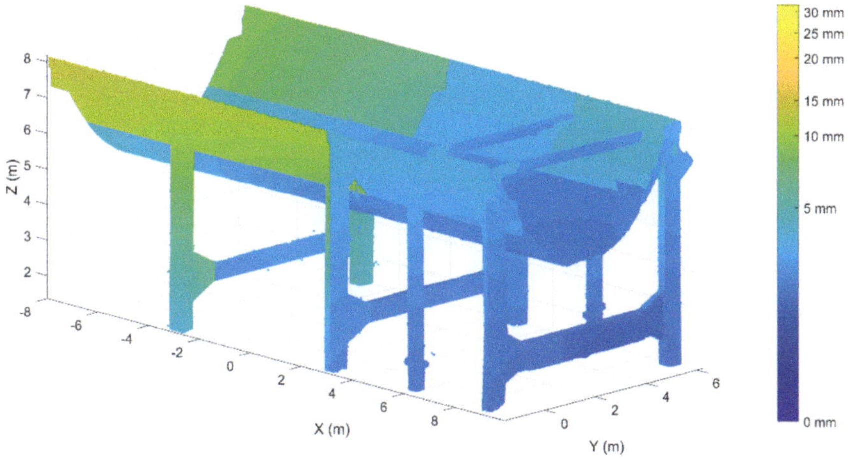

Figure 6.6. Point precision, additional parameter component.

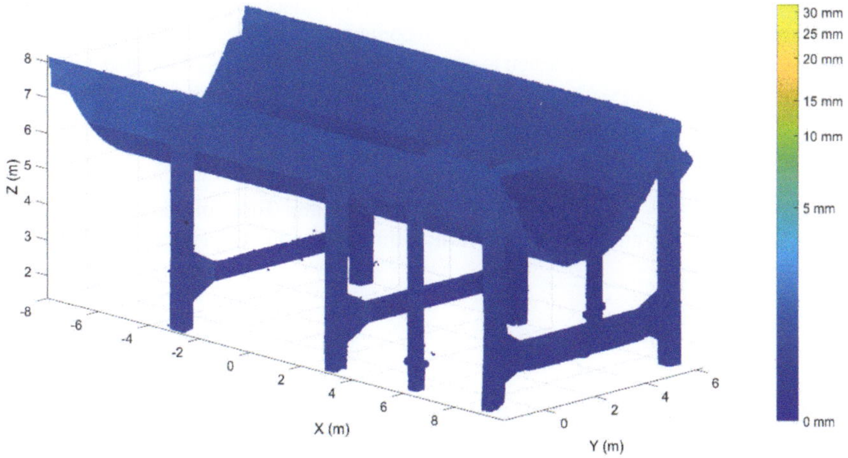

Figure 6.5. Point precision, registration component.

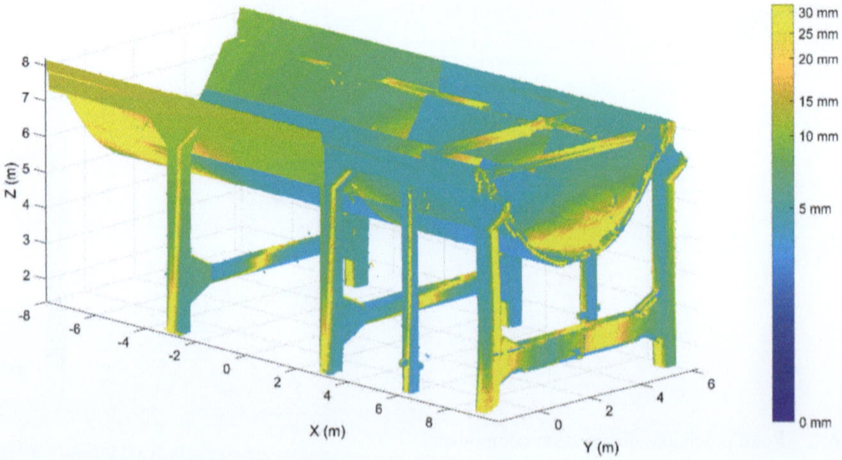

Figure 6.7. Point precision, overall.

the scanner, such as at the west end. The additional parameter component is the least significant in this example (±2 mm maximum) since the AP precision was very high. Many points were observed at high elevation angles, most notably the bottom of the flume, but the uncertainty was not large in these locations. Although high elevation angles lead to large uncertainty in direction due to the collimation axis and trunnion axis error models, the ranges are short, so these effects are not significant in comparison to the observation component.

10 SUMMARY

This chapter has presented an overview of the many factors that contribute to the quality of point clouds captured by terrestrial laser scanning instruments. The example is an attempt to illustrate just how large realistic precision estimates can be. Considering that the stated rangefinder precision is ±1 mm – often the most eye-catching of many TLS specifications – the RMS precision of ±13 mm may seem alarmingly large. The results must, of course, be interpreted carefully since they are specific to this example, which has been presented in a way so as to accentuate certain effects. However, they do demonstrate that point cloud precision can be much worse than one specification may suggest when all factors are rigorously considered in the stochastic model. The recommended practice is to perform this analysis for every project as part of the network design to ensure project specifications can be met.

11 ACKNOWLEDGEMENTS

The author acknowledges the support of Alberta Culture and Tourism and the assistance of Dr Peter Dawson, Adam Jahraus, Jeremy Steward, Zahra Hadavandsiri, Kaleel Al-Durgham, Fengman Jia and Shu Zhang.

REFERENCES

Ahn, J. & Wohn, K. (2016) Interactive scan planning for heritage recording. *Multimedia Tools and Applications*, 75, 3655–3675. https://doi.org/10.1007/s11042-015-2473-0.

Alba, M., Fregonese, L., Prandi, F., Scaioni, M. & Valgoi, P. (2006) Structural monitoring of a large dam by terrestrial laser scanning. ISPRS Archives – Volume XXXVI, Part 5. H.-G. Maas, D. Schneider (eds) *Proceedings of the ISPRS Commission V Symposium 'Image Engineering and Vision Metrology', September 25–27, 2006*, Dresden, Germany, p.6. Available from: https://www.isprs.org/proceedings/XXXVI/part5/

Al-Manasir, K. & Lichti, D.D. (2015) Self-calibration of a Lecia HDS7000 scanner. *Presented at the FIG Working Week 2015*. Sofia, Bulgaria. 12 pp.

Amann, M.-C., Bosch, T.M., Lescure, M., Myllylae, R.A. & Rioux, M. (2001) Laser ranging: A critical review of unusual techniques for distance measurement. *Optical Engineering*, 40, 10–20.

Barbarella, M., D'Amico, F., De Blasiis, M., Di Benedetto, A. & Fiani, M. (2017) Use of terrestrial laser scanner for rigid airport pavement management. *Sensors*, 18, 44. https://doi.org/10.3390/s18010044.

Besl, P. & McKay, N. (1992) A method for registration of 3-D shapes. *IEEE Transactions on Pattern Analysis and Machine Intelligence*, 14, 239–256.

Caspary, W.F., Haen, W. & Borutta, H. (1990) Deformation analysis by statistical methods. *Technometrics*, 32, 49. https://doi.org/10.2307/1269844.

Chen, X., Yu, K., Zhang, G., Hua, X., Wu, H. & An, Q. (2017) Precision estimation of the angular resolution of terrestrial laser scanners. *The Photogrammetric Record*, 32, 276–290. https://doi.org/10.1111/phor.12200.

Chen, Y. & Medioni, G. (1992) Object modelling by registration of multiple range images. *Image and Vision Computing*, 10, 145–155.

Fraser, C.S. (1982) Optimization of precision in close-range photogrammetry. *Photogrammetric Engineering and Remote Sensing*, 48, 561–570.

Fraser, C.S. (1984) Network design considerations for non-topographic photogrammetry. *Photogrammetric Engineering and Remote Sensing*, 50, 1115–1126.

Gordon, S.J. & Lichti, D.D. (2007) Modeling terrestrial laser scanner data for precise structural deformation measurement. *Journal of Surveying Engineering*, 133, 72–80.

Heidari Mozaffar, M. & Varshosaz, M. (2016) Optimal placement of a terrestrial laser scanner with an emphasis on reducing occlusions. *The Photogrammetric Record*, 31, 374–393. https://doi.org/10.1111/phor.12162.

Holst, C., Schunck, D., Nothnagel, A., Haas, R., Wennerbäck, L., Olofsson, H., . . . Kuhlmann, H. (2017) Terrestrial laser scanner two-face measurements for analyzing the elevation-dependent deformation of the onsala space observatory 20-m radio telescope's main reflector in a bundle adjustment. *Sensors*, 17, 1833. https://doi.org/10.3390/s17081833.

Hullo, J.-F., Thibault, G., Boucheny, C., Dory, F. & Mas, A. (2015) Multi-sensor as-built models of complex industrial architectures. *Remote Sensing*, 7, 16339–16362. https://doi.org/10.3390/rs71215827.

Jia, F. & Lichti, D. (2019) A model-based design system for terrestrial laser scanning networks in complex sites. *Remote Sensing*, 11, 1749. https://doi.org/10.3390/rs11151749

Lichti, D.D. & Jamtsho, S. (2006) Angular resolution of terrestrial laser scanners. *The Photogrammetric Record*, 21, 141–160.

Lichti, D.D. & Skaloud, J. (2010) Registration and calibration. In: *Airborne and Terrestrial Laser Scanning*. Caithness, UK: Whittles Publishing. pp.83–133.

Lichti, D.D. (2007) Error modelling, calibration and analysis of an AM–CW terrestrial laser scanner system. *The ISPRS Journal of Photogrammetry and Remote Sensing*, 61, 307–324. https://doi.org/10.1016/j.isprsjprs.2006.10.004.

Lichti, D.D. (2010) Terrestrial laser scanner self-calibration: Correlation sources and their mitigation. *The ISPRS Journal of Photogrammetry and Remote Sensing*, 65, 93–102. https://doi.org/10.1016/j.isprsjprs.2009.09.002.

Manz, D., Loov, R. & Webber, J. (1989) Brooks aqueduct. *Canadian Journal of Civil Engineering*, 16, 684–692.

Medić, T., Holst, C. & Kuhlmann, H. (2017) Towards system calibration of panoramic laser scanners from a single station. *Sensors*, 17, 1145. https://doi.org/10.3390/s17051145.

Mikhail, E. & Ackermann, F. (1976) *Observations and Least Squares*. New York, NY: IEP.

Montagu, J. (2004) Galvanometric and resonant scanners. In: *Handbook of Optical and Laser Scanning*. New York, NY: Marcel Dekker. pp.417–476.

Mukupa, W., Roberts, G.W., Hancock, C.M. & Al-Manasir, K. (2016) A review of the use of terrestrial laser scanning application for change detection and deformation monitoring of structures. *Survey Review*, 1–18. https://doi.org/10.1080/00396265.2015.1133039.

Muralikrishnan, B., Ferrucci, M., Sawyer, D., Gerner, G., Lee, V., Blackburn, C., . . . Palmateer, J. (2015) Volumetric performance evaluation of a laser scanner based on geometric error model. *Precision Engineering*, 40, 139–150. https://doi.org/10.1016/j.precisioneng.2014.11.002.

Olsen, M.J., Kuester, F., Chang, B.J. & Hutchinson, T.C. (2009) Terrestrial laser scanning-based structural damage assessment. *Journal of Computing in Civil Engineering*, 24, 264–272.

Park, H.S., Lee, H.M., Adeli, H. & Lee, I. (2007) A new approach for health monitoring of structures: Terrestrial laser scanning. *Computer-Aided Civil and Infrastructure Engineering*, 22, 19–30.

Pesci, A., Teza, G. & Bonali, E. (2011) Terrestrial laser scanner resolution: Numerical simulations and experiments on spatial sampling optimization. *Remote Sensing*, 3, 167–184. https://doi.org/10.3390/rs3010167.

Reshetyuk, Y. (2010) A unified approach to self-calibration of terrestrial laser scanners. *ISPRS Journal of Photogrammetry and Remote Sensing*, 65, 445–456. https://doi.org/10.1016/j.isprsjprs.2010.05.005.

Shakarji, C.M. (1998) Least-squares fitting algorithms of the NIST algorithm testing system. *Journal of Research of the National Institute of Standards and Technology*, 103, 633.

Soudarissanane, S.S. & Lindenbergh, R.C. (2011) Optimizing terrestrial laser scanning measurement set-up. *ISPRS Workshop Laser Scanning 2011, 29–31 August 2011*. Calgary, Canada; IAPRS, XXXVIII (5/W12), 2011. International Society for Photogrammetry and Remote Sensing (ISPRS).

Soudarissanane, S.S., Lindenbergh, R., Menenti, M. & Teunissen, P. (2011) Scanning geometry: Influencing factor on the quality of terrestrial laser scanning points. *ISPRS Journal of Photogrammetry and Remote Sensing*, 66, 389–399. https://doi.org/10.1016/j.isprsjprs.2011.01.005.

Teunissen, P.J.G. (2003) *Adjustment Theory*. VSSD, Delft, The Netherlands.

Teunissen, P.J.G. (2006) *Network Quality Control*. VSSD, Delft, The Netherlands.

Wang, J., Kutterer, H. & Fang, X. (2016) External error modelling with combined model in terrestrial laser scanning. *Survey Review*, 48, 40–50. https://doi.org/10.1080/00396265.2015.1097589

Wujanz, D., Burger, M., Mettenleiter, M. & Neitzel, F. (2017) An intensity-based stochastic model for terrestrial laser scanners. *ISPRS Journal of Photogrammetry and Remote Sensing*, 125, 146–155. https://doi.org/10.1016/j.isprsjprs.2016.12.006.

Wujanz, D. & Neitzel, F. (2016) Model based viewpoint planning for terrestrial laser scanning from an economic perspective. *ISPRS – The International Archives of the Photogrammetry, Remote Sensing and Spatial Information Sciences*, XLI-B5, 607–614. https://doi.org/10.5194/isprsarchives-XLI-B5-607-2016.

Laser Scanning – Riveiro and Lindenbergh
© 2020 ISPRS, ISBN 978-1-138-49604-0

Chapter 7

Semantic segmentation of dense point clouds

Martin Weinmann

ABSTRACT: This chapter addresses the semantic segmentation of dense point clouds. For this purpose, different strategies may be applied. On the one hand, the rather classic strategy may be applied, which relies on the use of hand-crafted features serving as input to a standard classification technique, which, in turn, delivers a semantic labeling with respect to defined object classes. On the other hand, the strategy relying on the interplay between traditional classification and segmentation techniques allows assigning both class-aware labels with respect to defined object classes and instance-aware labels with respect to objects in the considered scene. Beyond these strategies, the strategy of involving modern deep learning techniques is nowadays commonly applied, as it tends to outperform the other strategies on recent benchmark datasets. After revisiting the fundamentals of the three strategies, a variety of commonly used benchmark datasets for evaluating the performance of a semantic segmentation approach are presented, whereby most of these datasets have been acquired via terrestrial or mobile laser scanning within urban areas.

1 INTRODUCTION

For us humans, the visualization of a dense point cloud such as the one depicted in Figure 7.1 already allows reasoning about specific properties of a considered scene. On the one hand, a diversity of objects such as buildings, ground, road inventory, cars or vegetation can easily be detected by solely considering the spatial arrangement of 3D points. On the other hand, the human capability to detect a diversity of objects is rather robust to occlusions, strongly varying point density and an irregular point sampling. Accordingly, it seems desirable to transfer such a capability to automated systems. In this regard, each 3D point of the point cloud should be uniquely assigned with a semantic label. At the same time, a respective approach should also be capable of coping with local variations in point density and irregularly distributed 3D points, which are for instance given when considering different types of point clouds derived via terrestrial or mobile laser scanning, airborne laser scanning or multi-view stereo reconstruction.

To achieve a semantic point cloud labeling via automated systems, different concepts have been presented. In this regard, many scientific and application-oriented investigations focus on assigning a semantic class label to each 3D point of a considered point cloud by involving a standard classification technique. Thereby, the classes of interest need to be defined in advance and may for instance be given by *Wire*, *Pole/Trunk*, *Façade*, *Ground* and *Vegetation* as shown in Figure 7.2. Note that the class labels thus represent object categories and are not specific for each single object in the scene. When focusing on such a semantic classification task, the number of defined object classes and the similarity between the classes typically play important roles, and the involved features should be sufficiently distinctive to allow for distinguishing between the defined object classes. In contrast, there are also numerous investigations which aim at providing a meaningful partitioning of a considered point cloud into smaller, connected subsets corresponding to objects of interest or to parts of these. For this purpose, a variety of clustering techniques may be applied which deliver numerous segments exhibiting a homogeneous behavior with respect to a pre-defined criterion. Based on the corresponding characteristics, the derived segments are then assigned a respective

Figure 7.1. Point cloud acquired within an urban area via terrestrial laser scanning: this visualization reveals occlusions, a strongly varying point density and an irregular point sampling. Reprinted/adapted from Weinmann (2016) by permission from Springer, ©2016.

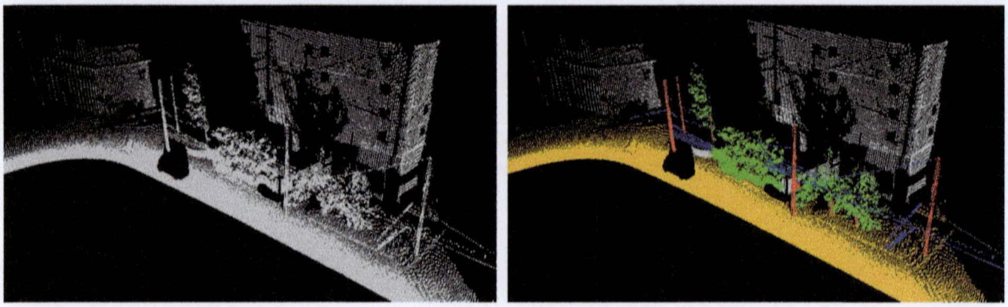

Figure 7.2. Visualization of a point cloud (left) and a corresponding reference labeling (right) which refers to five semantic classes represented by *Wire* (blue), *Pole/Trunk* (red), *Façade* (gray), *Ground* (orange) and *Vegetation* (green).

semantic label, whereby a labeling with respect to object categories is typically pursued. However, it may also be desirable to assign point-wise labels that are both class-aware and instance-aware (i.e. each 3D point is assigned a semantic class label indicating one of the defined object categories and an instance label indicating the respective object in the scene). This task is commonly referred to as an *instance-level segmentation* (Zhang *et al.*, 2016).

In the scope of this chapter, the focus is put on the semantic segmentation of dense point clouds with hundreds or thousands of 3D points per m^2 and different strategies that can be followed to achieve such a semantic segmentation. First, the rather classic strategy for semantic point cloud segmentation is considered, which is based on the use of hand-crafted features provided as input to a supervised classification technique. Subsequently, the strategy for semantic point cloud segmentation relying on the interplay between traditional classification and segmentation techniques to assign point-wise labels that are both class-aware and instance-aware is presented. Beyond such conventional strategies, particularly the strategy of involving modern deep learning techniques for semantic point cloud segmentation has recently been followed. Consequently, this chapter also provides a brief overview of approaches involving deep learning techniques. The last part of this chapter is dedicated to a variety of benchmark datasets that have been published so far and allow an objective comparison of proposed approaches for semantic point cloud segmentation. Most of these datasets have been acquired via terrestrial or mobile laser scanning within urban areas, which allows a comparative conclusion with respect to advantages and limitations of a proposed approach for different semantic segmentation tasks. Finally, the contents of this chapter are briefly summarized.

2 THE CLASSIC STRATEGY FOR POINT CLOUD SEGMENTATION AT POINT LEVEL

This section is dedicated to the classic strategy for point cloud segmentation. Here, local point cloud characteristics are described by first defining a suitable local neighborhood for each 3D point and subsequently using all other 3D points within the defined local neighborhood to derive a set of hand-crafted geometric (and optionally also radiometric) features. The extracted features, in turn, are provided as input to a classifier that has been trained on representative training data and thus allows assigning an appropriate label to the considered 3D point.

2.1 *Definition of local neighborhoods*

When focusing on the consideration of local point cloud characteristics around a considered 3D point $\mathbf{X}_0$, geometric features are typically extracted from the spatial arrangement of $\mathbf{X}_0$ and its neighboring points. Hence, it is of utmost interest to involve a suitable neighborhood definition which allows assessing the neighbors of $\mathbf{X}_0$. In this regard, different neighborhood definitions may be taken into account. Thereby, the difference may be given by the neighborhood type for which a spherical neighborhood (Lee and Schenk, 2002; Linsen and Prautzsch, 2001) or a cylindrical neighborhood (Filin and Pfeifer, 2005; Niemeyer *et al.*, 2014) is most commonly used. Another difference may be given by the parameterization of the selected neighborhood type, which may be specified by either a radius (Lee and Schenk, 2002; Filin and Pfeifer, 2005) or the k nearest neighbors with respect to the Euclidean 3D distance in case of spherical neighborhoods (Linsen and Prautzsch, 2001) or the k nearest neighbors with respect to the Euclidean 2D distance in case of cylindrical neighborhoods (Niemeyer *et al.*, 2014). Note that, for a cylindrical neighborhood, the height of the cylinder is often set as infinite, while the orientation is typically defined along the vertical direction. Accordingly, only one scale parameter represented by either a radius or the number of nearest neighbors has to be determined like in the case of a spherical neighborhood. To select an appropriate value for this scale parameter, prior knowledge about the scene and/or the data is typically involved, and it is often assumed that the same value of the scale parameter applies for each 3D point of the point cloud. Intuitively, however, such a strong assumption might seem less appropriate, as a suitable value of the scale parameter might depend on the local 3D structure and thus on local point cloud characteristics. Indeed, recent investigations reveal that structures related with different classes may favor a different neighborhood size (Weinmann *et al.*, 2015; Weinmann, 2016), which, in turn, motivates the use of data-driven approaches for optimal neighborhood size selection. Respective approaches for selecting the scale parameter for each 3D point individually are based on different criteria, e.g. the consideration of the local surface variation (Pauly *et al.*, 2003) or the combined consideration of curvature, point density and noise of normal estimation (Mitra and Nguyen, 2003; Lalonde *et al.*, 2005). Further techniques for optimal neighborhood size selection are represented by dimensionality-based scale selection (Demantké *et al.*, 2011) and eigenentropy-based scale selection (Weinmann *et al.*, 2015; Weinmann, 2016) and focus on minimizing a functional defined in a similar way as the Shannon entropy across different values of the scale parameter. The value minimizing the defined functional then indicates the optimal neighborhood size. To allow obtaining an impression about the characteristics of different neighborhood types, a visualization of the number of points within the respective local neighborhoods is provided in Figure 7.3, which clearly reveals the significantly different behavior of the considered local neighborhoods.

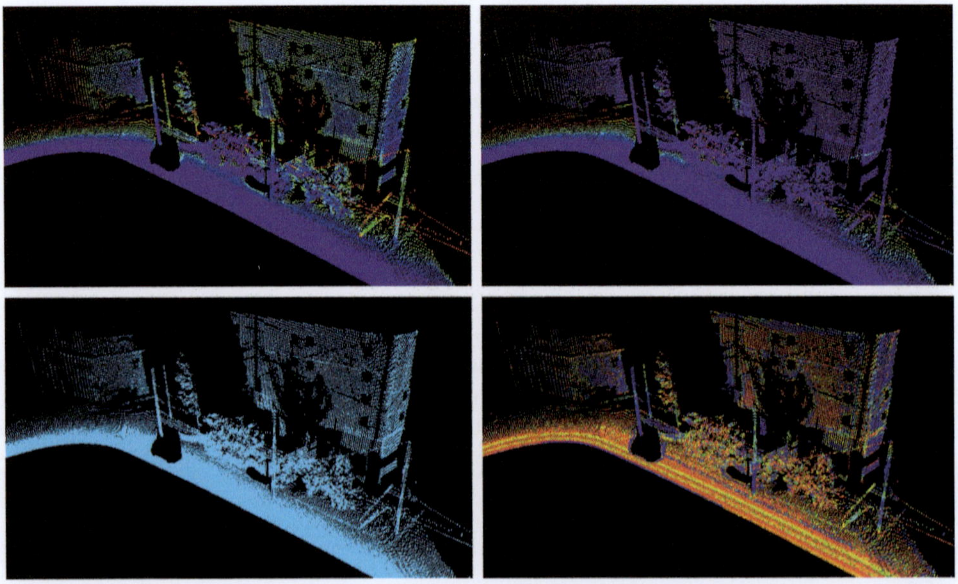

Figure 7.3. Number of points within the local neighborhood when using spherical neighborhoods with a radius of 1 m (top left), when using vertically-oriented cylindrical neighborhoods with a radius of 1 m and infinite height (top right), when using spherical neighborhoods comprising the 50 nearest neighbors with respect to the Euclidean 3D distance (bottom left) and when using locally-adaptive spherical neighborhoods determined via eigenentropy-based scale selection (bottom right): the color encoding indicates neighborhoods with 10 or fewer points in red and reaches via yellow, green, cyan and blue to violet for 100 and more points (Weinmann *et al.*, 2017a).

Instead of considering a single neighborhood as the basis for a subsequent extraction of geometric features, a combination of multiple neighborhoods can be used. Such a multi-scale neighborhood allows describing the local 3D structure at different scales and thus implicitly taking into account how the local 3D geometry changes across these scales. Among the most commonly used multi-scale neighborhoods are the combination of spherical neighborhoods with different radii (Brodu and Lague, 2012), the combination of cylindrical neighborhoods with different radii (Niemeyer *et al.*, 2014) or a multi-scale voxel neighborhood (Hackel *et al.*, 2016). Other alternatives are given by combining neighborhoods based on different entities such as voxels, blocks and pillars (Hu *et al.*, 2013) or such as spatial bins, planar segments and local neighborhoods (Gevaert *et al.*, 2016). The combination of multi-scale neighborhoods and different entities has been proposed with multi-scale, multi-type neighborhoods defined via a combination of both spherical and cylindrical neighborhoods with different scale parameters (Blomley and Weinmann, 2017). The more neighborhoods are used, the higher becomes the dimensionality of the data representation which is later derived to describe the local 3D structure around each 3D point.

2.2 *Extraction of hand-crafted features*

Once an appropriate local neighborhood has been defined for each 3D point $\mathbf{X}_0$, the spatial arrangement of all 3D points within this local neighborhood may be exploited to define a variety of geometric features, whereby the latter are concatenated to a respective feature vector. In this regard, the 3D coordinates of $\mathbf{X}_0$ and its k neighboring points $\mathbf{X}_i$ with $i = 1, \ldots, k$ are often used to derive the 3D structure tensor

$$\mathbf{S} = \frac{1}{k+1}\sum_{i=0}^{k}\left(\mathbf{X}_i - \bar{\mathbf{X}}\right)\left(\mathbf{X}_i - \bar{\mathbf{X}}\right)^T \tag{1}$$

with $\bar{\mathbf{X}}$ indicating the center of gravity. The eigenvalues λ_1, λ_2 and λ_3 of the 3D structure tensor with $\lambda_1 \geq \lambda_2 \geq \lambda_3 \geq 0$ allow reasoning about specific shape primitives (Jutzi and Gross, 2009; Dittrich *et al.*, 2017). Instead of such a reasoning, the distinctiveness of certain behaviors of the local 3D structure can be described by using the eigenvalues of the 3D structure tensor to define a set of hand-crafted features. This is of great interest for typical semantic segmentation tasks where no prior knowledge about expected shape primitives and their relation to the classes of interest is given. Respective features are for instance represented by the local 3D shape features of linearity L, planarity P, sphericity S, omnivariance O, anisotropy A, eigenentropy E and change of curvature C (West *et al.*, 2004; Pauly *et al.*, 2003), which allow a rather intuitive description of the local 3D structure with one value per feature and are therefore commonly used for semantic point cloud interpretation:

$$L = \frac{\lambda_1 - \lambda_2}{\lambda_1} \tag{2}$$

$$P = \frac{\lambda_2 - \lambda_3}{\lambda_1} \tag{3}$$

$$S = \frac{\lambda_3}{\lambda_1} \tag{4}$$

$$O = \sqrt[3]{\lambda_1 \lambda_2 \lambda_3} \tag{5}$$

$$A = \frac{\lambda_1 - \lambda_3}{\lambda_1} \tag{6}$$

$$E = -\sum_{i=1}^{3} \lambda_i \ln(\lambda_i) \tag{7}$$

$$C = \frac{\lambda_3}{\lambda_1 + \lambda_2 + \lambda_3} \tag{8}$$

Depending on the selected definition of local neighborhoods serving as the basis for extracting such features, different characteristics of the local neighborhood are described, which can for instance be verified by considering the behavior of the three dimensionality features of linearity L, planarity P and sphericity S for an exemplary scene as illustrated in Figure 7.4.

Further commonly used geometric features include angular characteristics (Munoz *et al.*, 2009), height and plane characteristics (Mallet *et al.*, 2011), a variety of low-level geometric 3D and 2D features (Weinmann *et al.*, 2015; Weinmann, 2016), moments and height features (Hackel *et al.*, 2016), or specific shape descriptors addressing surface properties, slope, height characteristics, vertical profiles and 2D projections (Guo *et al.*, 2015).

As an alternative to the use of rather intuitive geometric features representing one single property of the local neighborhood by a single value, it has been proposed to sample specific properties within a local neighborhood, e.g. in the form of histograms. Such sampled features can be derived via spin images (Johnson and Hebert, 1999), shape distributions (Osada *et al.*, 2002; Blomley and Weinmann, 2017), 3D shape context descriptors (Frome *et al.*, 2004), point feature histograms (Rusu *et al.*, 2009) or SHOT descriptors (Tombari *et al.*, 2010). The resulting feature vectors describing the local structure around a point $\mathbf{X}_0$ are typically of high dimension, and single entries are hardly interpretable.

Besides a variety of geometric features, other types of features such as radiometric features, full-waveform features or echo-based features provide complementary information and can be used as additional entries of the feature vector (Mallet *et al.*, 2011; Niemeyer *et al.*, 2014). In this regard, radiometric features may for instance allow a filtering of noisy data, as noisy measurements typically correspond to low reflectance values (Djuricic and Jutzi, 2013; Weinmann and Jutzi, 2015). Furthermore, a consideration of full-waveform features might reveal different surface structures and properties (Jutzi and Stilla, 2006).

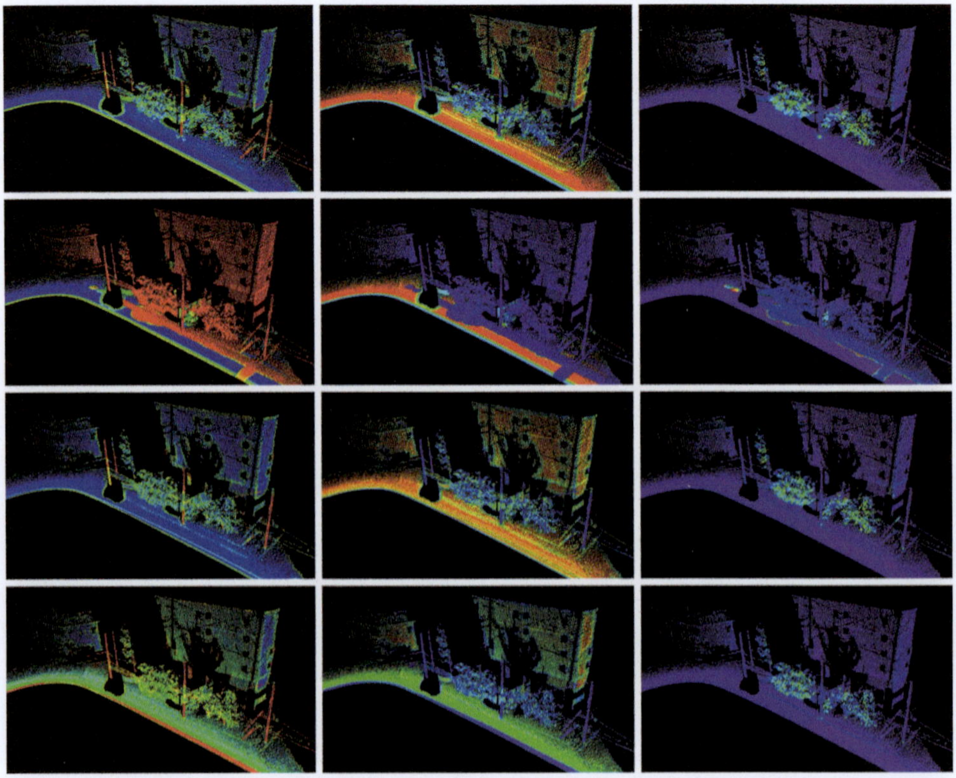

Figure 7.4. Behavior of the three dimensionality features of linearity L (left column), planarity P (center column) and sphericity S (right column) when using spherical neighborhoods with a radius of 1 m (first row), when using vertically-oriented cylindrical neighborhoods with a radius of 1 m and infinite height (second row), when using spherical neighborhoods comprising the 50 nearest neighbors with respect to the Euclidean 3D distance (third row) and when using locally-adaptive spherical neighborhoods determined via eigenentropy-based scale selection (fourth row): the color encoding indicates high values close to 1 in red and reaches via yellow, green, cyan and blue to violet for low values close to 0 (Weinmann *et al.*, 2017a).

2.3 *Supervised classification*

The semantic labeling of a point cloud relies on the extracted features which serve as input to a classification approach. Here, the focus is typically put on a supervised classification, which allows monitoring the progress during training and remains more efficient than unsupervised classification techniques. Among the most popular approaches for supervised classification are standard classifiers such as a support vector machine (Cortes and Vapnik, 1995) or a random forest classifier (Breiman, 2001). These classifiers are meanwhile available in a variety of software tools and rather easy-to-use for non-expert end-users. However, they treat each 3D point individually by solely considering the respectively derived feature vector as the basis for assigning a class label. Consequently, relations among neighboring 3D points are not directly taken into account, and a visualization of the derived semantic point cloud labeling thus tends to reveal a "noisy" behavior as shown in Figure 7.5 and in Figure 7.6, where classification results derived with a random forest classifier based on a set of hand-crafted geometric features are provided.

The classification results visualized in Figure 7.5 clearly indicate that the use of cylindrical neighborhoods is not that suitable for classifying terrestrial or mobile laser scanning data. This

Figure 7.5. Classification results obtained with a random forest classifier when using spherical neighborhoods with a radius of 1 m (top left), when using vertically-oriented cylindrical neighborhoods with a radius of 1 m and infinite height (top right), when using spherical neighborhoods comprising the 50 nearest neighbors with respect to the Euclidean 3D distance (bottom left) and when using locally-adaptive spherical neighborhoods determined via eigenentropy-based scale selection (bottom right): the color encoding refers to five semantic classes represented by *Wire* (blue), *Pole/Trunk* (red), *Façade* (gray), *Ground* (orange) and *Vegetation* (green) (Weinmann *et al.*, 2017a).

Figure 7.6. Classification result derived via geometric features serving as input to a random forest classifier: the color encoding addresses the classes *Façade* (gray), *Ground* (orange), *Cars* (blue), *2-Wheelers* (yellow), *Road Inventory* (red), *Pedestrians* (magenta) and *Vegetation* (green). Reprinted from Landrieu *et al.* (2017) with permission from Elsevier, ©2017.

is due to the consideration of vertically-oriented cylindrical neighborhoods with infinite height, which is not appropriate if different objects appear at different height levels as e.g. given for parts of the scene where wire is above ground or vegetation. Furthermore, Figure 7.5 reveals that such cylindrical neighborhoods deliver systematic misclassifications on the building façade. In contrast, the spherical neighborhood definitions yield appropriate classification results for almost all cases, although they have significantly different characteristics as shown in Figure 7.3. While the spherical neighborhoods with a radius of 1 m tend to comprise a large number of 3D points for most parts of the scene, the locally-adaptive spherical neighborhoods derived via eigenentropy-based scale selection tend to be comparably small. This, in turn, leads to a different behavior of respectively extracted geometric features as shown in Figure 7.4. Indeed, the corresponding quantitative classification results differ by less than two percent in overall accuracy (Weinmann *et al.*, 2017a), whereby the best results are obtained with the locally-adaptive spherical neighborhoods.

To take into account that the class labels of neighboring 3D points tend to be correlated and therefore enforce a spatially smooth labeling, statistical models of context are often involved such as associative Markov networks (Munoz *et al.*, 2009), non-associative Markov networks (Shapovalov *et al.*, 2010), conditional random fields (CRFs; Lafferty *et al.*, 2001; Niemeyer *et al.*, 2014) or structured regularization representing a more versatile alternative to the standard graphical model approach (Landrieu *et al.*, 2017). The latter relies on an initial (ideally probabilistic) labeling derived with a standard classifier and formulates the smooth labeling task as an optimization problem, where a wide range of fidelity functions and regularizers may be used. Thereby, the graphical model approach as used with a CRF can be expressed as a special instance of structured regularization, and a diversity of fast solving algorithms can be used instead of rather slow and memory-intensive message-passing algorithms. Furthermore, different cases are given under which the smoothed labeling remains probabilistic, which allows concluding about the uncertainty associated with each label. Whereas such advanced labeling techniques deliver a spatially regular and thus "smooth" labeling as depicted in Figure 7.7, they require

Figure 7.7. Classification result derived via structured regularization: the color encoding addresses the classes *Façade* (gray), *Ground* (orange), *Cars* (blue), *2-Wheelers* (yellow), *Road Inventory* (red), *Pedestrians* (magenta) and *Vegetation* (green). Reprinted from Landrieu *et al.* (2017) with permission from Elsevier, ©2017.

additional effort for inferring relations among neighboring 3D points from the training data, and, in most cases, more data is needed as the basis for training.

3 THE STRATEGY FOR POINT CLOUD SEGMENTATION AT OBJECT LEVEL

Instead of a standard point-wise classification and subsequent efforts for imposing spatial regularity on the labeling, the consideration of point cloud segments might also seem to be an appropriate approach. For this purpose, a segmentation of the point cloud into smaller, connected subsets corresponding to objects of interest or to parts of these is required which can be achieved with different approaches (Vosselman, 2013). While some approaches start with an oversegmentation of the considered point cloud and subsequently merge neighboring segments with similar characteristics, other approaches start with a selection of seed points and subsequently perform a region growing. In this regard, however, it has to be taken into account that the quality of achieved segmentation results strongly depends on the applied segmentation technique. Furthermore, most techniques delivering a generic, data-driven 3D segmentation reveal a high computational burden. Consequently, several approaches focus on the interplay between standard classification and segmentation techniques. On the one hand, many approaches start with an oversegmentation of the scene, e.g. by deriving supervoxels (Kim *et al.*, 2013; Aijazi *et al.*, 2013; Wolf *et al.*, 2015). Subsequently, hand-crafted features are extracted based on the derived segments and then provided as input to a classification technique. On the other hand, the result of a standard point-wise classification may serve as input for a subsequent segmentation in order to detect individual objects in the scene (Monnier *et al.*, 2012; Böhm *et al.*, 2016; Weinmann *et al.*, 2017b) or to improve the derived labeling with respect to the given object classes (Niemeyer *et al.*, 2015; Guignard and Landrieu, 2017). More specifically, the result of a point-wise classification derived via a CRF can be used as input for a subsequent region-growing algorithm which, in turn, connects points that are close to each other and meet certain conditions such as having been assigned the same label by the point-wise classification (Niemeyer *et al.*, 2015). Based on the segments corresponding to connected points, a further CRF-based classification delivers an improved semantic labeling. As an alternative to such a two-layer CRF, a non-parametric segmentation which partitions the considered point cloud into geometrically homogeneous segments may be integrated into a CRF in order to capture the high-level structure of the considered scene (Guignard and Landrieu, 2017).

In the following, the focus is put on the interplay between standard classification and segmentation techniques to extract individual objects from a considered point cloud. Exemplary objects of interest are represented by trees, which play an important role in urban areas since they provide measurable economic, environmental, social and health benefits (Kelly, 2011). Such an extraction of individual objects can be achieved in two steps (Weinmann *et al.*, 2017b). First, a point-wise classification is performed to achieve a class-aware labeling with respect to the classes *Tree* and *Background*. Subsequently, a segmentation based on those 3D points labeled as *Tree* is conducted to obtain an instance-aware labeling with respect to individual trees.

3.1 *Detection of tree-like structures in a point cloud*

To detect tree-like structures in a considered point cloud, a binary classification of 3D points with respect to the classes *Tree* and *Background* is performed. For this purpose, the classic strategy for point cloud segmentation is followed as shown in Figure 7.8. Accordingly, appropriate local neighborhoods are selected as the basis for extracting hand-crafted features, which, in turn, are provided as input to a trained classifier assigning an appropriate label to each 3D point. More specifically,

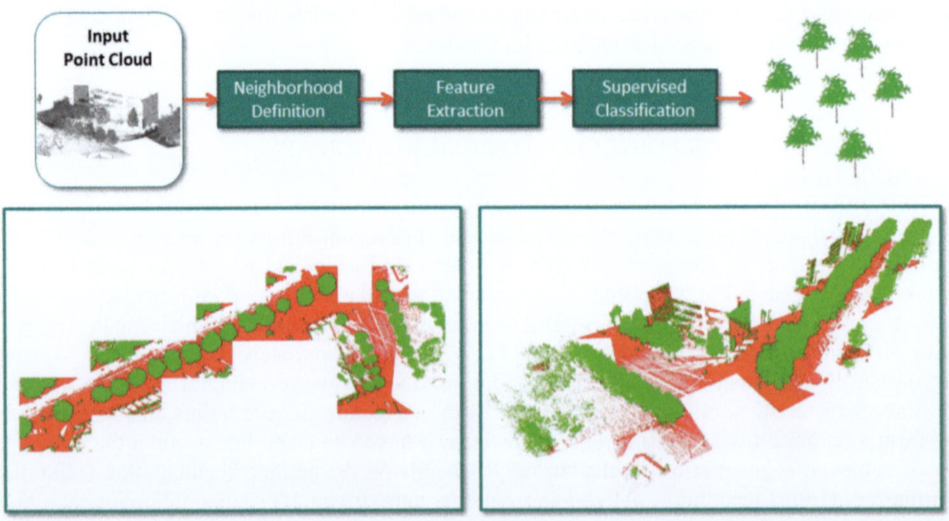

Figure 7.8. Result of a binary classification with respect to the classes *Tree* (green) and *Background* (red) when using a standard point-wise classification pipeline (top) as proposed by Weinmann *et al.* (2017b): nadir view (bottom left) and side view (bottom right).

locally-adaptive local neighborhoods derived via eigenentropy-based scale selection (Weinmann *et al.*, 2015; Weinmann, 2016) are used to extract the local 3D shape features of linearity, planarity, sphericity, omnivariance, anisotropy, eigenentropy and change of curvature (West *et al.*, 2004; Pauly *et al.*, 2003) as well as other 3D and 2D geometric features. If the given data also contain radiometric information, radiometric features may be used as well. The involved classifier is represented by a random forest classifier (Breiman, 2001).

3.2 *Separation of selected 3D points with respect to individual trees*

Once a binary point-wise classification with respect to the classes *Tree* and *Background* has been performed, those 3D points assigned to the class *Tree* are used for a subsequent separation into segments corresponding to individual trees in the scene. As generic segmentation methods tend to cause a high computational burden, several adaptations may be introduced to retain an efficient approach for individual tree extraction at point-level. In this regard, a recently proposed approach (Weinmann *et al.*, 2017b), summarized in Figure 7.9, introduces a downsampling of the 3D points assigned to the class *Tree* and this is followed by a 2D projection of the downsampled point cloud onto a horizontally-oriented plane. The latter is motivated by the fact that, for individual trees within urban environments, a larger spacing and less overlap is given than for trees in forested areas. On the basis of the derived 2D projections, applying a mean shift segmentation (Fukunaga and Hostetler, 1975; Cheng, 1995) allows deriving a meaningful partitioning with respect to individual trees in the considered scene without introducing strong assumptions on a specific geometric model or on the number of expected segments. As the derived segmentation results only refer to the downsampled point cloud, a respective upsampling to those 3D points corresponding to the class *Tree* is performed. Subsequently, a segment-based shape analysis based on semantic rules is carried out to derive plausible segments corresponding to individual trees. For all plausible tree segments, the location of the respective tree is estimated via the corresponding mode derived during the mean shift segmentation.

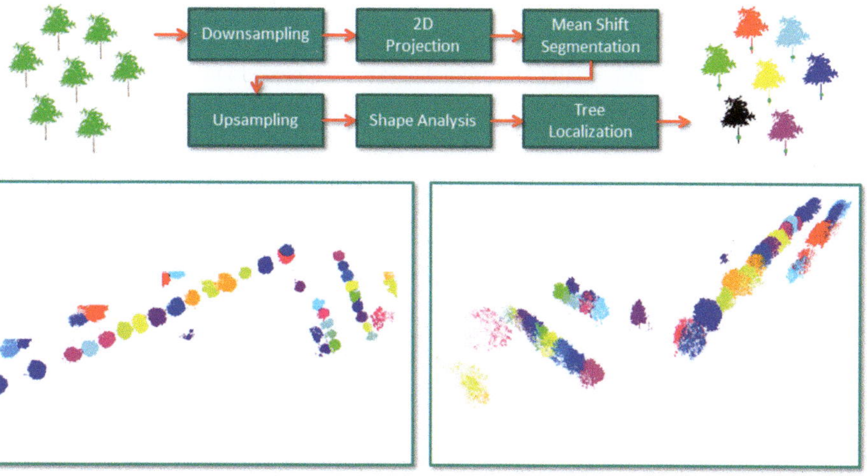

Figure 7.9. Result of an instance-aware segmentation with respect to individual trees (indicated in different colors) when using a straightforward segmentation pipeline (top) as proposed by Weinmann *et al.* (2017b): nadir view (bottom left) and side view (bottom right).

4 THE STRATEGY FOR POINT CLOUD SEGMENTATION BASED ON DEEP LEARNING

In contrast to those strategies relying on a set of hand-crafted features, recent approaches for point cloud segmentation focus on the use of modern deep learning techniques (Ioannidou *et al.*, 2017). Given a sufficiently large amount of training data, such techniques are capable of learning features which are relevant for the semantic segmentation task and thus allow deriving a semantic labeling for new data. Among a diversity of approaches for deep semantic point cloud segmentation that have been presented so far, two approaches have become very popular. On the one hand, standard deep networks can be adapted to receiving 3D data in the form of voxel-occupancy grids as input. On the other hand, 2D projections of the point cloud may be exploited which allow the subsequent use of standard deep image segmentation approaches.

4.1 *Deep semantic segmentation based on voxel-occupancy grids*

The straightforward approach to deep semantic point cloud segmentation is represented by the transfer of the considered point cloud to a regular 3D voxel grid and the direct adaptation of convolutional neural networks (CNNs) to 3D data. In this regard, several promising approaches focus on the classification of each 3D point $\mathbf{X}_0$ of a considered point cloud by exploiting a transformation of all points within its local neighborhood to a voxel-occupancy grid (with either a binary occupancy value per voxel or the point density as occupancy value per voxel) serving as input to a 3D-CNN (Huang and You, 2016; Tchapmi *et al.*, 2017; Hackel *et al.*, 2017; Roynard *et al.*, 2018a; Wang *et al.*, 2018). The 3D-CNN, in turn, may for instance be derived from a standard 2D-CNN by replacing the 2D convolutional layers and 2D pooling layers with 3D convolutional layers and 3D pooling layers, respectively (Huang and You, 2016).

4.2 *Deep semantic segmentation based on 2D projections*

As an alternative to the adaptation of a deep network for handling 3D input data, another intuitive approach relies on the use of a 2D-CNN designed for semantic image segmentation. This is particularly promising since much expertise on 2D-CNNs and a variety networks (partially even pretrained on image datasets, which significantly reduces the amount of required training data) are already available. To allow for applying such a 2D-CNN, 2D projections of the point cloud (e.g. in the form of a virtual color image, a virtual depth map or an elevation map) are derived, which serve as input for the 2D-CNN and, finally, the predicted labeling needs to be backprojected to 3D space to obtain the semantically labeled point cloud (Boulch *et al.*, 2017; Lawin *et al.*, 2017). Thereby, it seems appropriate to render multiple 2D projections from different viewpoints as done for 3D object classification and retrieval, where only a 3D shape model is considered as input data (Ioannidou *et al.*, 2017). Accordingly, an adequate fusion of the projections of the predicted labelings to 3D space is required to obtain an appropriate labeling of the considered point cloud.

4.3 *Deep semantic image segmentation*

Deep learning techniques have been shown to be good at learning suitable mid-level and high-level features from given input imagery, which can efficiently be used for the task of semantic image segmentation. In this regard, most of the presented techniques have been adapted from deep networks originally proposed for image classification, for which different network architectures may be used such as the VGG networks (Simonyan and Zisserman, 2014) and the residual network (ResNet; He *et al.*, 2016). Besides such adaptations, the fully convolutional network (FCN; Long *et al.*, 2015) and encoder-decoder architectures (Volpi and Tuia, 2017; Badrinarayanan *et al.*, 2017) are widely used to achieve a pixel-wise semantic labeling of input imagery. For more details about such networks, the reader is referred to (Zhu *et al.*, 2017) and references therein, while the idea of using an ensemble of CNNs has been presented by Marmanis *et al.* (2016).

Given derived 2D projections of radiometric and/or geometric information in the form of a color and/or a depth image, the segmentation task becomes fairly the same as e.g. addressed in the *ISPRS Benchmark on 2D Semantic Labeling* (Rottensteiner *et al.*, 2012; Gerke, 2014), where true orthophotos and the corresponding digital surface models (DSMs) as illustrated in Figure 7.10 are

Figure 7.10. The *ISPRS Benchmark on 2D Semantic Labeling*: given a true orthophoto (top left) and the corresponding DSM (top center), the objective is to derive a semantic labeling close to the reference labeling (top right). Based on the given data, other radiometric or geometric features may be extracted such as the NDVI (bottom left) or the nDSM (bottom center). Using a variety of radiometric and geometric features, an exemplary labeling (bottom right) derived with a modern deep learning framework is depicted as well (Chen *et al.*, 2018).

considered as the basis for semantic image segmentation. Among the proposed approaches, several ones also rely on the additional use of hand-crafted features derived from the true orthophotos and/or their corresponding DSMs as input to a modern deep learning technique. In particular, the normalized difference vegetation index (NDVI) and the normalized digital surface model (nDSM) as provided in Figure 7.10 are commonly used (Audebert *et al.*, 2016; Liu *et al.*, 2017; Kampffmeyer *et al.*, 2016). Furthermore, the use of a diversity of hand-crafted radiometric and geometric features reveals promising results (Chen *et al.*, 2018), which holds particularly for those cases where the amount of training data available to tune the deep network might not necessarily be sufficient.

5 BENCHMARK DATASETS FOR POINT CLOUD SEGMENTATION

This section is dedicated to a variety of benchmark datasets that have been presented for evaluating the performance of approaches for the semantic segmentation of dense point clouds.

5.1 *Oakland 3D Point Cloud Dataset*

The Oakland 3D Point Cloud Dataset (Munoz *et al.*, 2009) was acquired with a mobile laser scanning system in the vicinity of the CMU campus in Oakland, PA, USA. More specifically, a vehicle was equipped with side-looking SICK LMS laser scanners used in push-broom mode, and this vehicle drove within an urban environment with a speed of up to 20 km/h. The complete dataset has been split into a training set comprising about 36.9k 3D points, a validation set comprising about 91.5k 3D points and a test set comprising about 1.3M 3D points. For each 3D point, a reference labeling with respect to three structural classes represented by *Linear Structures*, *Planar Structures* and *Volumetric Structures* is available as well as a reference labeling with respect to five semantic classes represented by *Wire*, *Pole/Trunk*, *Façade*, *Ground* and *Vegetation*. The latter reference labeling is visualized in Figure 7.2 for the validation set.

5.2 *Paris-rue-Madame Database*

The Paris-rue-Madame Database (Serna *et al.*, 2014) was acquired in February 2013 with the mobile laser scanning system *L3D2* (Goulette *et al.*, 2006) in the city of Paris, France. This system uses a Velodyne HDL-32 to capture geometric information about the local environment. The complete dataset comprises 20M 3D points and corresponds to a street section with a length of about 160 m. In addition to the spatial 3D coordinates and the corresponding reflectance values, a point-wise reference labeling derived via manual annotation is provided. This labeling is both class-aware and instance-aware, and it addresses point-wise labels with respect to 26 object classes and IDs corresponding to a total number of 642 objects in the scene.

As several of the object classes comprise less than 0.05% of the complete dataset, it may be assumed that these object classes are not covered in a representative way. Accordingly, it has been proposed to only address the six dominant object classes represented by *Façade*, *Ground*, *Cars*, *Motorcycles*, *Traffic Signs* and *Pedestrians*, which comprise 99.81% of the Paris-rue-Madame Database (Weinmann *et al.*, 2015; Weinmann, 2016).

5.3 *IQmulus/TerraMobilita Benchmark Dataset*

The IQmulus/TerraMobilita Benchmark Dataset (Vallet *et al.*, 2015) was acquired in January 2013 with the mobile mapping system *Stereopolis II* (Paparoditis *et al.*, 2012) in the city of Paris, France. This system is equipped with two plane sweep lidar sensors of type Riegl LMS-Q120i which are placed on each side of the vehicle and mainly used to observe the building façades with centimeter accuracy. Furthermore, the system involves a 3D lidar sensor of type Velodyne HDL-64E, which is used to observe the bottom part in between the building façades. In total, the acquired dataset consists of more than 100M 3D points corresponding to 10 km of streets within the 6th district of Paris. In addition to the spatial 3D coordinates, reflectance information about the backscattered intensity is provided as well as the number of the respective echo. For the complete dataset, a point-wise reference labeling derived via manual annotation is given, whereby the ground truth is class-aware with respect to object classes organized in a very detailed hierarchy of semantic classes and instance-aware with respect to objects in the scene.

For testing purposes, a learning dataset including the reference labeling is publicly available, which is referred to as the Paris-rue-Cassette Database. This part of the complete dataset comprises 12M 3D points which correspond to a street section with a length of about 200 m. However, several of the defined classes comprise less than 0.05% of the Paris-rue-Cassette Database and may therefore be considered as not being representatively covered. Accordingly, it has been proposed to discard 3D points belonging to such classes and define the object classes of interest with respect to seven dominant object classes represented by *Façade*, *Ground*, *Cars*, *2-Wheelers*, *Road Inventory*, *Pedestrians* and *Vegetation*, which cover about 99.56% of the Paris-rue-Cassette database (Weinmann *et al.*, 2015; Weinmann, 2016).

5.4 *IQPC'15 Benchmark Dataset*

The IQPC'15 Benchmark Dataset (Gorte *et al.*, 2015) was acquired with the Fugro DRIVE-MAP system in the vicinity of the university campus in Delft, the Netherlands. Besides spatial 3D coordinates, this system also delivers both reflectance and color information. The complete dataset is organized in 509 tiles, each covering an area of 25 m × 25 m.

For testing purposes, a smaller part of the dataset corresponding to 26 tiles and comprising about 10.13M 3D points is available which also includes the respective reference labeling with respect to the classes *Tree* and *Background*. This allows for performance evaluation with respect to a binary classification of 3D points, while a separation of those 3D points assigned to the class *Tree* with respect to individual trees in the scene has to be verified via visual inspection.

5.5 *MLS 1 – TUM City Campus Dataset*

The MLS 1 – TUM City Campus Dataset (Gehrung *et al.*, 2017; Xu *et al.*, 2018) was acquired in April 2016 with the mobile mapping system MODISSA of Fraunhofer IOSB. This system uses four lidar sensors, two of type Velodyne HDL-64E and two of type Velodyne VLP-16 Puck, to acquire spatial 3D information. In addition, the system contains eight cameras which allow texturing the acquired 3D measurements. To obtain georeferenced data, the navigation data provided by an Applanix POS LV navigation system was used. Given these capabilities, the system was moved on the city campus of the TU München and along public buildings in the vicinity. In total, the acquired dataset comprises 1735M 3D points and, for a part of the dataset, a reference labeling obtained via manual annotation is available which refers to the classes *Unlabeled*, *Artificial Terrain*, *Natural Terrain*, *High Vegetation*, *Low Vegetation*, *Building*, *Hardscape*, *Artefact* and *Vehicle*.

5.6 *Paris-Lille-3D Dataset*

The Paris-Lille-3D Dataset (Roynard *et al.*, 2018b) was acquired with the MLS prototype *L3D2* (Goulette *et al.*, 2006) which involves a lidar sensor of type Velodyne HDL-32E. The complete dataset comprises 143.1M 3D points corresponding to street sections with a length of about 1.9 km within the cities of Paris and Lille in France. In addition to the spatial 3D coordinates and the corresponding reflectance values, a point-wise reference labeling derived via manual annotation is provided. On the one hand, this labeling is class-aware with respect to object classes organized in a very detailed class hierarchy similar to the one used for the IQmulus/TerraMobilita Benchmark Dataset (Vallet *et al.*, 2015). On the other hand, this labeling is instance-aware and given in the form of object labels corresponding to more than 2k objects in the scene.

5.7 *Semantic3D.net Dataset*

The Semantic3D.net Dataset (Hackel *et al.*, 2017) was acquired with surveying-grade terrestrial laser scanners within various different natural and man-made scenes. For each 3D point, the given information comprises the spatial 3D coordinates, an intensity value and three color values referring to the RGB color space. The complete dataset comprises more than four billion manually labelled 3D points, whereby the classes of interest are represented by *Man-made Terrain*, *Natural Terrain*, *High Vegetation*, *Low Vegetation*, *Buildings*, *Clutter (hard scape)*, *Scanning Artefacts* and *Cars*.

6 SUMMARY AND RECENT TRENDS

In this chapter, the focus has been put on the semantic segmentation of dense point clouds as typically acquired with terrestrial or mobile laser scanning devices or as generated from image collections. On the one hand, the classic strategy of a semantic point cloud segmentation relying on the use of hand-crafted features as input to a standard classification technique has been considered in detail. On the other hand, the strategy focusing on the interplay of traditional classification and segmentation techniques has been addressed to derive a point-wise labeling which is both class-aware with respect to defined object categories and instance-aware with respect to the respective objects in the scene. Beyond these strategies for semantic point cloud segmentation, the strategy of involving modern deep learning techniques has been revisited. Finally, an overview of state-of-the-art benchmark datasets for the semantic segmentation of dense point clouds has been provided.

Recent trends addressing the semantic segmentation of dense point clouds are related to the combined use of an efficient organization of point clouds and modern deep learning techniques. In this regard, a promising approach first partitions a considered point cloud into geometrically homogeneous simple shapes called superpoints and subsequently constructs a superpoint graph by linking nearby superpoints by superedges with rich edge features encoding the contextual relationship between object parts in point clouds (Landrieu and Simonovsky, 2018). The derived superpoints are finally transformed into compact embeddings which, in turn, are processed with graph convolutions to make use of contextual information and classified by a graph convolutional network. This approach is scalable toward the processing of large point clouds and significantly outperforms other approaches on recent benchmark datasets.

REFERENCES

Aijazi, A.K., Checchin, P. & Trassoudaine, L. (2013) Segmentation based classification of 3D urban point clouds: A super-voxel based approach with evaluation. *Remote Sensing*, 5(4), 1624–1650.

Audebert, N., Le Saux, B. & Lefèvre, S. (2016) Semantic segmentation of Earth observation data using multimodal and multi-scale deep networks. *Proceedings of the 13th Asian Conference on Computer Vision, 20–24 November 2016*. Taipei, Taiwan. pp.180–196.

Badrinarayanan, V., Kendall, A. & Cipolla, R. (2017) SegNet: A deep convolutional encoder-decoder architecture for image segmentation. *IEEE Transactions on Pattern Analysis and Machine Intelligence*, 39(12), 2481–2495.

Blomley, R. & Weinmann, M. (2017) Using multi-scale features for the 3D semantic labeling of airborne laser scanning data. *ISPRS Annals of the Photogrammetry, Remote Sensing and Spatial Information Sciences*, IV-2/W4, 43–50.

Böhm, J., Brédif, M., Gierlinger, T., Krämer, M., Lindenbergh, R., Liu, K., . . . Sirmacek, B. (2016) The IQmulus urban showcase: Automatic tree classification and identification in huge mobile mapping point clouds. *The International Archives of the Photogrammetry, Remote Sensing and Spatial Information Sciences*, XLI-B3, 301–307.

Boulch, A., Le Saux, B. & Audebert, N. (2017) Unstructured point cloud semantic labeling using deep segmentation networks. *Proceedings of the Eurographics Workshop on 3D Object Retrieval, 23–34 April 2017*. Lyon, France. pp.17–24.

Breiman, L. (2001) Random forests. *Machine Learning*, 45(1), 5–32.

Brodu, N. & Lague, D. (2012) 3D terrestrial lidar data classification of complex natural scenes using a multi-scale dimensionality criterion: Applications in geomorphology. *ISPRS Journal of Photogrammetry and Remote Sensing*, 68, 121–134.

Chen, K., Weinmann, Mi., Gao, X., Yan, M., Hinz, S., Jutzi, B. & Weinmann, M. (2018) Residual shuffling convolutional neural networks for deep semantic image segmentation using multi-modal data. *ISPRS Annals of the Photogrammetry, Remote Sensing and Spatial Information Sciences*, IV-2, 65–72.

Cheng, Y. (1995) Mean shift, mode seeking, and clustering. *IEEE Transactions on Pattern Analysis and Machine Intelligence*, 17(8), 790–799.

Cortes, C. & Vapnik, V. (1995) Support-vector networks. *Machine Learning*, 20(3), 273–297.

Demantké, J., Mallet, C., David, N. & Vallet, B. (2011) Dimensionality based scale selection in 3D lidar point clouds. *The International Archives of the Photogrammetry, Remote Sensing and Spatial Information Sciences*, XXXVIII-5/W12, 97–102.

Dittrich, A., Weinmann, M. & Hinz, S. (2017) Analytical and numerical investigations on the accuracy and robustness of geometric features extracted from 3D point cloud data. *ISPRS Journal of Photogrammetry and Remote Sensing*, 126, 195–208.

Djuricic, A. & Jutzi, B. (2013) Supporting UAVs in low visibility conditions by multiple-pulse laser scanning devices. *The International Archives of the Photogrammetry, Remote Sensing and Spatial Information Sciences*, XL-1/W1, 93–98.

Filin, S. & Pfeifer, N. (2005) Neighborhood systems for airborne laser data. *Photogrammetric Engineering & Remote Sensing*, 71(6), 743–755.

Frome, A., Huber, D., Kolluri, R., Bülow, T. & Malik, J. (2004) Recognizing objects in range data using regional point descriptors. *Proceedings of the European Conference on Computer Vision, 11–14 May 2004*. Prague, Czech Republic. pp. III:224–III:237.

Fukunaga, K. & Hostetler, L. (1975) The estimation of the gradient of a density function, with applications in pattern recognition. *IEEE Transactions on Information Theory*, 21(1), 32–40.

Gehrung, J., Hebel, M., Arens, M. & Stilla, U. (2017) An approach to extract moving objects from MLS data using a volumetric background representation. *ISPRS Annals of the Photogrammetry, Remote Sensing and Spatial Information Sciences*, IV-1/W1, 107–114.

Gerke, M. (2014) *Use of the Stair Vision Library Within the ISPRS 2D Semantic Labeling Benchmark (Vaihingen)*. Technical Report, ITC, University of Twente.

Gevaert, C.M., Persello, C. & Vosselman, G. (2016) Optimizing multiple kernel learning for the classification of UAV data. *Remote Sensing*, 8(12), 1025:1–1025:22.

Gorte, B., Oude Elberink, S., Sirmacek, B. & Wang, J. (2015) IQPC 2015 track: Tree separation and classification in mobile mapping lidar data. *The International Archives of the Photogrammetry, Remote Sensing and Spatial Information Sciences*, XL-3/W3, 607–612.

Goulette, F., Nashashibi, F., Abuhadrous, I., Ammoun, S. & Laurgeau, C. (2006) An integrated onboard laser range sensing system for on-the-way city and road modelling. *The International Archives of the Photogrammetry, Remote Sensing and Spatial Information Sciences*, XXXVI-1, 1–6.

Guignard, S. & Landrieu, L. (2017) Weakly supervised segmentation-aided classification of urban scenes from 3D lidar point clouds. *The International Archives of the Photogrammetry, Remote Sensing and Spatial Information Sciences*, XLII-1/W1, 151–157.

Guo, B., Huang, X., Zhang, F. & Sohn, G. (2015) Classification of airborne laser scanning data using JointBoost. *ISPRS Journal of Photogrammetry and Remote Sensing*, 100, 71–83.

Hackel, T., Savinov, N., Ladicky, L., Wegner, J.D., Schindler, K. & Pollefeys, M. (2017) SEMANTIC3D.NET: A new large-scale point cloud classification benchmark. *ISPRS Annals of the Photogrammetry, Remote Sensing and Spatial Information Sciences*, IV-1/W1, 91–98.

Hackel, T., Wegner, J.D. & Schindler, K. (2016) Fast semantic segmentation of 3D point clouds with strongly varying point density. *ISPRS Annals of the Photogrammetry, Remote Sensing and Spatial Information Sciences*, III-3, 177–184.

He, K., Zhang, X., Ren, S. & Sun, J. (2016) Deep residual learning for image recognition. *Proceedings of the IEEE Conference on Computer Vision and Pattern Recognition, 26 June – 1 July 2016*. Las Vegas, NV, USA. pp.770–778.

Hu, H., Munoz, D., Bagnell, J.A. & Hebert, M. (2013) Efficient 3-D scene analysis from streaming data. *Proceedings of the IEEE International Conference on Robotics and Automation, 6–10 May 2013*. Karlsruhe, Germany. pp.2297–2304.

Huang, J. & You, S. (2016) Point cloud labeling using 3D convolutional neural network. *Proceedings of the International Conference on Pattern Recognition, 4–8 December 2016*. Cancun, Mexico. pp.2670–2675.

Ioannidou, A., Chatzilari, E., Nikolopoulos, S. & Kompatsiaris, I. (2017) Deep learning advances in computer vision with 3D data: A survey. *ACM Computing Surveys*, 50(2), 20:1–20:38.

Johnson, A.E. & Hebert, M. (1999) Using spin images for efficient object recognition in cluttered 3D scenes. *IEEE Transactions on Pattern Analysis and Machine Intelligence*, 21(5), 433–449.

Jutzi, B. & Gross, H. (2009) Nearest neighbour classification on laser point clouds to gain object structures from buildings. *The International Archives of the Photogrammetry, Remote Sensing and Spatial Information Sciences*, XXXVIII-1-4-7/W5, 1–6.

Jutzi, B. & Stilla, U. (2006) Range determination with waveform recording laser systems using a Wiener filter. *ISPRS Journal of Photogrammetry and Remote Sensing*, 61(2), 95–107.

Kampffmeyer, M., Salberg, A.-B. & Jenssen, R. (2016) Semantic segmentation of small objects and modeling of uncertainty in urban remote sensing images using deep convolutional neural networks. *Proceedings of the IEEE Conference on Computer Vision and Pattern Recognition Workshops, 26 June – 1 July 2016*. Las Vegas, NV, USA. pp.680–688.

Kelly, M. (2011) Urban trees and the green infrastructure agenda. *Proceedings of the Urban Trees Research Conference, 13–14 April 2011*. Birmingham, UK. pp.166–180.

Kim, B.S., Kohli, P. & Savarese, S. (2013) 3D scene understanding by voxel-CRF. *Proceedings of the IEEE International Conference on Computer Vision, 1–8 December 2013*. Sydney, Australia. pp.1425–1432.

Lafferty, J.D., McCallum, A. & Pereira, F.C.N. (2001) Conditional random fields: Probabilistic models for segmenting and labeling sequence data. *Proceedings of the International Conference on Machine Learning, 28 June – 1 July 2001*. Williamstown, MA, USA. pp.282–289.

Lalonde, J.-F., Unnikrishnan, R., Vandapel, N. & Hebert, M. (2005) Scale selection for classification of point-sampled 3D surfaces. *Proceedings of the International Conference on 3-D Digital Imaging and Modeling, 13–16 June 2005*. Ottawa, Canada. pp.285–292.

Landrieu, L., Raguet, H., Vallet, B., Mallet, C. & Weinmann, M. (2017) A structured regularization framework for spatially smoothing semantic labelings of 3D point clouds. *ISPRS Journal of Photogrammetry and Remote Sensing*, 132, 102–118.

Landrieu, L. & Simonovsky, M. (2018) *Large-scale Point Cloud Semantic Segmentation With Superpoint Graphs*. arXiv preprint arXiv:1711.09869. Available from: https://arxiv.org/pdf/1711.09869.pdf [accessed May 9, 2018].

Lawin, F.J., Danelljan, M., Tosteberg, P., Bhat, G., Khan, F.S. & Felsberg, M. (2017) Deep projective 3D semantic segmentation. *Proceedings of the 17th International Conference on Computer Analysis of Images and Patterns, 22–24 August 2017*. Ystad, Sweden. pp.95–107.

Lee, I. & Schenk, T. (2002) Perceptual organization of 3D surface points. *The International Archives of the Photogrammetry, Remote Sensing and Spatial Information Sciences*, XXXIV-3A, 193–198.

Linsen, L. & Prautzsch, H. (2001) Local versus global triangulations. *Proceedings of Eurographics, 5–7 September 2001*. Manchester, UK. pp.257–263.

Liu, Y., Piramanayagam, S., Monteiro, S.T. & Saber, E. (2017) Dense semantic labeling of very-high-resolution aerial imagery and lidar with fully-convolutional neural networks and higher-order CRFs. *Proceedings of the IEEE Conference on Computer Vision and Pattern Recognition Workshops, 21–26 July 2017*. Honolulu, HI, USA. pp.1561–1570.

Long, J., Shelhamer, E. & Darrell, T. (2015) Fully convolutional networks for semantic segmentation. *Proceedings of the IEEE Conference on Computer Vision and Pattern Recognition, 7–12 June 2015*. Boston, MA, USA. pp .3431–3440.

Mallet, C., Bretar, F., Roux, M., Soergel, U. & Heipke, C. (2011) Relevance assessment of full-waveform lidar data for urban area classification. *ISPRS Journal of Photogrammetry and Remote Sensing*, 66(6), S71–S84.

Marmanis, D., Wegner, J.D., Galliani, S., Schindler, K., Datcu, M. & Stilla, U. (2016) Semantic segmentation of aerial images with an ensemble of CNNs. *ISPRS Annals of the Photogrammetry, Remote Sensing and Spatial Information Sciences*, III-3, 473–480.

Mitra, N.J. & Nguyen, A. (2003) Estimating surface normals in noisy point cloud data. *Proceedings of the Annual Symposium on Computational Geometry, 8–10 June 2003*. San Diego, CA, USA. pp.322–328.

Monnier, F., Vallet, B. & Soheilian, B. (2012) Trees detection from laser point clouds acquired in dense urban areas by a mobile mapping system. *ISPRS Annals of the Photogrammetry, Remote Sensing and Spatial Information Sciences*, I-3, 245–250.

Munoz, D., Bagnell, J.A., Vandapel, N. & Hebert, M. (2009) Contextual classification with functional max-margin Markov networks. *Proceedings of the IEEE Conference on Computer Vision and Pattern Recognition, 20–25 June 2009*. Miami, FL, USA. pp.975–982.

Niemeyer, J., Rottensteiner, F. & Soergel, U. (2014) Contextual classification of lidar data and building object detection in urban areas. *ISPRS Journal of Photogrammetry and Remote Sensing*, 87, 152–165.

Niemeyer, J., Rottensteiner, F., Soergel, U. & Heipke, C. (2015) Contextual classification of point clouds using a two-stage CRF. *The International Archives of the Photogrammetry, Remote Sensing and Spatial Information Sciences*, XL-3/W2, 141–148.

Osada, R., Funkhouser, T., Chazelle, B. & Dobkin, D. (2002) Shape distributions. *ACM Transactions on Graphics*, 21(4), 807–832.

Paparoditis, N., Papelard, J.-P., Cannelle, B., Devaux, A., Soheilian, B., David, N. & Houzay, E. (2012) Stereopolis II: A multi-purpose and multi-sensor 3D mobile mapping system for street visualization and 3D metrology. *Revue Française de Photogrammétrie et de Télédétection*, 200, 69–79.

Pauly, M., Keiser, R. & Gross, M. (2003) Multi-scale feature extraction on point-sampled surfaces. *Computer Graphics Forum*, 22(3), 81–89.

Rottensteiner, F., Sohn, G., Jung, J., Gerke, M., Baillard, C., Benitez, S. & Breitkopf, U. (2012) The ISPRS benchmark on urban object classification and 3D building reconstruction. *ISPRS Annals of the Photogrammetry, Remote Sensing and Spatial Information Sciences*, I-3, 293–298.

Roynard, X., Deschaud, J.-E. & Goulette, F. (2018a) *Classification of Point Cloud Scenes With Multiscale Voxel Deep Network*. arXiv preprint arXiv:1804.03583. Available from: https://arxiv.org/pdf/1804.03583.pdf [accessed May 9, 2018].

Roynard, X., Deschaud, J.-E. & Goulette, F. (2018b) *Paris-Lille-3D: A Large and High-Quality Ground Truth Urban Point Cloud Dataset for Automatic Segmentation and Classification*. arXiv preprint arXiv:1712.00032. Available from: https://arxiv.org/pdf/1712.00032.pdf [accessed May 9, 2018].

Rusu, R.B., Blodow, N. & Beetz, M. (2009) Fast Point Feature Histograms (FPFH) for 3D registration. *Proceedings of the IEEE International Conference on Robotics and Automation, 12–17 May 2009*. Kobe, Japan. pp.3212–3217.

Serna, A., Marcotegui, B., Goulette, F. & Deschaud, J.-E. (2014) Paris-rue-Madame database: A 3d mobile laser scanner dataset for benchmarking urban detection, segmentation and classification methods. *Proceedings of the IEEE International Conference on Robotics and Automation, 6–8 March 2014*. Angers, France. pp.819–824.

Shapovalov, R., Velizhev, A. & Barinova, O. (2010) Non-associative Markov networks for 3D point cloud classification. *The International Archives of the Photogrammetry, Remote Sensing and Spatial Information Sciences*, XXXVIII-3A, 103–108.

Simonyan, K. & Zisserman, A. (2014) *Very Deep Convolutional Networks for Large-Scale Image Recognition*. arXiv preprint arXiv:1409.1556. Available from: https://arxiv.org/pdf/1409.1556.pdf [accessed May 9, 2018].

Tchapmi, L.P., Choy, C.B., Armeni, I., Gwak, J.Y. & Savarese, S. (2017) SEGCloud: Semantic segmentation of 3D point clouds. *Proceedings of the International Conference on 3D Vision, 10–12 October 2017*. Qingdao, China.

Tombari, F., Salti, S. & Di Stefano, L. (2010) Unique signatures of histograms for local surface description. *Proceedings of the European Conference on Computer Vision, 5–11 September 2010*. Heraklion, Greece. pp.III:356–III:369.

Vallet, B., Brédif, M., Serna, A., Marcotegui, B. & Paparoditis, N. (2015) TerraMobilita/iQmulus urban point cloud analysis benchmark. *Computers & Graphics*, 49, 126–133.

Volpi, M. & Tuia, D. (2017) Dense semantic labeling of subdecimeter resolution images with convolutional neural networks. *IEEE Transactions on Geoscience and Remote Sensing*, 55(2), 881–893.

Vosselman, G. (2013) Point cloud segmentation for urban scene classification. *The International Archives of the Photogrammetry, Remote Sensing and Spatial Information Sciences*, XL-7/W2, 257–262.

Wang, L., Huang, Y., Shan, J. & He, L. (2018) MSNet: Multi-scale convolutional network for point cloud classification. *Remote Sensing*, 10(4), 612:1–612:19.

Weinmann, M. (2016) *Reconstruction and Analysis of 3D Scenes – From Irregularly Distributed 3D Points to Object Classes*. Springer International Publishing, Cham, Switzerland.

Weinmann, M. & Jutzi, B. (2015) Geometric point quality assessment for the automated, markerless and robust registration of unordered TLS point clouds. *ISPRS Annals of the Photogrammetry, Remote Sensing and Spatial Information Sciences*, II-3/W5, 89–96.

Weinmann, M., Jutzi, B., Hinz, S. & Mallet, C. (2015) Semantic point cloud interpretation based on optimal neighborhoods, relevant features and efficient classifiers. *ISPRS Journal of Photogrammetry and Remote Sensing*, 105, 286–304.

Weinmann, M., Jutzi, B., Mallet, C. & Weinmann, Mi. (2017a) Geometric features and their relevance for 3D point cloud classification. *ISPRS Annals of the Photogrammetry, Remote Sensing and Spatial Information Sciences*, IV-1/W1, 157–164.

Weinmann, M., Weinmann, Mi., Mallet, C. & Brédif, M. (2017b) A classification-segmentation framework for the detection of individual trees in dense MMS point cloud data acquired in urban areas. *Remote Sensing*, 9(3), 277:1–277:28.

West, K.F., Webb, B.N., Lersch, J.R., Pothier, S., Triscari, J.M. & Iverson, A.E. (2004) Context-driven automated target detection in 3-D data. *Proceedings of SPIE*, 5426, 133–143.

Wolf, D., Prankl, J. & Vincze, M. (2015) Fast semantic segmentation of 3D point clouds using a dense CRF with learned parameters. *Proceedings of the IEEE International Conference on Robotics and Automation, 26–30 May 2015*, Seattle, WA, USA. pp.4867–4873.

Xu, Y., Sun, Z., Boerner, R., Koch, T., Hoegner, L. & Stilla, U. (2018) Generation of ground truth datasets for the analysis of 3D point clouds in urban scenes acquired via different sensors. *The International Archives of the Photogrammetry, Remote Sensing and Spatial Information Sciences*, XLII-3, 2009–2015.

Zhang, Z., Fidler, S. & Urtasun, R. (2016) Instance-level segmentation for autonomous driving with deep densely connected MRFs. *Proceedings of the IEEE Conference on Computer Vision and Pattern Recognition, 27–30 June 2016*, Las Vegas, NV, USA. pp.669–677.

Zhu, X.X., Tuia, D., Mou, L., Xia, G.-S., Zhang, L., Xu, F. & Fraundorfer, F. (2017) Deep learning in remote sensing: A comprehensive review and list of resources. *IEEE Geoscience and Remote Sensing Magazine*, 5(4), 8–36.

© 2020 ISPRS, ISBN 978-1-138-49604-0

Chapter 8

Laser scanning for operational multiscale structural monitoring

Roderik Lindenbergh, Sylvie Soudarissanane, Jinhu Wang, Abdul Nurunnabi, Adriaan van Natijne and Cserép Máté

ABSTRACT: Our hard infrastructure is key to the functioning of our current society. It comprises roads and railroads, tunnels and bridges, but also dams and notably public and private buildings. As soon as such structure is delivered, its state starts to deteriorate because of natural wear. Structural deficiencies or extreme external forces may accelerate this state deterioration. Structural monitoring aims at detecting possibly dangerous deformations in a structure, resulting in either intervention or normal maintenance. In recent years, laser scanning emerged as a suitable technique to efficiently measure structures at scales between millimeters to hundreds of kilometers. Laser scanning samples a surface at intervals regular to the sensor platform, resulting in huge point clouds. Still it is challenging to operationally extract useful geometric information from such point clouds. This chapter reviews recent methodological developments in the full processing chain from data acquisition to reporting the quality of the results. It will shortly discuss different laser scanning systems and corresponding acquisition issues, alignment issues, and different existing and emerging methodologies to extract changes in structures at different scales.

1 INTRODUCTION

With increasing population and urbanization, also the share amount of infrastructure increases, including its associated costs for maintenance. Traditionally, infrastructure monitoring is mainly based on in situ inspection by structural experts. This practice has several disadvantages, however. In situ inspection is expensive and often dangerous, as people have to directly access locations that are either difficult to reach, like the top of a bridge, or in heavy use, like a railway yard or highway. In addition, inspection by individual operators introduces subjectivity in the inspection results.

As an alternative, in recent years, different close- and long-range remote sensing techniques have become available (Rees, 2012; Marsella and Scaioni, 2018; Vaghefi *et al.*, 2011). Long-range remote sensing refers to in general nonspecific observations obtained from satellites or airborne platforms. Close-range remote sensing comprises a number of different techniques that are directly employed on or around the object of interest. The most well-known satellite remote sensing technique is spectral remote sensing, as implemented in e.g. the Landsat missions and in Sentinel-2. Spectral remote sensing is suitable for mapping and monitoring land cover, but its spatial resolution of 10 m and lower (for freely available data), in combination with its sensitivity to cloud cover, only makes it fit for larger urban or structural inventories, e.g. (Zhang *et al.*, 2002; Tapete and Cigna, 2018). Reversely, measurements from global navigation satellite systems (GNSS) enable the monitoring of fixed points positions at mm level and find many applications in structural monitoring, e.g. (Im *et al.*, 2011). The main drawback of GNSS is its lack of spatial coverage. It measures the position of a limited number of fixed points at high quality and at high temporal resolution. When monitoring e.g. a dam of a reservoir or a big wall of a large historical building, GNSS will not be able to identify subtle spatial patterns of variations in change.

Radar techniques are able to combine the quality of GNSS with the spatial coverage of spectral measurements. Synthetic aperture radar (SAR) operates in the microwave part of the electromagnetic spectrum, which implies that SAR signals penetrate clouds. SAR images display the strength of the radar return signal. As buildings are in general good scatterers, SAR is suitable to monitor the presence of or damage to buildings (Matsuoka and Yamazaki, 2004), at pixel resolution. As SAR pixel resolution is, at best ~5 meter, pinpointing the location of damage will in general also have a precision in the order of ~5m. Except for the signal strength or amplitude, SAR receivers also measure the phase of the employed radar waves. By combining phase returns from different viewpoints and after correction for atmospheric and topographic effects (Hanssen, 2001), a powerful interferometric SAR (InSAR) analysis is able to estimate deformations in the order of mm, provided that the deforming location is scattering consistently through time (Ferretti *et al.*, 2001). InSAR is suitable for large-scale structural monitoring, e.g. (Chang *et al.*, 2017) but currently misses the geolocation quality to reliably pinpoint deformations of mm rate (Van Natijne *et al.*, 2018), that is, the quality of the deformation estimates is high, but the geolocation quality is low, relative to other sensor solutions.

The techniques discussed so far all rely on satellite systems. The spectral techniques and InSAR are in general unspecific in the sense that no particular object to monitor is fixed before the measurements take place. GNSS is specific, however, as monitoring by GNSS typically involves the placement of GNSS infrastructure (from simple measurement bold to full receiver) at a particular, carefully selected position on or near an object of interest. Note that InSAR can be made site specific by using so-called corner reflectors. These are objects placed in a scene that are constructed such that they dominate the natural SAR backscatter signal and therefore are guaranteed to be picked up in the post-processing (Qin *et al.*, 2013).

Traditional surveying is either close-range or in situ. A method that is still unprecedented in quality is traditional levelling, possibly enhanced by automatic total station measurements (Anderson and Mikhail, 1997). But again, like GNSS, this type of specific close-range measurement is strong in considering a limited number of points of interest. On these points a bold or prism is placed, and its (possibly relative) position is assessed over and over again by either a human operator or a robotic instrument. The main drawback of these kind of measurements is that a surface or object is not fully sampled, such that spatial variations in change might be missed.

Close-range techniques that have surface sampling capacities are basically ground-based SAR, 2D imaging techniques and 3D laser scanning. The principles of ground-based SAR are the same as for satellite-based SAR (Monserrat *et al.*, 2014), and as for satellite-based measurement, either the SAR data are processed as amplitude images, enabling the use of image processing techniques, or the SAR data are processed using interferometric techniques. For the latter, coherent data is required, spatial resolution is limited, but deformations in the line of sight towards the sensor at the submillimetre to millimetre level are achievable.

Close-range imaging techniques use a spectral camera, a normal RGB camera or a thermal camera to sample the surface of an object. Photogrammetry exploits intersection in overlapping images to estimate camera positions and position individual pixel from images on the objects in the scene (Förstner and Wrobel, 2016). Photogrammetry is a widely accepted method for structural assessment, and extracting 3D spatial information up to submillimetre level is possible (Maas and Hampel, 2006). But achieving this is not straightforward, and requires expert knowledge on measurement setup, camera calibration and image processing (Valença *et al.*, 2012). Image-based techniques require sufficient light and matching images are problematic in areas that are dark or have low texture.

Image based analysis can be enhanced by using a wide range of available computer vision and image sequence analysis techniques. For example, object tracking enables users to assess the structural response of a bridge to different loading schemes (Zaurin and Catbas, 2009). Adding additional spectral bands, notably in the thermal infrared part of the spectrum, makes it possible to extract cracks in structures or identify parts of walls that are affected by moisture (Orbán and Gutermann, 2009).

In this short overview, different close- and long-range surveying techniques were discussed. Key in selecting a suitable technique are the requirements on geometric quality and on sampling density. Point-based techniques like leveling and GNSS have high geometric quality but lack sampling density. InSAR has good possibilities to estimate deformation but lacks in locating these deformations. Photogrammetry is the only technique discussed so far that is able to acquire dense and high-quality geometric change, but obtaining high-quality results is in general not straightforward and is impossible in case of poor lightning and texture conditions.

The focus of this chapter is on using laser scanning as a technique to obtain dense and high-quality geometric change information on structures. Laser scanning is little sensitive to surface texture or light conditions, as it provides its own laser illumination, while, with current equipment, obtaining faithful 3D information is relatively simple. Still, obtaining reliable geometric change information at different scales is difficult: there are many factors influencing both the quality of a laser point cloud acquisition and of consecutive processing steps, combining several such point clouds (Lindenbergh and Pietrzyk, 2015) for further comparison, Kim *et al.* (2019) gives a recent review of both image and laser based quality assessment of buildings and civil structures.

In section 2 an overview is given of issues related to point cloud acquisition using laser scanning. Systems are discussed, typical acquisition errors and some strategies to limit acquisition errors. Section 3 discusses some methodology to extract objects in point clouds, while section 4 focusses on object based change detection. Alternatively, section 5 discusses how to detect changes at point level, while section 6 discusses how scanned point clouds can be fused to other surveying techniques to profit from the strengths of both.

2 DATA ACQUISITION

This chapter will review recent developments in laser point cloud acquisition methodology. We will first discuss sensor systems and typical issues in acquired point clouds. Next we will discuss developments in measurement planning, which is a way to mitigate some of these issues. Finally, examples of fusion with complementary sensors will be discussed.

2.1 *Laser scanning systems*

A typical laser scanning system consists of a light detection and ranging (lidar) unit, an opto-mechanical scanner, a positioning and orientation unit and a control unit (Vosselman and Maas, 2010). The basic principle of lidar is to record the two way travel time of a laser pulse to object and back. The scanner mechanism ensures that pulses are emitted in different directions, while the positioning and orientation unit, if available, monitors the location of the scanner, e.g. using GNSS, and its orientation, using an initial measurement unit (IMU). The control unit manages the scan settings and stores and combines the different observations.

There are two different setups of laser scanning systems, static and kinetic. Static scanners don't move location while scanning. 3D terrestrial laser scanners or panoramic scanners sample the 3D environment of their scan location by combining a rotating mirror with a rotating head. The rotating head ensures that the scanner is able to acquire points in different azimuth directions, while the rotating mirror enables scanning at different elevation angles. Therefore, panoramic scanners acquire scans in a spherical coordinate system. The alignment of different static scans into a common coordinate systems is often referred to as registration, and the process of registration will introduce inevitably some registration errors (Vosselman and Maas, 2010). On the other hand, the static way of scanning avoids IMU and GNSS errors. Static scanners have in general the highest scan quality, due to this lack of additional sensors with corresponding errors.

Kinetic scan systems use the motion of the scanning platform to sample their 3D environment. That is, in its most simple form, the scanner itself is only rotating in one direction, while the platform, by covering a certain trajectory, moves in a direction perpendicular to the rotation direction. In this way, kinetic scanners acquire data in a cylindrical coordinate system. Scanners have been implemented on a variety of platforms, starting from manned airborne platforms, via vehicles (Puente *et al.*, 2013), including notably cars, boats and trains, towards, more recently, trolleys and backpack systems (Kukko *et al.*, 2012), drones (Kasturi *et al.*, 2016) and handheld scanners (Sirmacek *et al.*, 2016). Often kinetic systems combine several laser scanners or profilers with cameras and positioning and orientation sensors.

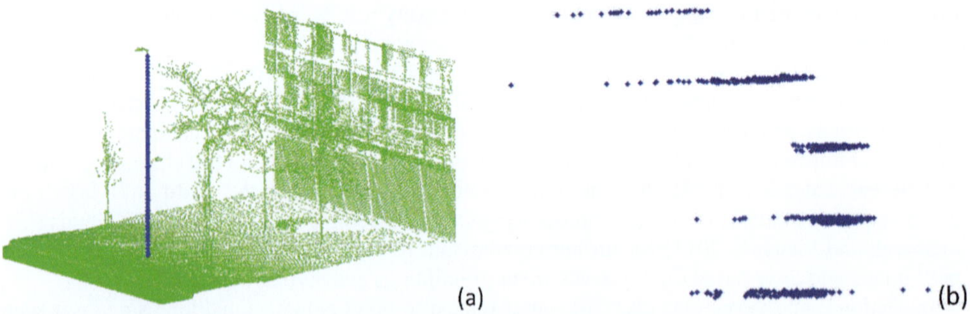

Figure 8.1. Part of a mobile mapping point cloud sampling the side of a road. In subfigure (a) the pole appears to be fully sampled. Projecting the blue points on the ground results in subfigure (b). Apparently, scanning at discrete azimuth angles results in a point cloud that samples the cylindrical pole at only five different circular positions. Further processing this point clouds, e.g. by fitting a cylinder, will be difficult (Nurunnabi *et al.*, 2019).

2.2 *Point cloud issues*

All spatial point clouds have issues. A first problem is incomplete acquisition. This is also illustrated in Figure 8.1. Car-based mobile mapping acquires data from the road. As a consequence, only that part of objects is sampled that is visible from the road. In Figure 8.1b, only the right part of the lamp pole in Figure 8.1a is sampled. Figure 8.1b also illustrates the effect of the discrete sampling nature of laser scanners. A pole of about 10 cm diameter is sampled by in this case five scan lines of one of several scanners on a mobile mapping system. By adding data from other scanners on the same platform, data becomes more dense, but also alignment issues between the different scanners may be introduced.

Acquisition geometry has a direct effect on point density and individual point cloud quality (Soudarissanane *et al.*, 2011). If the pole in Figure 8.1 would be positioned further from the road, it would be hit by fewer scan lines. In addition, the footprint of individual laser pulses hitting the pole will increase in area, corresponding to a drop in the signal-to-noise ratio, which again results in an increased noise level. Additional primary error sources are sensor calibration issues (Vosselman and Maas, 2010) and atmospheric conditions during scanning (Friedli *et al.*, 2019). Finally also the material has effect on the amount of backscatter and therefore on the noise levels.

2.3 *Measurement acquisition*

To mitigate point cloud errors, like incompleteness, noise and systematic offsets, a specific measurement acquisition strategy can be applied. the goal of such strategy is first to ensure that a scene is fully sampled after acquisition, without any gaps due to e.g. occlusions. In addition, it is often required that the acquired point cloud is of sufficient quality. That is, certain requirements on point density or maximum noise level should be met. As for point cloud positioning, static laser scanning leads to different methods than kinetic scan systems.

The case of static scanning is illustrated in Figure 8.2 in a simplified, 2D, setting, depicting a large indoor space consisting of a number of rooms. In the figure, a first scan position is indicated by O. From this scan position part of the indoor scene is visible for the scanner, as indicated by the green polygon. To obtain a complete point cloud of the indoor scene, a number of scans is required, such that the union of all their visibility polygons covers the complete scene. An acquisition is considered optimal when a minimal number of scan positions is used to fully cover the scene. This problem is closely connected to two NP hard problems in computational geometry, first the Art Gallery Problem, which searches the minimum number of cameras to monitor every position in an art gallery (Berg *et al.*, 2008) and the Next Best View problem, which searches an additional camera position that covers maximal yet unseen space (Surmann *et al.*, 2003). NP-hard implies that for practical applications, algorithms aim at finding an approximately optimal solution, Different algorithmic

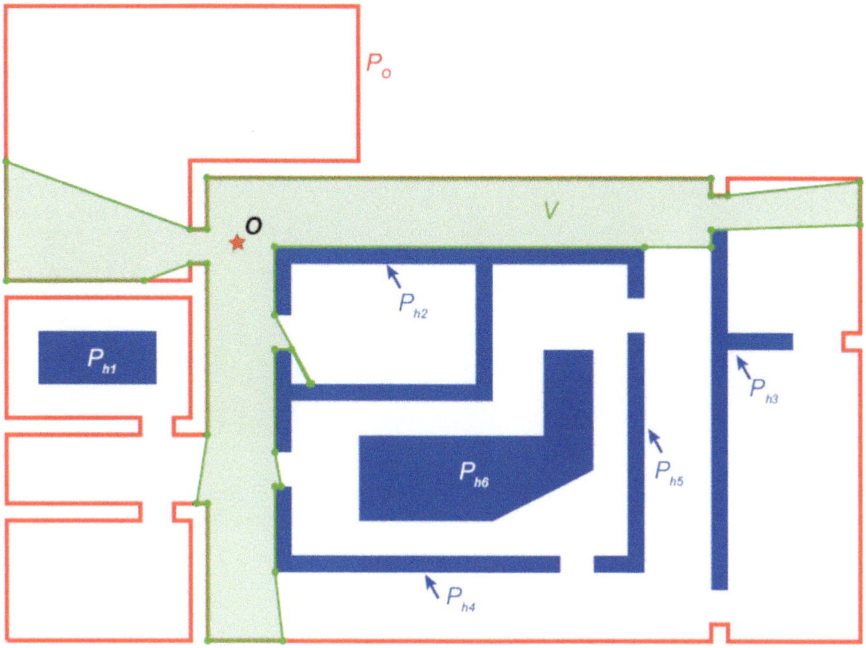

Figure 8.2. Visibility analysis. A simulated 2D complex building interior is shown, consisting of exterior walls, in red, and interior walls, in blue. A terrestrial scanner located at position O is only able to view visibility polygon V, which is indicated by a green outline. For a complete scan of the building interior, several additional scan positions are required.

solutions have been proposed in the setting of static laser scanning. Surmann *et al.* (2003) propose a greedy approach starting from a first scan; Soudarissanane and Lindenbergh (2011) propose a greedy approach given a rough outline of the scene, compare Figure 8.2, where Jia and Lichti (2017) compare different optimization methods, while Karaszewski *et al.* (2012) consider the use of a similar device for close-range scanning of sculpture-like objects. Some literature also explicitly incorporates additional constraints, e.g. on maximum incidence angle (Soudarissanane and Lindenbergh, 2011), or overlap between scans, to enable registration (Wujanz and Neitzel, 2016).

Literature on optimal 3D mapping strategies for kinetic scan systems is sparse. In robotics, a breakthrough was the first simultaneous localization and mapping (SLAM) algorithm (Dissanayake *et al.*, 2001), but SLAM aims at the simultaneous positioning of the acquisition system, i.e. a robot with laser or camera sensor and the mapping of the space around the acquisition system. In recent years, the focus is on how to use SLAM related solutions to acquire good quality point cloud data in areas without GNSS coverage like mines (Zlot and Bosse, 2014) or underground forts (Nocerino *et al.*, 2017). How challenging point cloud acquisition can be in practical settings is illustrated in Figure 8.3. Static scanning is only partially possible, as the many beams obstruct scanner placement. Handheld scan data could complement or replace static data but lacks methodology for evaluating scan completeness.

3 OBJECT DETECTION

Once a point cloud is acquired, a next step is to extract, in our case, geometric information on specific points or objects. In Figure 8.3, objects could be individual beams, while points could be the 'center of the top end of the beam'. As point cloud acquisition is not specific, in the sense that it is not pinpointing specific 3D locations, an algorithm or human operator is required extract such points and objects.

Figure 8.3. Part of the wooden tower construction of the Bavo Church, Haarlem, The Netherlands. Both complete point cloud acquisition and segmentation of an acquired point cloud are still challenging.

3.1 *Segmentation and decomposition*

A first step to organize a point cloud is often segmentation. Segmentation is the process of decomposing an image or point cloud into segments, groups of geometrically connected points, such that a certain homogeneity criterion holds either for the segment as a whole or locally, e.g. for local patches within the segment. Examples of such criteria are planarity in 3D or color in 2D. Segmentation is well studied for point cloud processing (Vosselman and Maas, 2010). A classic method is region growing, where starting from a seed point, points are added to growing segments as long as homogeneity criterion holds (Vosselman and Maas, 2010). Also density based clustering is successful in 3D, where points are connected to a growing cluster if they are sufficiently close according to a distance criterion as implemented in the DBSCAN algorithm (Ester *et al*., 1996; Aljumaily *et al*., 2017).

Disadvantages of these methods is that all individual points are processed, at the cost of computational efficiency. To avoid this, often point clouds are organized in a voxel or octree structure. A voxel structure uses regular 3D boxes of the same size, while in octree processing, boxes or initial voxels are subdivided into eight smaller voxels if more detail is required. Vo *et al*. (2015) combine octrees and region growing to speed up segmentation of urban point clouds. Voxels have also been used to assess the topology of a 3D scene (Bucksch *et al*., 2010). In Figure 8.3, the topology can be described as the configuration of beams and which beams meet. To describe the topology just the skeleton or the center line of the beams suffices, and efficient algorithms to acquire such skeletons, even for noisy 3D data, are available (Peters, 2018). Another upcoming way to explicitly integrate structure is semantic segmentation (Landrieu and Simonovsky, 2018). Here a graph structure, connecting e.g. geometric primitives is used to assign class labels like 'beam', 'pole' or 'road' to initial segments.

3.2 *Detecting and assessing objects*

The last method, i.e. Landrieu and Simonovsky (2018), already interprets a point cloud by assigning class labels to combinations of segments. In case of road environment monitoring by a municipality, a problem might be to detect all lamp poles or traffic signs of a certain type, compare Figure 8.4, from laser and possibly image mobile mapping data (Ma *et al*., 2018). Once traffic signs are detected, it is possible to assess if these are still in good shape or are bent, rusty or needing maintenance for other reasons. To do so, compare Soilán *et al*. (2016) or Wang *et al*. (2017), first terrain points are identified and removed using a suitable segmentation strategy. Off-terrain points are clustered and traffic signs

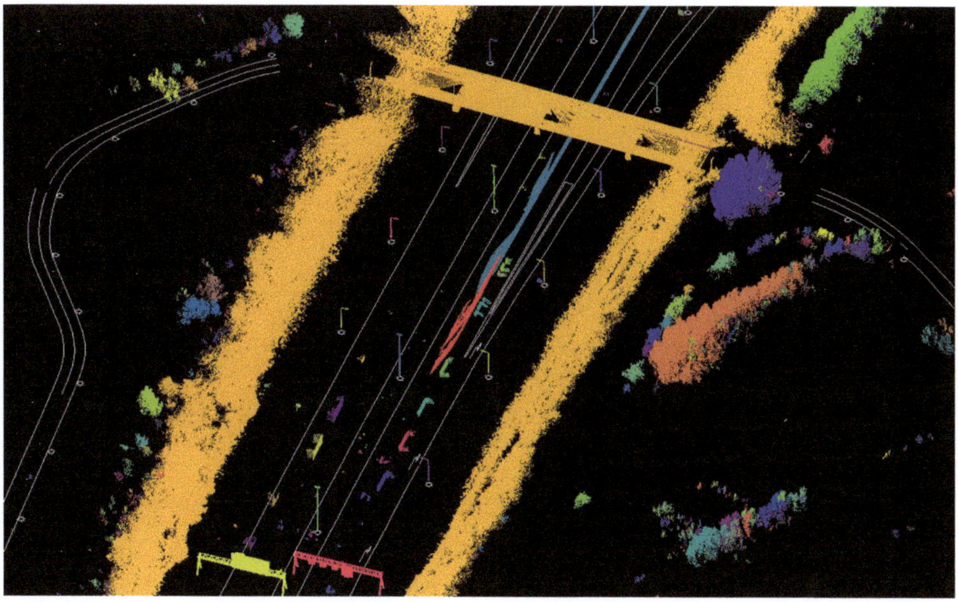

Figure 8.4. Automatic object extraction from a laser mobile mapping point cloud sampling a highway. After removing ground points, points sampling vegetation and roadside elements remain. These points are clustered using connected component analysis. Clusters of suitable size are consecutively tested to match template point clouds sampling a certain type of pole.

are identified based on intensity, that is laser backscatter (Soilán *et al.*, 2016), or by 3D shape matching (Wang *et al.*, 2017). Soilán *et al.* (2016) integrates image data for final traffic sign recognition.

After segmentation, classification and object detection, a scene sampled by a point cloud may be assumed to be divided into single objects. This could for example be roads, single bridge components, a traffic sign or single wooden beams; compare Figure 8.3. Next the fine geometric information, up to individual point level, can be used to assess the geometric state of these objects. For example Cabaleiro *et al.* (2017) describe 3D methodology to detect and assess cracks from terrestrial laser scan data sampling wooden beams like in Figure 8.3. On the other hand, Cho *et al.* (2018) use 2D imaging techniques to identify cracks in concrete walls after projecting their 3D point cloud data to 2D depth images. Also Jiang *et al.* (2018) convert 3D wall point to 2D images to detect earthquake-induced wall cracks. Of big interest is also the use of laser and other sensors to automatically identify road damage, using a variety of methods using profile and panoramic laser scanning (Coenen and Golroo, 2017; Ragnoli *et al.*, 2018). Possible damage of masonry arch bridges is detected in Ye *et al.* (2018) by first decomposing a point cloud into parts like piers and barrels, and second, assessing the state of these parts by evaluating the fit of geometric primitives like planes or cylinders (Nurunnabi *et al.*, 2014). Finally, deviations are linked to possible past movement scenarios.

4 DETECTING CHANGES AT OBJECT LEVEL

Supercomputing facilities make it possible these days to perform change detection at large scale. Here we report on the large-scale detection of changes at building level. This is illustrated in Figure 8.5, which shows results from the massive processing of Dutch archived Actueel Hoogtebestand Nederland (AHN) data. AHN is an online, open-source lidar data set sampling The Netherlands at 6 to 10 samples per square meter, resulting in 640 billion points for one complete

Figure 8.5. Detecting changes in building height by comparing two releases of the Dutch national airborne laser scan archive "Actueel Hoogtebestand Nederland'. Positive changes in yellow to red colors may indicate new or heightened buildings. Negative changes may indicate removed buildings. The figure shows the surroundings of Amsterdam Central Station, The Netherlands. False detections, corresponding to ships, are recognizable in the IJ river, in the middle of the picture.

acquisition (Van Oosterom *et al.*, 2015). Currently the third version is being acquired, which also allows one to detect changes between acquisitions. Here we focus on AHN2 and AHN3. AHN2 was acquired in different years around 2010, is organized in 1370 tiles and covers about 36 000 km^2. AHN3 was acquired between 2014 and 2019. Here, the 426 AHN3 tiles were considered that were made available first. These tiles together cover an area of about 13.000 km^2 corresponding roughly to the Dutch provinces of South-Holland, Zeeland, Utrecht, Friesland and the city of Amsterdam. The maximal reported errors in elevation in both AHN2 and AHN3 at point level are 5 cm systematic and 5 cm random errors.

The goal of the study is the detection of changes in the built-up area. For this purpose the 0.5 m AHN raster product was used, as this will limit storage and evaluation time and simplify computations. As both AHN2 and AHN3 are classified into a terrain and off-terrain class, a first step is to restrict the study to areas that where off-terrain in at least one epoch. Indeed, these areas should include buildings that stayed buildings, removed buildings and new buildings. Differencing at raster cell level and only keeping locations with bigger change is still far from the product shown in Figure 8.5: the resulting change set is strongly affected by noise, caused e.g. by vegetation present before a building was constructed or after a building was removed. Additionally, systematic effects, like small geolocation errors, may result in large vertical differences near building edges. To remove such effects, a noise filter followed by a filtering operation were performed. To restore building roofs, boundaries were expanded by morphological dilation, while artificial holes were filled using majority filtering.

The sketched workflow was performed both on a desktop and on a supercomputer. Processing of all 2 × 426 times on a desktop with 3 distributed processes took almost 2 days. The same workflow performed on a super computer using 30 nodes and 7 processes per node could reduce the computation time to 1 hour, which makes it possible for example to efficiently tune parameters. Results comparing the demo website (Cserép, 2016), were aggregated at different spatial scales, from individual raster cells, via neighborhoods, electoral districts to municipalities. Figure 8.5 shows an example of the Central Station area, of Amsterdam, The Netherlands. Results show that Amsterdam is the municipality that gained the largest building volume per hectare at almost 2500 m^3. Zooming in shows that Amsterdam city center gained a slightly lower building volume of 2140 m^3/ha, while the district Burgwallen-Nieuwe Zijde containing the Central Station, gained 2705 m^3/ha. Changes at segment level as shown in Figure 8.5 correspond to well-known new buildings, like the Eye Filmmuseum (Maro Kiris, 2018) or removed buildings. At the same time, some false detections, e.g. caused by ships in the water, require further improvement of the workflow.

5 DETECTING CHANGES AT POINT LEVEL

The problem of lack of fixed points in point clouds can be solved by different strategies. Notably in point clouds sampling human-made structures, human recognizable features can be pinpointed using an algorithm profiting from the measurement redundancy in point clouds. This approach is used in (Shen *at al.*, 2017), where, first, the centers of bricks are identified in individual epochs, second, brick centers are connected in single epochs by virtual baselines and, third, corresponding baseline lengths are compared between different epochs to identify geometric change in a certain, a priori unknown, part of a structure. A similar approach has already been used in (Lindenbergh *et al.*, 2011) to assess the stability of a target network in a TLS point cloud.

Alternatively, a 3D feature descriptor algorithm can be used to estimate 3D descriptors at arbitrary point locations (Rusu *et al.*, 2009). Consecutively, these can be applied to extract either global or local matches between point cloud parts, to detect e.g. displacement vectors of objects, like rocks (Gojcic *et al.*, 2018). In (Wujanz *et al.*, 2016) matching of point clouds using first global and next local least squares matching is used to identify areas affected by deformation. The latter two approaches consider non-structural deformation applications, but it would be interesting to apply similar methodology to detect deformation in buildings.

Another issue, notably with the use of terrestrial laser scanning for change detection, for example after an earthquake (Pesci *et al.*, 2013), is the local variations in point density corresponding to local variations in acquisition geometry (Soudarissanane *et al.*, 2011). As a consequence, the potential for change detection is not homogeneous, but smaller changes are detectable in well-sampled areas in a scene. Pesci *et al.* (2013) tackle this problem by suggesting a deformation analysis workflow that takes the *expected distortions* into account by actively incorporating the measurement geometry. By additionally incorporating geometric primitives, like planes, the authors were able to identify a small but significant change in inclination in a tower in an earthquake affected area.

6 IMPROVED CHANGE DETECTION BY SENSOR FUSION

Although laser scanning has several unique advantages above geospatial data obtained by other sensor systems, it also has problems or is simply overruled by superior possibilities of other sensor systems. Lidar data acquisition is relatively expensive, and point cloud data sampling lifted planes, as are common on building roofs, often suffer from jagged edges as a combination of the laser acquisition pattern and the so-called mixed-pixel effect. In recent years, several studies were reported in which lidar data was acquired simultaneously with very-high-resolution (VHR) image data, enabling a direct comparison (Widyaningrum and Gorte, 2017) and even a fusion, where strengths of both methods could be optimally exploited (Mandlburger *et al.*, 2017). As VHR image data is more frequently acquired, it is also interesting to assess the validity of older airborne laser scan data using newly acquired VHR imagery (Zhou *et al.*, 2018), either by a comparison of derived 3D products, which suffers from the weak points of the applied processing chains, or by an integrated approach, more profiting from the strengths of each sensor system (Zhou *et al.*, 2018).

While the above fusion approach considers two strongly related data inputs, images and laser scanning, there are also less obvious geospatial data to fuse with laser scan data. Radar interferometry is a technique that extracts deformations at the millimeter range from overlapping synthetic aperture radar (SAR) satellite images. Obtaining deformation information of a comparable quality at country scale using laser scanning is currently still beyond imagination. However, InSAR has the drawback that pinpointing the derived deformations has a geolocation accuracy that is at best in the order of several meters (Dheenathayalan *et al.*, 2016). Recently, researchers also started to combine InSAR and laser point cloud data, e.g. (Anghel *et al.*, 2016). In Anghel *et al.* (2016), specific reflectors were placed on a reservoir dam and a landslide area, that would give a distinguishable reflection in SAR images obtained by passing SAR satellite missions. At the same time, the two sites were also monitored using terrestrial laser scanning, enabling a direct comparison of the deformation results of both sensors.

Explicit fusion of InSAR deformations and airborne laser scan point clouds is described in Van Natijne *et al.* (2018). In this research, about 3 million deformation trends, estimated from 72 consecutive SAR images from the TerraSAR-X mission in 2016 and 2017, were potentially relocated by linking them to 3 billion airborne lidar points from the Dutch national archive AHN. As the error budget of the estimated InSAR trends is rather specific, the quality description of the trends was explicitly incorporated in the linking procedure. Indeed, the location quality of the trends is much higher in the line of sight towards the directions of the radar signal, resulting in a 3D error ellipse with an axis ratio of 1/2/22 in range direction, azimuth direction and cross-range direction, respectively. The largest error, in the cross-range direction, equals almost 3 m at 1σ. Given these errors, the vertical AHN errors at 0.05 m systematic and 0.05 m random were ignored. Compared to the AHN data, some InSAR trend locations are situated in open space. To improve their geolocation, an InSAR trend location is relocated to the closest AHN point, relative to the InSAR quality description, using nearest neighbors. To enable the use of ordinary nearest neighbors, first both InSAR trend locations and AHN locations are transformed by a so-called Whitening transform (Kessy *et al.*, 2018). This boils down to a change of basis, to a basis defined by the eigenvectors (orthogonal directions) and eigenvalues (length of the basis vectors) of the covariance matrix of the InSAR trend locations.

Some results of these procedures are shown in Figure 8.6. InSAR estimation points are visualized in both pictures by dots, colored by the corresponding deformation trend in mm/yr. Points in

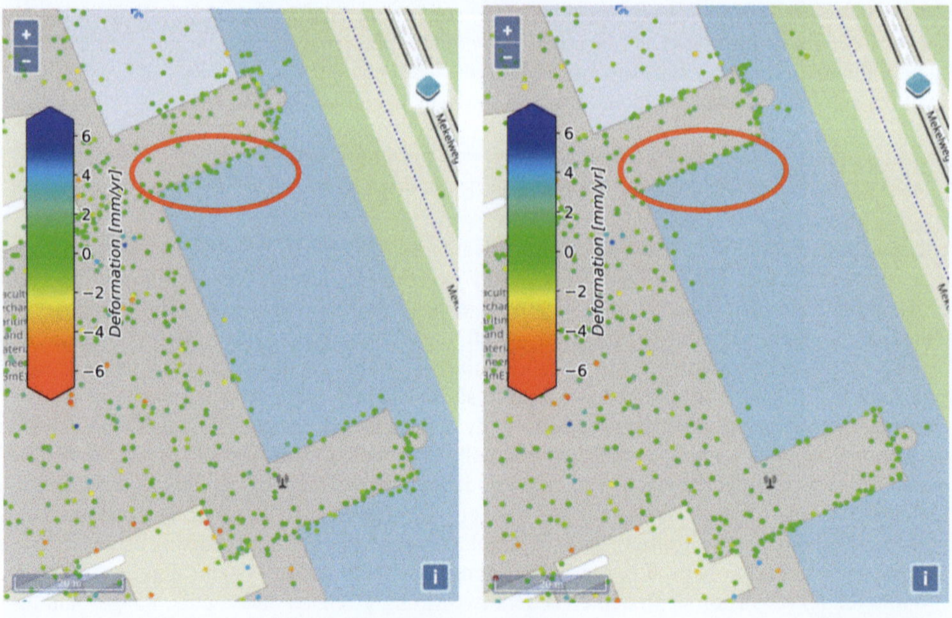

Figure 8.6. Geolocation of PS-InSAR–derived linear deformation trends, in the order between −6 to +6 mm/yr. Left: dots are located at geolocations directly estimated from the SAR satellite products. Right: dots are clipped to the AHN-3 point cloud, from the Dutch national airborne laser scan archive. Notably improvements are obtained at facades, for example within the red ellipse.

the red ellipse are still floating in the left image but are clipped to the close-by building façade by the procedure of above. In total, 85% of the points was clipped to a nearby surface, where 'nearby' is defined in terms of the InSAR covariance matrix. Results can even be further improved by not clipping to the nearest AHN point, but to the nearest AHN segment, but this requires additional processing of the AHN point cloud data.

Alternative ways of fusing or combining laser point cloud data and InSAR estimation trends are also thinkable: InSAR could be used to identify areas, like building parts, that are deforming at a suspicious trend, or where suspicious changes in trends occur. Next, such areas could be automatically classified or geometrically characterized using co-located airborne point cloud data.

7 SUMMARY

In this chapter an overview has been presented of the use of laser scanning for the purpose of structural monitoring, illustrated by a number of examples. First, laser scanning as a sensor technique has been positioned within the different available close- and long-range remote sensing methods. Laser scanning is not cheap, but as an active technique it has some advantages over photogrammetry, as data acquisition in a way is straightforward and not hampered by lack of texture or light. As for photogrammetry, it has dense spatial coverage and quality in the order of cm or better. InSAR is a complementary technique that has the possibility to estimate deformation trends at mm/yr but at lower point density and more dependent on object geometry and material characteristics.

Still, laser scanning is already hampered by errors at the acquisition phase, notably incompleteness and larger noise combined with lower point density. Specific measurement acquisition strategies have been or are being developed to mitigate such errors, but there is certainly room for improvements of such methodology. Once a point cloud is available, it is decomposed by grouping nearby points that are somehow similar, segmentation or by identifying objects in the point clouds, using a classification or matching method. Once objects are detected, the state of an object can be assessed in more detail by zooming in towards the object surface to identify for example cracks, or variations in deviation from a geometric primitive like a plane can be used to assess the tilting of the face of a tower at different heights. Methodology to do so is partly available but is expected to develop further in the coming years given upcoming machine learning methodology like deep learning, upcoming cloud computing facilities and a wish for stakeholders to automatically monitor the state of objects in management.

A final topic considered in this chapter was structural monitoring using laser scan point clouds fused with somehow complementary geospatial data. Obvious complementary data is image data, and several approaches exist to simultaneously extract information from either image and point cloud data acquired at the same time or at different epochs. Similarly, spectral images could be fused to point clouds. Also multispectral lidar is upcoming but so far was not assesses for structural monitoring purposes to the best of our knowledge. Promising are fusion approaches with completely different geospatial data sources, like the combination of nationwide airborne lidar data with operationally acquired satellite data from SAR missions or Spectral missions.

To conclude, this chapter gave a partial overview of recent developments in structural monitoring at multiple scales, from individual cracks to nationwide building changes. These developments show that after being around as a technique for several decades, laser scanning is still very much alive and in development as an active sensor technique to monitor our infrastructure.

REFERENCES

Aljumaily, H., Laefer, D.F. & Cuadra, D. (2017) Urban point cloud mining based on density clustering and MapReduce. *Journal of Computing in Civil Engineering*, 31(5), 04017021.

Anderson, J. & Mikhail, E. (1997) *Surveying, Theory and Practice*. 7th ed. McGraw-Hill Education, Boston.

Anghel, A., Vasile, G., Boudon, R., d'Urso, G., Girard, A., Boldo, D. & Bost, V. (2016) Combining spaceborne SAR images with 3D point clouds for infrastructure monitoring applications. *ISPRS Journal of Photogrammetry and Remote Sensing*, 111, 45–61.

Berg, M.D., Cheong, O., Kreveld, M.V. & Overmars, M. (2008) *Computational Geometry: Algorithms and Applications*. Springer-Verlag, Berlin, Heidelberg.

Bucksch, A., Lindenbergh, R. & Menenti, M. (2010) SkelTre – Robust skeleton extraction from imperfect point clouds. *The Visual Computer*, 26(10), 1283–1300.

Cabaleiro, M., Lindenbergh, R., Gard, W.F., Arias, P. & Van de Kuilen, J.W.G. (2017) Algorithm for automatic detection and analysis of cracks in timber beams from LiDAR data. *Construction and Building Materials*, 130, 41–53.

Chang, L., Dollevoet, R.P. & Hanssen, R.F. (2017) Nationwide railway monitoring using satellite SAR interferometry. *IEEE Journal of Selected Topics in Applied Earth Observations and Remote Sensing*, 10(2), 596–604.

Cho, S., Park, S., Cha, G. & Oh, T. (2018) Development of image processing for crack detection on concrete structures through terrestrial laser scanning associated with the octree structure. *Applied Sciences*, 8(12), 2373.

Coenen, T.B. & Golroo, A. (2017) A review on automated pavement distress detection methods. *Cogent Engineering*, 4(1), 1374822.

Cserép, M. (2016) *AHN 2–3 Altimetry Websites in the Built-Up Area in The Netherlands*, demo website [Online]. Available from: http://skynet.elte.hu/tudelft/ahn_urban_nl.html [accessed 1st February 2019].

Dheenathayalan, P., Small, D., Schubert, A. & Hanssen, R.F. (2016) High-precision positioning of radar scatterers. *Journal of Geodesy*, 90(5), 403–422.

Dissanayake, M.G., Newman, P., Clark, S., Durrant-Whyte, H.F. & Csorba, M. (2001) A solution to the simultaneous localization and map building (SLAM) problem. *IEEE Transactions on Robotics and Automation*, 17(3), 229–241.

Ferretti, A., Prati, C. & Rocca, F. (2001) Permanent scatterers in SAR interferometry. *IEEE Transactions on Geoscience and Remote Sensing*, 39(1), 8–20.

Förstner, W. & Wrobel, B.P. (2016) *Photogrammetric Computer Vision*. Springer International Publishing, Cham, Switzerland.

Friedli, E., Presl, R. & Wieser, A. (2019) Influence of atmospheric refraction on terrestrial laser scanning at long range. *Proceedings of the 4th Joint International Symposium on Deformation Monitoring (JISDM), 15–17 May 2019*, Athens, Greece.

Gojcic, Z., Zhou, C. & Wieser, A. (2018) Learned compact local feature descriptor for TLS-based geodetic monitoring of natural outdoor scenes. *ISPRS Annals*, IV-2, 113–120.

Hanssen, R.F. (2001) *Radar Interferometry: Data Interpretation and Error Analysis*. Vol. 2. Springer Science & Business Media, Dordrecht and Boston.

Im, S.B., Hurlebaus, S. & Kang, Y.J. (2011) Summary review of GPS technology for structural health monitoring. *Journal of Structural Engineering*, 139(10), 1653–1664.

Jia, F. & Lichti, D. (2017) A comparison of simulated annealing, genetic algorithms and particle swarm optimization in optimal first-order design of indoor TLS networks. *ISPRS Annals*, 4(2/W4), 75–82.

Jiang, H., Li, Q., Jiao, Q., Wang, X. & Wu, L. (2018) Extraction of wall cracks on earthquake-damaged buildings based on TLS point clouds. *IEEE Journal of Selected Topics in Applied Earth Observations and Remote Sensing*, 11(9), 3088–3096.

Karaszewski, M., Sitnik, R. & Bunsch, E. (2012) On-line, collision-free positioning of a scanner during fully automated three-dimensional measurement of cultural heritage objects. *Robotics and Autonomous Systems*, 60(9), 1205–1219.

Kasturi, A., Milanovic, V., Atwood, B.H. & Yang, J. (2016) UAV-borne lidar with MEMS mirror-based scanning capability. *Laser Radar Technology and Applications*, XXI(9832), 98320M. International Society for Optics and Photonics.

Kessy, A., Lewin, A. & Strimmer, K. (2018) Optimal whitening and decorrelation. *The American Statistician*, 72(4), 309–314.

Kim, M.K., Wang, Q. & Li, H. (2019) Non-contact sensing based geometric quality assessment of buildings and civil structures: A review. *Automation in Construction*, 100, 163–179.

Kukko, A., Kaartinen, H., Hyyppä, J. & Chen, Y. (2012) Multiplatform mobile laser scanning: Usability and performance. *Sensors*, 12(9), 11712–11733.

Landrieu, L. & Simonovsky, M. (2018) Large-scale point cloud semantic segmentation with superpoint graphs. *Proceedings, IEEE Conference on Computer Vision and Pattern Recognition*, The Computer Vision Foundation, pp.4558–4567. Available from: http://www.thecvf.com

Lindenbergh, R.C. & Pietrzyk, P. (2015) Change detection and deformation analysis using static and mobile laser scanning. *Applied Geomatics*, 7(2), 65–74.
Lindenbergh, R.C., Soudarissanane, S.S., De Vries, S., Gorte, B.G. & De Schipper, M.A. (2011) Aeolian beach sand transport monitored by terrestrial laser scanning. *The Photogrammetric Record*, 26(136), 384–399.
Ma, L., Li, Y., Li, J., Wang, C., Wang, R. & Chapman, M. (2018) Mobile laser scanned point-clouds for road object detection and extraction: A review. *Remote Sensing*, 10(10), 1531.
Maas, H.G. & Hampel, U. (2006) Photogrammetric techniques in civil engineering material testing and structure monitoring. *Photogrammetric Engineering & Remote Sensing*, 72(1), 39–45.
Mandlburger, G., Wenzel, K., Spitzer, A., Haala, N., Glira, P. & Pfeifer, N. (2017) Improved topographic models via concurrent airborne LIDAR and dense image matching. *ISPRS Annals*, IV-2/W4, 259–266.
Maro Kiris, I. (2018) Biology and architecture: Two buildings inspired by the anatomy of the visual system. *Turkish Neurosurgery*, 28(5).
Marsella, M. & Scaioni, M. (2018) Sensors for deformation monitoring of large civil infrastructures. *Sensors*, 18(11), 3941.
Matsuoka, M. & Yamazaki, F. (2004) Use of satellite SAR intensity imagery for detecting building areas damaged due to earthquakes. *Earthquake Spectra*, 20(3), 975–994.
Monserrat, O., Crosetto, M. & Luzi, G. (2014) A review of ground-based SAR interferometry for deformation measurement. *ISPRS Journal of Photogrammetry and Remote Sensing*, 93, 40–48.
Nocerino, E., Menna, F., Remondino, F., Toschi, I. & Rodríguez-Gonzálvez, P. (2017) Investigation of indoor and outdoor performance of two portable mobile mapping systems. *Videometrics, Range Imaging, and Applications*, XIV(10332), 103320I. International Society for Optics and Photonics.
Nurunnabi, A., Belton, D. & West, G. (2014) Robust statistical approaches for local planar surface fitting in 3D laser scanning data. *ISPRS Journal of Photogrammetry and Remote Sensing*, 96, 106–122.
Nurunnabi, A., Sadahiro, Y., Lindenbergh, R. & Belton, D. (2019) Robust cylinder fitting in laser scanning point cloud data. *Measurement*, 138, 632–651.
Orbán, Z. & Gutermann, M. (2009) Assessment of masonry arch railway bridges using non-destructive in-situ testing methods. *Engineering Structures*, 31(10), 2287–2298.
Pesci, A., Teza, G., Bonali, E., Casula, G. & Boschi, E. (2013) A laser scanning-based method for fast estimation of seismic-induced building deformations. *ISPRS Journal of Photogrammetry and Remote Sensing*, 79, 185–198.
Peters, R.Y. (2018) *Geographical Point Cloud Modelling With the 3D Medial Axis Transform*. PhD thesis, Delft University of Technology.
Puente, I., González-Jorge, H., Martínez-Sánchez, J. & Arias, P. (2013) Review of mobile mapping and surveying technologies. *Measurement*, 46(7), 2127–2145.
Qin, Y., Perissin, D. & Lei, L. (2013) The design and experiments on corner reflectors for urban ground deformation monitoring in Hong Kong. *International Journal of Antennas and Propagation*, 2013.
Ragnoli, A., De Blasiis, M. & Di Benedetto, A. (2018) Pavement distress detection methods: A review. *Infrastructures*, 3(4), 58.
Rees, W.G. (2012) *Physical Principles of Remote Sensing*. 3rd ed. Cambridge University Press, Cambridge, UK.
Rusu, R.B., Blodow, N. & Beetz, M. (2009) Fast Point Feature Histograms (FPFH) for 3D registration. *Robotics and Automation, 2009. ICRA'09. IEEE International Conference on*. IEEE. pp.3212–3217.
Shen, Y., Lindenbergh, R. & Wang, J. (2017) Change analysis in structural laser scanning point clouds: The baseline method. *Sensors*, 17(1), 26.
Sirmacek, B., Shen, Y., Lindenbergh, R., Zlatanova, S. & Diakite, A. (2016) Comparison of Zeb1 and Leica C10 indoor laser scanning point clouds. *ISPRS Annals*, 3, 143–149.
Soilán, M., Riveiro, B., Martínez-Sánchez, J. & Arias, P. (2016) Traffic sign detection in MLS acquired point clouds for geometric and image-based semantic inventory. *ISPRS Journal of Photogrammetry and Remote Sensing*, 114, 92–101.
Soudarissanane, S.S. & Lindenbergh, R.C. (2011) Optimizing terrestrial laser scanning measurement set-up. *ISPRS Archives*, XXXVIII (5/W1). Available from: http://www.isprs.org/publications/archives.aspx
Soudarissanane, S.S., Lindenbergh, R.C., Menenti, M. & Teunissen, P. (2011) Scanning geometry: Influencing factor on the quality of terrestrial laser scanning points. *ISPRS Journal of Photogrammetry and Remote Sensing*, 66(4), 389–399.
Surmann, H., Nüchter, A. & Hertzberg, J. (2003) An autonomous mobile robot with a 3D laser range finder for 3D exploration and digitalization of indoor environments. *Robotics and Autonomous Systems*, 45(3–4), 181–198.
Tapete, D. & Cigna, F. (2018) Appraisal of opportunities and perspectives for the systematic condition assessment of heritage sites with Copernicus sentinel-2 high-resolution multispectral imagery. *Remote Sensing*, 10(4), 561.

Vaghefi, K., Oats, R.C., Harris, D.K., Ahlborn, T.T.M., Brooks, C.N., Endsley, K.A., . . . Dobson, R. (2011) Evaluation of commercially available remote sensors for highway bridge condition assessment. *Journal of Bridge Engineering*, 17(6), 886–895.

Valença, J., Julio, E.N.B.S. & Araújo, H.J. (2012) Applications of photogrammetry to structural assessment. *Experimental Techniques*, 36(5), 71–81.

Van Natijne, A.L., Lindenbergh, R.C. & Hanssen, R.F. (2018) Massive linking of PS-INSAR deformations to a national airborne laser point cloud. *ISPRS Archives*, 42(2), 1137–1144.

Van Oosterom, P., Martinez-Rubi, O., Ivanova, M., Horhammer, M., Geringer, D., Ravada, S., . . . Gonçalves, R. (2015) Massive point cloud data management: Design, implementation and execution of a point cloud benchmark. *Computers & Graphics*, 49, 92–125.

Vo, A.V., Truong-Hong, L., Laefer, D.F. & Bertolotto, M. (2015) Octree-based region growing for point cloud segmentation. *ISPRS Journal of Photogrammetry and Remote Sensing*, 104, 88–100.

Vosselman, G. & Maas, H.G. (2010) *Airborne and Terrestrial Laser Scanning*. CRC Press, Dunbeath.

Wang, J., Lindenbergh, R. & Menenti, M. (2017) SigVox–A 3D feature matching algorithm for automatic street object recognition in mobile laser scanning point clouds. *ISPRS Journal of Photogrammetry and Remote Sensing*, 128, 111–129.

Widyaningrum, E. & Gorte, B.G.H. (2017) Comprehensive comparison of two image-based point clouds from aerial photos with airborne LIDAR for large-scale mapping. *ISPRS Archives*, XLII-2/W7, 557–565.

Wujanz, D., Krueger, D. & Neitzel, F. (2016) Identification of stable areas in unreferenced laser scans for deformation measurement. *The Photogrammetric Record*, 31(155), 261–280.

Wujanz, D. & Neitzel, F. (2016) Model based viewpoint planning for terrestrial laser scanning from an economic perspective. *ISPRS Archives*, XLI(B5), 607–614.

Ye, C., Acikgoz, S., Pendrigh, S., Riley, E. & DeJong, M.J. (2018) Mapping deformations and inferring movements of masonry arch bridges using point cloud data. *Engineering Structures*, 173, 530–545.

Zaurin, R. & Catbas, N. (2009) Integration of computing imaging and sensor data for structural health monitoring of bridges. *Smart Material and Structures*, 19(1), 1–15.

Zhang, Q., Wang, J., Peng, X., Gong, P. & Shi, P. (2002) Urban built-up land change detection with road density and spectral information from multi-temporal Landsat TM data. *International Journal of Remote Sensing*, 23(15), 3057–3078.

Zhou, K., Gorte, B., Lindenbergh, R. & Widyaningrum, E. (2018) 3D building change detection between current VHR images and past LIDAR data. *ISPRS Archives*, XLII-2, 1229–1235.

Zlot, R. & Bosse, M. (2014) Efficient large-scale 3D mobile mapping and surface reconstruction of an underground mine. In: *Field and Service Robotics*. Springer, Berlin, Germany and Heidelberg, Germany. pp.479–493.

Laser Scanning – Riveiro and Lindenbergh
© 2020 ISPRS, ISBN 978-1-138-49604-0

Chapter 9

A Smart Point Cloud Infrastructure for intelligent environments

Florent Poux and Roland Billen

ABSTRACT: 3D point cloud data describes our physical world spatially. Knowledge discovery processes including semantic segmentation and classification are a great way to complement this information by leveraging analytic or domain knowledge to extract semantics. Combining efficiently this information is an opening on intelligent environments and deep automation. This chapter provides a conceptual data model to structure 3D point data, semantics and topology proficiently. It aims at creating an interactive clone of the real world usable by cognitive decision systems. A multi-modal infrastructure integrating this data model is presented that includes knowledge extraction, knowledge integration and knowledge representation for automatic agent-based decision-making over enriched point cloud data. A knowledge base processing with ontologies is provided for extended interoperability.

1 INTRODUCTION

Knowledge extraction (KE), also known as knowledge discovery process (Fayyad *et al.*, 2002; Kurgan and Musilek, 2006) is the process to mine[1] information and create new knowledge from structured/unstructured data. Specifically, KE-oriented processes such as semantic segmentation and classification permit users to extract relevant information regarding an application domain. However, KE is only one step from a global pipeline that aims at creating an interactive space for autonomous decision-making: to represent the world in a form that a computer can use to reason (Figure 9.1).

Looking at this modular framework from raw data to intelligent environment (Novak, 1997), the processes of integrating, structuring, reasoning and interacting with the data are very challenging, especially when dealing with massive datasets from heterogeneous sources. In this chapter, we explore a solution driven by a need in automation (Figure 9.1) to progressively achieve fully autonomous cognitive decision-making based on 3D digital data.

State of the art in KE-automation applied to point clouds permits nowadays to efficiently extract different information by means of feature-based algorithms (Weinmann *et al.*, 2015), by using Knowledge-based inference (Poux *et al.*, 2017b) or more actively through machine learning[2] with promising results using neural networks (Boulch *et al.*, 2018; Guerrero *et al.*, 2018; Maturana and Scherer, 2015; Qi *et al.*, 2017, 2016; Riegler *et al.*, 2016; Zhou and Tuzel, 2017) and decision trees (Weinmann *et al.*, 2017). Extracted knowledge often comprises new patterns, rules, associations or classifications and is ultimately useful for one application. But adapting this extra information to be usable within various domains is a great interoperable challenge. A transversal field coverage demands bijective communications through a great generalization and normalization effort.

The processing module defined as knowledge integration (KI) in Figure 9.1 specifically addresses this task of synthetizing multiple knowledge sources, representations and perspectives often layering multiple domains. This can be decomposed into (1) information integration (i.e. merging information that is based on different schemas and representation models) and (2) synthetizing the understanding of one domain into a common index that keep track of the variance within perspectives. KI specifically addresses integration and structuration of the data (often within database systems), which may resolve conflicts with hitherto assumed knowledge.

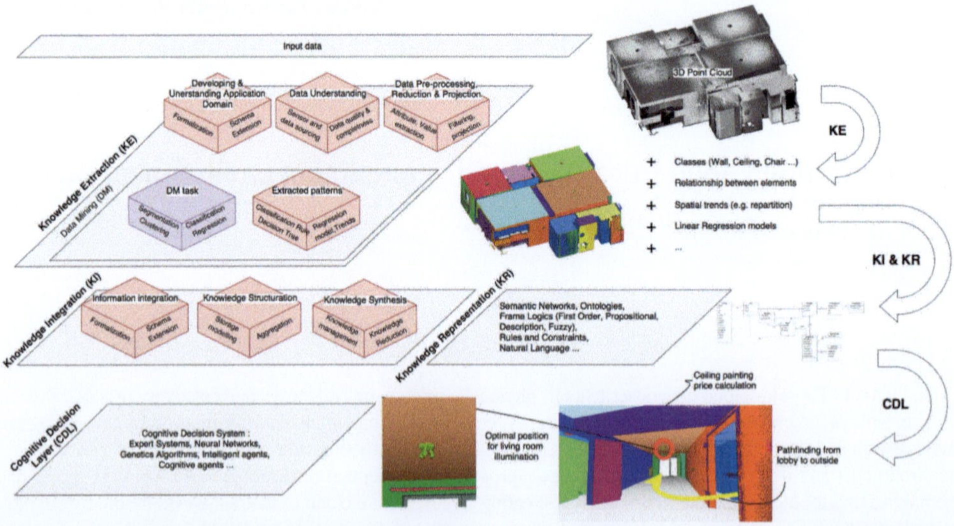

Figure 9.1. Modular framework for the creation of an intelligent virtual environment, illustrated over an indoor point cloud. (1) KE; (2) knowledge integration; (3) knowledge representation; (4) reasoning from cognitive decisions layer (CDL)

Thus, the extended data structure must be translated into an explicit knowledge representation (KR) to permit a computer to achieve intelligent behaviour and access knowledge reasoning through a set of logical and inference rules. This is very motivating if you want to deepen the operations made by the computer rather than interpreting on the fly (brain work). Once metadata is attached to 3D data, then you can more easily grasp or even make calculations impossible before (e.g. in Figure 9.1 for 3D indoor pathfinding, indoor lighting simulations, reasoning for optimal positioning, extraction of surfaces per room for digital quotations and inventories).

While such a universal solution to decision-making situations is very attractive, KE, KI and KR highly depend on the initial application domain definition, data understanding and its integration within a complete infrastructure. This strains that the underlying data structure must synthetize knowledge through pertinent KR while permitting inference reasoning based on an efficient cognitive decision layer (CDL).

In the first part, we study the attempts, standards and existing reflections to define a common scheme for exchanging relevant 3D information while situating the current integration state of point clouds within these normalizations. Second, we present a data model for point cloud structuration that retains knowledge. We then propose a point cloud infrastructure integrating this data model to allow KE, KI and KR following the modular framework presented in Figure 9.1. Finally, we benchmark such a global solution against several datasets to test its response to established requirements and identify limitations for new research directions.

2 SEMANTICS AND 3D DATA

In the context of 3D data, the wide array of applications implies a vast diversity in how the data is used/conceived. This research environment adds important complexities to the integration within any generalized workflow, of which 3D point cloud data are quasi-inexistent. This explicitly demands that data-driven applications enable targeted information extraction specific to each use case. While this is rather convenient for one use case, making a general rule that applies to all models is a daunting task. In this section, we review existing attempts and standards favouring interoperability through well-established 3D spatial information systems (2.1), KI reflexions over 3D data (2.2) and point clouds (2.3).

2.1 *3D spatial information systems*

Datasets that explicitly include spatial information are typically distinguished regarding the data models and structures used to create, manage, process and visualize the data. In a 3D context, we, analogically to Poux *et al.* (2017a) and Ross (2010), differentiate three main categories:

- 3D GIS: GIS systems usually model the world itself, retaining information about networks, conductivity, connectivity, topology and associativity. This enables geospatial analysis, often carried on large collections of 3D instances stored in data warehouses with coordinates expressed in a frame of reference.
- 3D CAD (computer-aided design): CAD/CAM techniques model objects from the real world through parametric and triangular modelling tools. The topology is often partial or planar (although vendors extend functions to include semantics and higher descriptive topology (Zlatanova and Rahman, 2002), and the data usually plays on a visual scale. The distinction between visualization and storage is not as clear as in 3D GIS systems, and one file generally describes one complex 3D object. CAD files carry visualization information that is not relevant to the data itself. A simplistic difference consists in thinking of 3D GIS systems as 3D spatial databases, whereas 3D CAD models are rather related to 3D drawings. The coordinate system is therefore linked to a defined point of interest (often the centroid) in the scene.
- BIM: It constitutes working methods and a 3D parametric digital model that contains "intelligent" and structured data initially for planning and management purposes. It is often studied for its integration with 3D GIS systems with an extensive review in Liu *et al.* (2017), but their parallel evolution (conditioned by temporal and hermetic domain research) and fundamentally different application scopes are slowing down their common assimilation. BIM models share many properties with 3D CAD models, including their expression of coordinates in a local system, but benefit of a higher semantic integration.

The emergence of new data sources and evolution in data models constantly put in question the suitability of these categorizations. Established and emerging data types and their integration/characterization can become difficult for meeting the characteristics of one of these categories. For example, more primal spatial data from a more direct data source such as 3D point clouds could benefit from their own category. Indeed, they have a very small direct integration in these groups but rather serve as a support for the creation of CAD/CAM models, BIM models or 3D GIS systems. In some advanced cases, the information included in 3D point clouds can help extract metadata for the future data model.

However, it is important to note that while barriers between each category were well defined five years ago, the improvement and added functionalities to each category as well as interoperability and integration research and standardization play a major part in blurring the respective frontiers.

2.2 *Knowledge integration solutions for 3D semantically-rich data*

"*Semantic interoperability is the technical analogue to human communication and cooperation*" (Kuhn, 2005). This sentence pertinently summarizes the drive in GIS research to formalize semantics in order to facilitate the communication of data among different communities. Different levels of interoperability exist, and we are looking at the technical parts without looking at societal issues raised by enterprise-oriented information sharing (Harvey *et al.*, 1999). However, the conceptualization of interoperability in our computerized environment remains a challenge at different levels:

- The nature of concepts that defines interoperability should not arise from simplistic assumptions as notions evolve with time;
- The ever-growing use of 3D data makes it very hard to define a common "language" to be spoken by all professionals;
- The knowledge involved is sparse enough to constrain natural language extension in a computerized formalism;

- Standardization efforts need international cooperation to represent as thoroughly as possible the reality and benefit of effective coordination;
- Retaining semiotic relationships between concept, symbol and entity, as in the semantic triangle.[3]

In a narrower context, 3D data as 3D models are largely used for a high number of applications, which vary in scope, scale, elaboration and representativity. Therefore, semantic schemes as generic as possible provide a potential solution for interoperability.

Ontologies are a good way to explicitly define knowledge in order to address semantic heterogeneity problematics arising from this large variety. However, independent work and research limit their extension to a broader audience especially looking at 3D content. But the rise in usage demands that specific solutions allow 3D data to be exchanged and used as thoroughly as possible. Independent development and uncoordinated actions in the research field of ontologies applied to GIS are addressed by entities such as the World Wide Web Consortium (W3C), the International Organization for Standardization (ISO), the Open Geospatial Consortium (OGC), the International Alliance for Interoperability (IAI) and the rise of open-source developments and repositories. Clarifying standardization processes over 3D data is especially important, with issues arising at both a technical level and a consideration level (how is 3D data considered by the community?).

In general, a standard defines a data model at two levels: properties and geometry. A well-known example is the standard GML3 issued by the OGC, which is used by the CityGML data model describing the geometrical, topological and semantic aspects of 3D city models (Kolbe *et al.*, 2005). The specification and the decomposition in level of details (LOD) as well as the current 2.0 version allowing to define semantic concepts has made the integration of city models easier and applicable to a wider range of use cases (Biljecki *et al.*, 2015). Indeed, this gives the possibilities for decision makers to impose a specific "abstraction figure" (LOD1, LOD2 . . .) that characterizes the granularity level of the wanted geometry and semantic concepts. This interoperability "tool" is a leap forward in the democratization of the standardized data model CityGML. However, its integration with other standards or ontologies is still being discussed and studied, where a discrete number of LOD with "unconnected" (potentially uneven) levels could be a concern (Karim *et al.*, 2017). This illustrates the need to find interoperable systems between already established standards to benefit of higher semantics and topology integration that enhances our comprehension and usability of 3D data.

The semantic web is a great tool standardized through Semantic Web 3.0 that can create links between already established standards, which encourages the use of web-based data formats and exchange protocols, with the resource description framework (RDF) as the basic format. Indeed, this has the potential to greatly reduce the gap/frontier between each previously defined category in 2.1 and better integrate knowledge within 3D spatial data. This is especially efficient if we better integrate 3D point clouds, from which we today derive so many systems and data models. Indeed, initially it could serve as transition data, but given time it could provide all the necessary information if correctly integrated. Point cloud data's large volume and high resolution make it suitable for LOD management and rendering.

Finding hidden patterns and information for knowledge discovery requires complex multimodal systems as presented in Figure 9.1. Data interaction needs flexibility and scalability for different tasks: processing, data management and visualization. To solve these challenges, both the data model and the storage model must be investigated.

2.3 *Point cloud solutions for knowledge integration*

Few solutions exist for managing semantics and geometry directly on the point cloud, demanding new data model to capture key logical aspects of the data structure. On top, the large datasets that point clouds constitute cannot directly fit in the main memory, demanding adapted systems that can exploit efficiently the information regarding a storage model.

We can identify two ways that point cloud can integrate semantics, geometry and topology: through file-based solutions or through database management systems (DBMS). The storage model used will condition how efficiently the variability and redundancy of the amount of observations is handled. In available DBMS the data store is either as a Block model (i.e. points are grouped in blocks, usually neighbourhoods, which are stored in a database table, one row per block) or a flat table model (i.e. points are directly stored in a database table, one row per point, resulting in tables with many rows). All file-based solutions use a type of blocks model, where points are stored in files in a certain format and processed by solution-specific software. While they are common point cloud storing systems managed through hierarchical-like database models, sharing, compatibility, query efficiency and data retrieval are the main limitations in these solutions. For example, the standard LAS[4] format allows only one byte for user data, making it a very limited choice for managing highly variable knowledge with different perspectives for categorization. Therefore, a fixed schema will quickly become obsolete when trying to cope with thousands of applications.

Dobos *et al.* (2014) introduced the concept of point cloud database for scientific applications, based on relational tables. Classical relational DBMS for such application exists, but binary trees limited scalability that struggle with huge datasets size and non-adapted vectorization and indexation schemes often specific for one usage are hard to exploit on many different servers. Building on this, they point requirements of the structure for analysis of point clouds mainly filtering capabilities, key look-up and nearest-neighbour search, cluster analysis, outlier identification, histogram and density estimation, random sampling, interactive visualization, data loading, insert and updates. van Oosterom *et al.* (2015) extend the concept by defining point cloud data as the third type of spatial representation (the first one being vector data – row-like single feature specification – and the second raster data – multipoint object). Their extensive work focus on benchmarking several available commercial point cloud data management systems (PCDMS; block model and flat model of PostgresQL-PostGIS, block model and flat table model of Oracle, flat model of MonetDB, file-based LAStools) to define which one is the most fitted for point cloud management. While some improvements need to be implemented to fix issues in available solutions, each provides a benefit compared to the others, but none can answer efficiently combined queries, data I/O and real-time visualization. The interoperability stays essential to combine point cloud data with vector data and raster data. They also show in a brilliant way the need of linking user needs, user type with user experience to define a standard in point cloud design and implementation. The NoSQL database robustness to massive data with weak relationship can scale up to many computers, but functionalities are today very limited.

Other research work by Ben Hmida *et al.* (2012), Cura *et al.* (2015) and Otepka *et al.* (2013) presents some solution to the integration of domain knowledge through a priori or a posteriori KR in ontologies, but the efficiency and extensibility to production processes depend on the underlying structure for efficient processing, analysis and visualization. As stated by Otepka *et al.* (2013), naïve strategies especially considering query complexity of neighbour search $O(n^2)$ are unrealistic for industrial applications. Indexing techniques provide a solution to storing, compressing and managing the data (Dobos *et al.*, 2014; Richter and Döllner, 2013; van Oosterom *et al.*, 2015), but efficiency and extensibility to dynamic semantic update and ontological reasoning stays limited. Queries over octree derived indexing techniques can provide an efficient solution for out-of-core rendering and parallel processing, but data structuration cannot efficiently include context adaptation and inference reasoning.

While these constitute pertinent examples of knowledge and semantic enhancing capabilities, no clear and defined structure is developed. Identifying links and relations within objects of interest becomes essential to truly understand how each spatial entity relates to its surroundings and connecting GIS, CAD and BIM concepts (as seen in section 2.1) to 3D point clouds through contextual segmentation and object storage. The work of Poux *et al.* (2017a, 2017b, 2016) is a first step in this direction: it proposes a global framework to classify, organize, structure and validate objects detected through a flexible and highly contextual structure that can adapt to three identified knowledge sources: domain, device and analytic knowledge. This lays the groundwork for the development of a new data model – the smart point cloud – that can address previously identified issues while retaining a high level of interoperability with existing standards and ensuing the Figure 9.1 outline.

3 THE SMART POINT CLOUD (SPC) DATA MODEL

The smart point cloud concept gives the conceptual tools and definitions for a point cloud knowledge-based structure contextually subdivided according to KE results (namely segmentation and/or classification). It presents a broad framework for the semantic enrichment and structuration of point clouds for intelligent agents and decision-making systems. Our approach infers initial relationships/topology and separate spatial/attribute information to provide efficient data mining capabilities. The domain specialization relies on ontologies to allow high interoperability and specialization through derived semantic enrichment layers (Janowicz *et al.*, 2010). The structuration of the in-base knowledge relies on a categorization first introduced in Poux *et al.* (2016) is extended in section 3.1, and a conceptual model is proposed in section 3.2 and described in sections 3.3, 3.4 and 3.5.

3.1 *Knowledge categorization*

KR and reasoning are the areas of artificial intelligence (AI) in which we study how knowledge can be represented symbolically and manipulated automatically by reasoning engines.[5] Mainly, the use of logic in this context will study entailment relations languages, truth conditions and rules of inference to enable reasoning. This demands that part of the knowledge be explicitly represented (knowledge base) and constitute the foundation of what the system believes. In order to structure this knowledge base (KB) for 3D point clouds, we propose a simple yet efficient categorization of knowledge for deriving implicit conclusions from our explicitly represented KB.

The first step toward knowledge integration considers knowledge categorization. In order to cope with the heterogeneity within information's and perspectives, Knowledge for point cloud processing was categorized in three branches: device knowledge, analytic knowledge and domain knowledge. The latter constitutes what is the closest to domain applications and thus is attached to ontologies of specialization.

3.2 *Conceptual SPC (smart point cloud) model*

The overview of current practices showed a need to improve automation, data management and interaction. The semantization process relies on geometrical descriptors as well as a domain analogy integrated in a new structuration of the point cloud data through correct indexing techniques. At a higher conceptual level, the creation of an intelligent virtual environment from point clouds is inspired by our cognitive system: recognizing an object means accessing symbolic units stored in a semantic memory and which are abstract from our previous experiences while being independent from any context. Disposing of either digital copies of the real world, invention/conception of "things" to be integrated in the world or a combination of both, we refer to geometries from the "physical space" and "fictional space" (immaterial, concept-based) as in Billen *et al.* (2012). In their paper, they propose an ontology of space in order to facilitate an explicit definition of CityGML. Extending the formalism, it constitutes a basis for semantic injection into point clouds. However, the study of ethno-physiography as well as human cognition of geospatial information is mandatory for defining information system ontologies. Indeed, the closer (and the richer) the model is to the domain concept, the better (and more extensible) the ontology will be. But the questions of how detailed an ontology should be depend on the levels of interoperability that are envisioned.

The purpose of the SPC characterization is to represent the real world spatially described by point clouds in a computerized form: a user-centered frame representation serving an intelligent environment. The definition of a generic model that applies to a general purpose is very complex, as opening on all domains that benefit from 3D semantically rich models and point clouds range from neuro-psychiatry to economics or geo-information. Our approach was thought to allow a maximum flexibility by defining a conceptualization on which different domain formalization can be attached (Figure 9.2).

We divided the characterization (KR and data modelling) in different hierarchical levels of abstraction to (1) avoid overlap to existing models and (2) enhance the flexibility and opening to all possible formalized structure. The core instruction is that the lower levels are closer to a domain

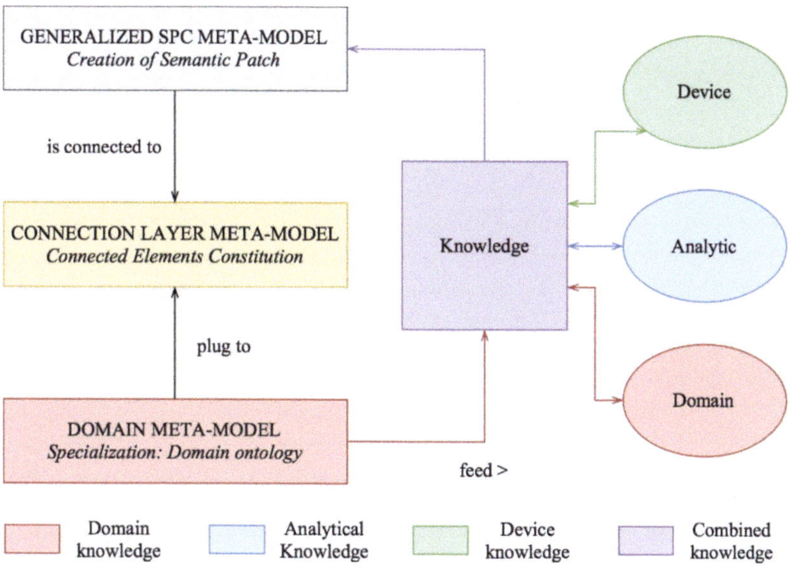

Figure 9.2. Meta-model articulation for the creation of an SPC

representation than higher levels (level 0 being the highest level), but they impose their constraints. The overall structure can be seen as a pyramidal assembly, allowing the resolution of thematic problems at lower levels with reference to constraints formally imposed by the higher levels.

KI is essential to the creation of the SPC structure, as it constitutes the necessary source for the meaning and adaptation of different entities within this pyramidal model. By default, we integrate a core algorithmic module that allows us to extract a raw relationship graph based on a voxel element mining routine inspired by Gorte and Pfeifer (2004 and Wang *et al.* (2017). This was established because it does not require any external semantic information other than pure spatial information which encourage flexibility and adaptation. However, more domain-verse classification modules such as Xiong *et al.* (2013) provide potentially enhanced workflows. As seen beforehand, while Relational DBMS are a great fit logically speaking, they do not perform well considering the very high number of rows. Clustering via indexing schemes is mandatory for interactive visualization as well as efficient data loading, inserts and updates. Building this spatial structure over an object-based binary host/guest structure enables powerful analysis and visualization exploitation. In parallel, the ontology-based KR allows inference reasoning/semantics retention and is directly linked to the spatial structure. Thus, it defines relationships and topology at both the point and object geometrical levels. The top conceptual level, called level 0, gathers data, information and knowledge about the core SPC components.

3.3 *Level 0: generalized SPC meta-model*

For clarity, we specifically target point clouds, but the model can be extended to all kinds of massive gathered data from our physical world and, in an extended version, provide an opening for 3D meshes or parametric model integration. The different meta-models are formalized in UML and provide a conceptual definition for implementations. We therefore modelled as a goal to provide a clear vision and comprehension of the underlying system, but the database creation slightly differs to privilege performances; therefore adaptations are made at the relation scheme modelling level.

The generalized SPC meta-model (Figure 9.3) formalizes the core components needed for constituting semantic point patches. It starts with the most primitive geometry: a point. It has a position defined by three coordinates in Euclidean space (R^3): X, Y and Z. Each point has a limited number

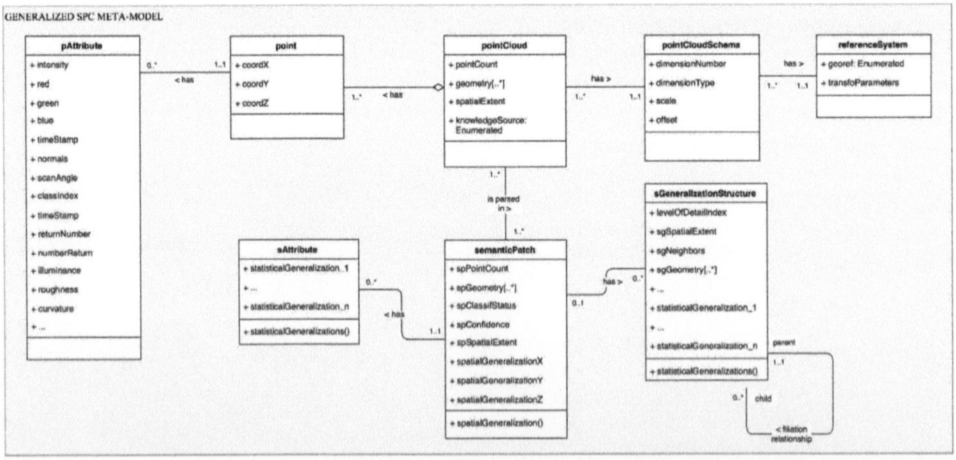

Figure 9.3. Level 0 generalized SPC meta-model (UML). A point cloud constituted of points is block-wise organized through semantic patches. These can be pure spatial conglomerate or retain a coherent semantic relationship between constituting points. Generalizations via different schemes are possible using the generalization structure to provide additional analysis flexibility.

of attributes, for an example in Figure 9.3, derived from three different sources: device knowledge (scan angle, intensity . . .), analytic knowledge (normal, curvature, roughness . . .) or domain knowledge (definition, representativeness . . .). While the UML model shows a one-to-many relationship, to avoid too many SQL joints and for performance's sake, the attributes can directly be integrated within the point table (the same applies to semanticPatch). However, it is important to note that one point can have many sets of attributes (consequently, as does a semanticPatch).

A collection of points sharing the same type of dimensions (spatial and semantic) constitutes a point cloud. This is a data-driven aggregation, as depending on the definition of the dataset, the point cloud object parameters will differ. However, one dataset often represents a coherent point aggregation which serves a domain purpose. This point cloud entity also benefits from a knowledge source pointer to identify which knowledge source it relates to (if multiple domain-specific ontologies are connected to the model). To cope with heterogeneity in point cloud sources, a schema is defined and attached to all point clouds that share a similar dimension number, dimension type, scale and offset, similarly to Cura *et al.* (2017). Each point cloud is then parsed in semantic patches, regarding available knowledge and an adapted subdivision technique. Arbitrarily, such a technique could be point-number related, geometry related, or position related. While the existing PostgreSQL plugin Pgpointcloud defining patches in an XML scheme provides spatial patches, we propose to greatly enhance such an approach by constructing semantic patches, which retain both spatial and semantic properties. It constitutes small spatial subsets of points that share a relationship based on available (and injected) knowledge. By default, our proposed voxel-based subdivision method groups point using geometrical and topological properties that implicitly relate to abstract conceptualization of our mind (such as geometric shapes to group points belonging to a plane, others floating above it . . .). As such, they are indirectly semantically enriched. "semanticPatch" retains many attributes, with an emphasis on two specifics: a classification status (which can be 0: unclassified, 1: one class only or 2: many classes) and a confidence level for the classification. These are computed through a segmentation and classification routine as described in section 4.1, which is independently developed from the proposed point cloud data model. In order to speed up computations, allow enhanced spatial and semantic searches and provide new generalization possibilities to better address our representation of the data, a LOD generalization structure definition is directly linked to the semantic patches (Figure 9.4).

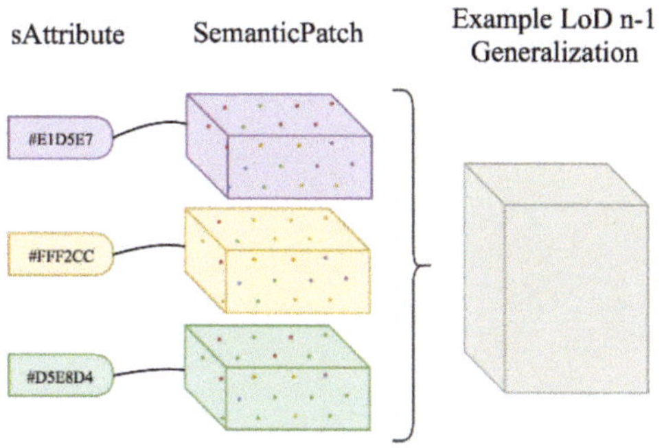

Figure 9.4. Example of a basic LOD n-1 generalization of 3 semanticPatches from a point cloud with colour attributes only.

It defines the indexation scheme used, the different levels (if any), its node spatial extent and neighbours, associated geometries (if any) and other generalized attributes derived from statistical computations (average, Gaussian mixture . . .). Poux *et al.* (2016) suggests a 3DOR-Tree as defined in Gong and Ke (2011) for improved performances, but hashing and implicit storage (Cura *et al.*, 2016) can also greatly improve the internal coherence.

3.4 *Level 1: connection-layer meta-model*

The connection-layer meta-model (i.e. the strict framework that drives the use of a formalism and resolves any ambiguities about the use of its concepts) plays the role of a plug system: an interface between the core SPC level 0 generalized meta-model and a domain ontology that formalizes the domain-specialization of a generic ontology. It is constituted of two sub-levels, L1-1 and L1-2 (Figure 9.5).

The core element in this meta-model is created by an aggregation of one or many patches that define a connected element (ConnectedElements, CEL). These are the entities that closely relate to classified objects, retaining both spatial and metadata coherence. Connected elements transparently describe a portion of the space that is by default indirectly influenced by analytical knowledge and device knowledge, from the underlying patch organization. Connected elements have a spatial extent computed from the aggregation of patches, as well as one or several geometries that can be obtained by topological calculations from the patches. Aside from geometrical attributes including a spatial generalization (which can be for example the barycentre of the spatial extent, but also more representative statistical generalization) they retain raw semantics from the underlying patch aggregation rule dependent of a CEL. The integration of domain knowledge gives the opportunity at this level to deepen the representativeness of a connected element. Nevertheless, one connected element regarding a variety of applications can have different spatio-semantic interests. Therefore, aggregated elements constitute an aggregation of connected elements which provide additional granularity and flexibility (a table, with 4 feet and one horizontal working area, which is either five connected elements or one aggregate element). Similarly, each connected element retains relationships with its surrounding environment: we detect and store host and guest relationship information (the table is the guest of the floor, and the floor is the host of the table). These strong concepts have an influence on how deep the selectivity can go. Retaining relations and organizing hierarchically through topological relations refers to mereology, applied on point clouds object generalization regarding DE-9IM (Clementini and Di Felice, 1997). The existing topological relations between 3D spatial objects with internal space are Disjoint, Meet, Overlap, Equal, Contain, ContainedBy, Cover, CoveredBy (Zlatanova *et al.*, 2004). Therefore, a double structural definition retaining generalization and point primitives (level 0) allows new analysis combining multi-LOD definitions. This pyramidal graph relationship formalization

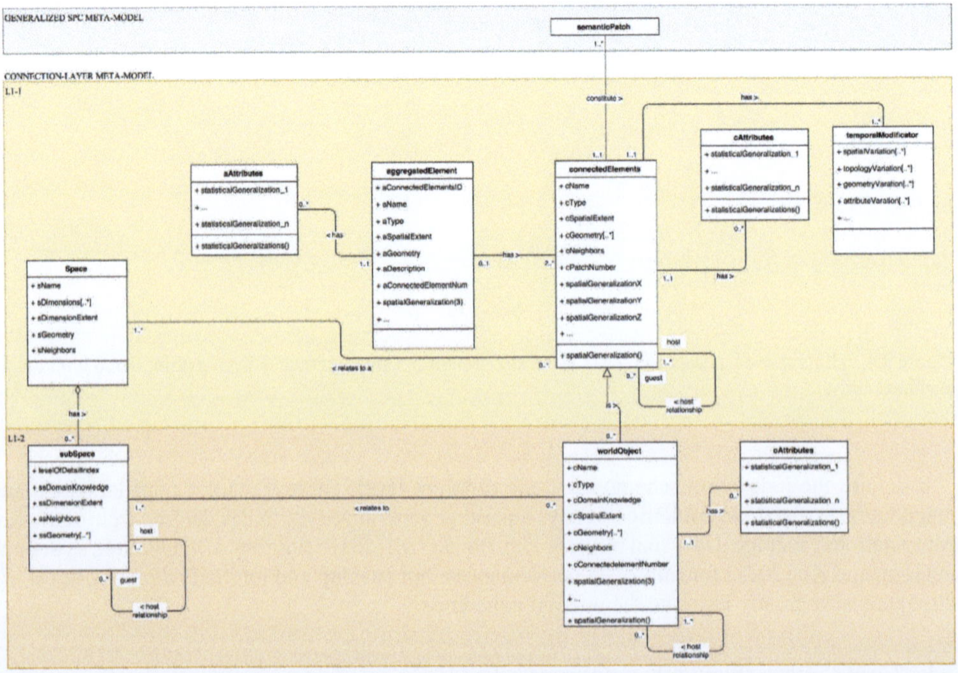

Figure 9.5. Level 1 connection-layer meta-model. It is directly linked to the level 0: semantic patches constitute ConnectedElements. AggregatedElements and topological notions gives flexibility to the deepness of an element characterization. ConnectedElements can relate to one or multiple spaces defined by their dimensions. These are subsequently divided in subspaces regarding a concept from a domain knowledge characterization, similarly to the world objects (being a specialization of ConnectedElements).

permits users to easily access a spatially connected graph for reasoning engines that interpret topological relations. These conditions can be used to infer a physical description and combine many possible analyses, for example the possibility to recreate occluded zones, reason about position in time and space, conduct structural investigation . . . Connected elements also have additional properties and specific attributes inherited from the patches that they relate to. While every point can retain a date stamp, a connected element can be influenced by temporal variations, but duplicating a physical description of the connected element at every discrete temporal interval would not be enough. Therefore, ConnectedElements can have a temporal modifier that will describe the different modifications from the in-base initial state. They can also relate to multiple spaces defined as a set of dimensions in $R, R^2 \dots R^n$ (for example X/Y/Z in R^3).

"Space" and "ConnectedElements" are related to a lower abstraction level L1-2 within the connection-layer. A space can have many subspaces defined in respect to the space dimensions. In a spatial context, it is interesting to note that they are mostly fiat subspaces in regard to Smith and Varzi (2000). Indeed, bona fide boundaries represent physical separators, whereas fiat boundaries will describe a fictional border, and most of subspaces for human cognition have a fictional border (e.g. a room with an open door). The topological inward relation allows use to constitute different subspace Levels of Detail (we can consider a building, or the first floor of that building, or the room 2/43 of that first floor . . .). "SubSpace" therefore retains a domain knowledge source pointer that can be dedicated to one or many specific domains (it can be a subspace regarding the ontology of buildings, to the archaeology temporal findings in Australia . . .). The concept of world objects results from the definition of Billen *et al.* (2012), which is a mind conceptualization of an object that also follows the categorization of Smith and Varzi (2000). "WorldObject" is a specialization of "ConnectedElements" retaining a domain related semantic pointer similarly to "SubSpace" (a knowledge source mirroring the domain conceptualization). Geometries attached to these entities are useful for topological calculations, and the direct link to "SubSpace" allows many possible queries for information extraction (testing the inclusion of a world object in a subspace, testing the intersection of two object's geometries with a fiat boundary from a subspace . . .). "SubSpace" and "WorldObject" constitute the entry points on which domain ontologies can be associated to adapt to a specific application.

3.5 *Level 2: domain adaptation*

As stated by Tangelder and Veltkamp (2007) "any fully-fledged system should apply as much domain knowledge as possible, in order to make shape retrieval effective". With the rise of online solutions, we have seen a great potential in using knowledge databases for classification to analogically associate shapes and groups of points with similar features. This association through analogy "is carried out by rational thinking and focuses on structural/functional similarities between two things and hence their differences. Thus, analogy helps us understand the unknown through the known and bridge gap between an image and a logical model" (Nonaka *et al.*, 1996). This introduces the concept of data association for data mining and relationships between seemingly unrelated data in a relational database or other information repositories. Enabling the use and analysis of domain knowledge through explicit domain assumptions while separating domain knowledge from operational knowledge refers to domain ontologies. This shares interoperability notions with our proposed SPC structure; while one domain meta-model formalization is suited for some applications, another can be more adapted for others and create different results that will be used differently. These will dictate how the final point cloud data model should be used (for which application).

Therefore, the level 2 meta-model is directly linked to different knowledge sources, which are specified in the level 1(2) meta-model interfaces: "SubSpace" and "WorldObject". Their conceptual abstraction in between pure spatial data (point clouds) and specific domain-verse data constitute a generic door for the potential connection to many level 2 domain specializations. This allows a great flexibility and a context adaptation to a very wide range of application, limited only by the underlying domain ontology. In fine, the domain meta-model attached to the connection-layer meta-model, and indirectly to the generalization meta-model, constitute the SPC model.

In a simple example (Figure 9.6), we illustrate over a basic indoor ontology the connection of a level 2 meta-model to the connection-layer meta-model. It contains two class elements

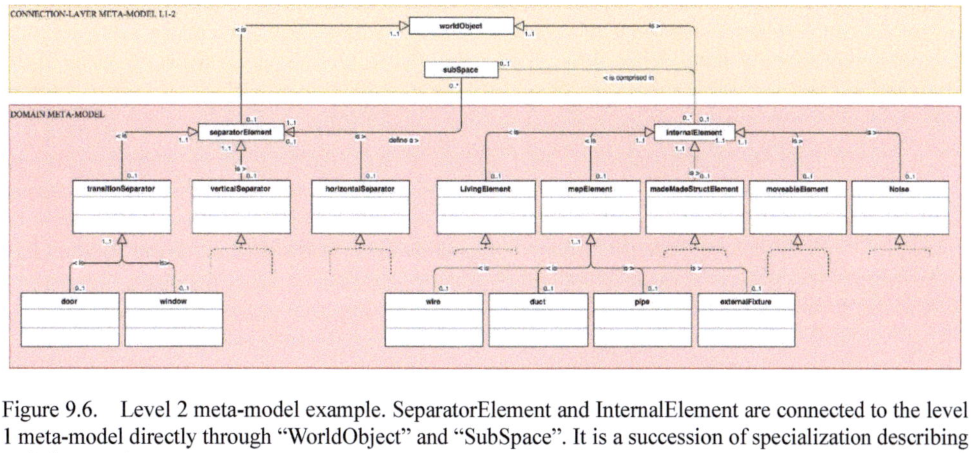

Figure 9.6. Level 2 meta-model example. SeparatorElement and InternalElement are connected to the level 1 meta-model directly through "WorldObject" and "SubSpace". It is a succession of specialization describing an indoor environment.

(SeparatorElement and InternalElement) specialized in eight classes (transitionSeparator, verticalSeparator, horizontalSeparator, livingElement, mepElement, madeMadeStructElement, moveableElement, noise) which can also be specialized in a refinement process to get as close possible from the abstract idea that the human mind has of a concrete or abstract object of thought. This crude example is inspired by already-established BIM standards and is used on a simple test case to enable rapid perception of natural language requests. As such, one selected "WorldObject" can be specialized and identified as an internalElement, a mepElement (mechanical, engineering, plumbing), specifically a duct of the "subspace" room 4 in the higher LOD level "subspace" building 7 and attached by an "externalFixture" next to the exit "door". The possibility to play on all possible scales is therefore an opening on a flexible system that can be adapted to many real-world applications within an infrastructure.

4 FRAMEWORK ARTICULATION AND AUTOMATION FOR INTELLIGENT ENVIRONMENTS

The SPC data model permits us to structure the information (3D geometry, semantics and topology) in order to leverage knowledge for accessing decision support tools and reasoning capabilities. Indeed, at the frontier between a point cloud GIS system and a spatial infrastructure for agent-based decision support systems, its flexibility allows its extension through new developments mainly in artificial intelligence and machine learning. As such, a modular SPC-based conception permits a high extensibility, with a few constraints described in section 4.1.

4.1 *Infrastructure modularity and extensibility*

The SPC data model integration within a computerized environment was thought to first allow an end-to-end usage even with limited automation. Thus, each module as seen in Figure 9.7 has been developed and implemented in a standard version that allows immediate usage but which can be upgraded when every defined constraint is met. The choice was oriented to provide an open, functional and evolutionary infrastructure that can easily be replicated/extended. It mainly addresses level 0 and level 1 SPC conceptual model, and for some datasets the level 2 is presented.

The Point Cloud is defined by its characteristics that must follow the prescriptions detailed in section 4.1.1. The parsing (4.1.2) and processing (4.1.3) modules are designed for KE and for organizing efficiently in SemanticPatches the data which is integrated with other spatio-semantic information in the point cloud database (4.1.4). This will constitute the main data repository for the knowledge processing engine (4.1.5) including classification, language and query processing, reasoning, which is directly linked to the knowledge management layer described in 4.1.7 (training data and ontologies). Finally, when through the GUI (4.1.9) an interaction necessitates an agent intervention, the query engine/reasoner and agent layer (4.1.8) permits AI-based decision-making.

4.1.1 Point cloud characteristics

To be processed and usable by the SPC infrastructure, a point cloud P must at least be constituted of n points with each three spatial attributes: X, Y, Z. This criterion is the minimal condition to be compatible with the parsing module. If the point cloud has one or multiple attributes a_j, these will be kept, and a specific schema will formally describe the format as an XML document stored in the pointcloud_formats table from the SPC data model level 0.

4.1.2 Point cloud parsing

In order to integrate both classified point clouds and unclassified point clouds, two methods are included in the SPC infrastructure. If knowledge information is available through classification metadata (e.g. .las attributes, ASCII with classification pointers) or file-on-disk organization (.txt

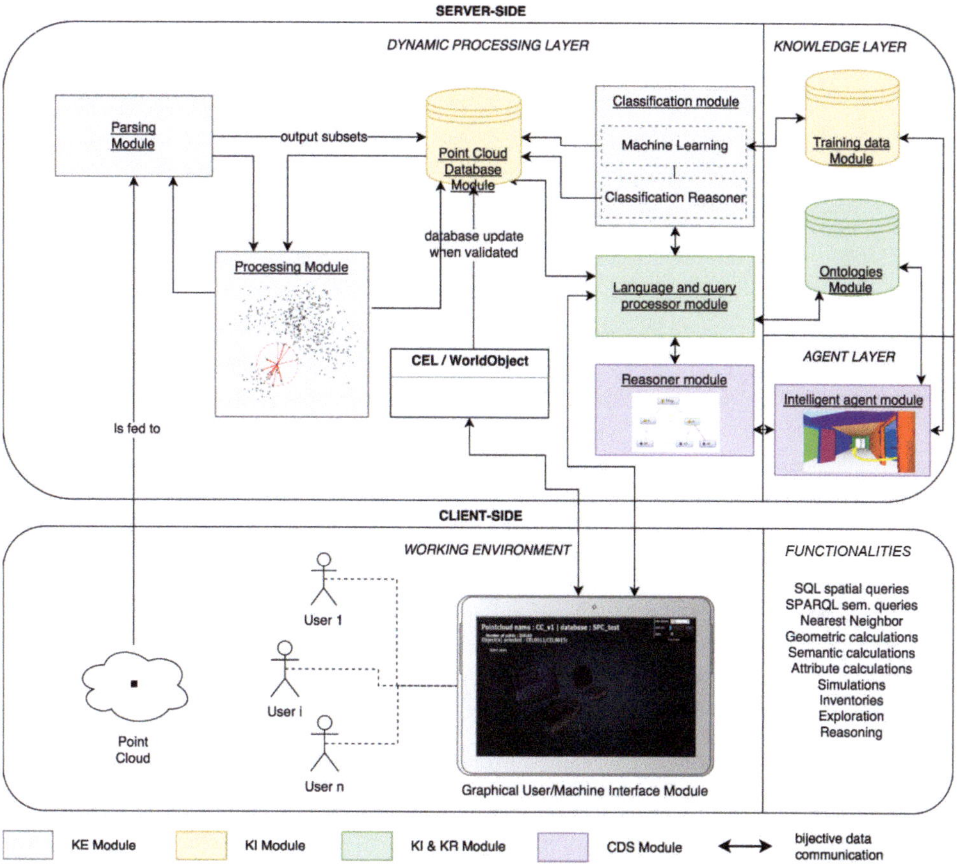

Figure 9.7. SPC modular framework. The point cloud is fed to the parsing module (KE) that is directly adjusted regarding the processing module (KE). The different point subsets are extracted and injected in the point cloud database module (KI). This module is central and influenced by the Classification module (KE) and the language and query processing module (KI & KR), which are themselves linked to the knowledge and agent layers.

file), each independent instance of each class is then divided into a point-based number of separate semantic patches pointing to each instance according to the hardware configuration of the server and the version of the used database. The point cloud parsing standard module for raw point cloud data is based on a semantic segmentation framework that groups points in a voxel-based space at a given octree level calculated automatically regarding device knowledge.[6] Each voxel is then studied by analytic featuring and similarity analysis to define a ConnectedElement. This process is conducted regarding an initial connected component from graph representation process after automatically detecting the main "ground" element and different perpendicular/parallel elements that are candidates to wall and ceiling. The voxels containing "edges" or multiple possible points that should belong to separate objects are further subdivided by studying the topology and features with their surrounding elements (Table 9.1). Thus, the indexation is defined both spatially and semantically to define SemanticPatches either "pure" voxel or leaf nodes from "hybrid" voxels (see Poux and Billen, 2019 for extended details). While this is very efficient, any decomposition can be though, for example regarding a 3D-OR Tree (Gong *et al.*, 2012), a Kd-tree (Kakde, 2005) or a Sparse Voxels Octree . . . On the implementation side, the parser module was developed in

Table 9.1 Point Cloud parsing methodology

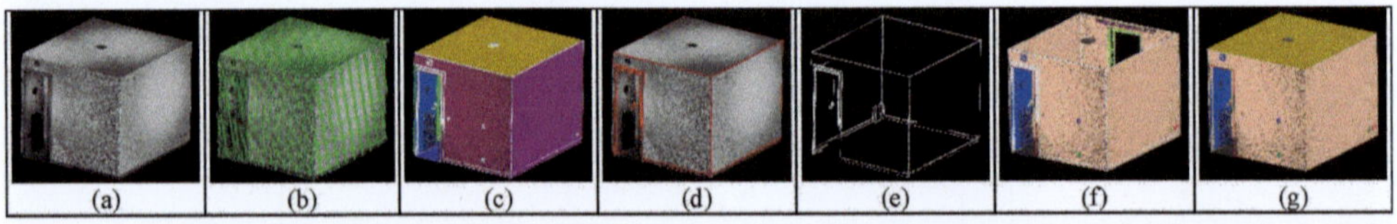

Source: (a) Raw point cloud from TLS; (b) voxelization at different Octree LOD levels; (c) segmentation: (d) voxel-based topology featuring; (e) extraction of highly representative points; (f) voxel classification; (g) connected element constitution and patch decomposition.

Python, including the following libraries: numpy, laspy, tensorflow, scikit-learn, math, networkx. The storage model is therefore of type Block.

4.1.3 Point cloud processing

The added functionalities play on multi-LOD (point-based or object-based) KE routines. By default, statistical generalizations are computed, as well as a topological skeleton based on voxel adjacency, object information (if available), space decomposition (if available) and in any pertinent space (XYZ, RGB . . .). This module then rearranges SemanticPatches to corresponding ConnectedElements and WorldObjects if applicable. Eigenvectors and eigenvalues are also extracted through principal component analysis, as well as parametric shapes, planarity estimators, shape regularity, concave/convex 3D estimator. This is especially important regarding the host/guest topological (8-relationships study) inference used for further reasoning and computations.

This module can directly be improved by joining classification procedures linked to the Knowledge processing engine for added feature computation or ConnectedElements specialization. The constraints are the update of WorldObjects table, the topology and internal relationships. The standard processing module was developed in Python and C++ and is directly connected to the Point Cloud Database through the psycopg/JSON Python library. Some calculations and extraction are directly in-Base (SQL).

4.1.4 Point cloud database

The decision support system is constructed over a point cloud DBMS which provides an interface for integrating, updating, and accessing 3D point clouds. In addition, a LOD data structure and indexing scheme are prepared for fast data access by hierarchically subdividing the spatial area or using existing indexes. It provides an access to every component as described in the conceptual model, permits topology featuring (Table 9.2) and allows various modules to be plugged, thus playing the role of a centralized KI module. In addition, point/objects attributes resulting from capture, analysis, simulation or processing stages can be stored. Efficient processing requires a certain data quality that can be ensured by applying knowledge-based filtering and registration methods to the input data. The module was implemented in the open-source PostgreSQL 9.6 DBMS, with the extensions PostGIS and Pgpointcloud activated.

4.1.5 The knowledge processing engine

For the training data and ontologies present in the knowledge management layer to be usable, a processing module makes use of this information to allow possible reasoning on point cloud data. As such, it is composed of a reasoner engine that can be used in the case of classification tasks for creating new knowledge (KE). It is also able to extract new information based on data stored in-base. First-order logic (FOL) is used for expressing logical conditions. The new knowledge is then stored as attributes and made available through a schema definition's modification. New KI is possible. A convolutional neural network based on PointNet Vanilla and trained using different datasets is developed in Python, and an indirect connection allows automatic classification based

Table 9.2 Example of point cloud in-base topology determinations for relationship extraction

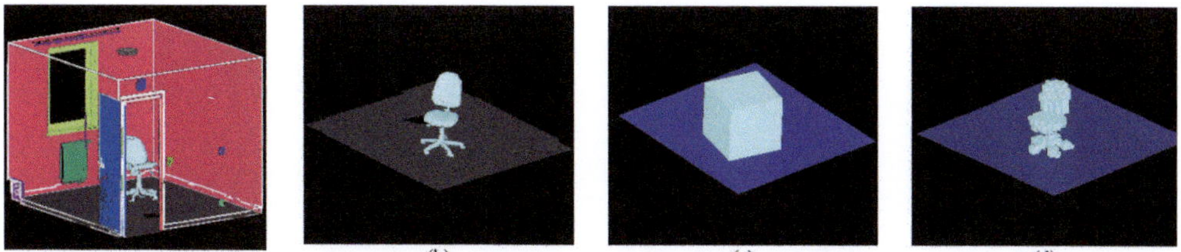

(a) (b) (c) (d)

Source: (a) Example of CEL block storage; (b) host/guest CEL/ground; (c) generalization and topology inference; (d) higher LOD topology inference.

on machine learning, using the trained model. Extending this provides a great opportunity for added representativity and automatic processing. The implementation was made using Protégé, the FOL reasoner Pellet and Python for the link between RDF/JSON and SQL statements.

4.1.6 The language and query processing engine

This module allows us to navigate between different languages, with the aim to provide a direct interface for natural language processing. The version in the SPC covers the ability to realize SQL and NoSQL queries. The language and query processors are developed in Python, and uses the following libraries: psycopg, json, SPARQLWrapper.

4.1.7 The knowledge management layer

The training data is structured to keep for each classified instance a link to every possible related domain formally expressed in an ontology of specialization such as provided in Poux *et al.* (2018) . The ontology also permits us to formalize rules and constraints for KR. The reasoner used for extracting new information is Pellet, and the developments are made in Protégé, linked to the database through SPARQL queries and connected through an EndPoint, or a server hosting. The main used language is OWL/RDF. KI and specifically knowledge structuration was constructed to allow efficient KR and reasoning while insuring interoperability with already established models.

4.1.8 The intelligent agent layer

The intelligent agent layer indirectly linked to a reasoner through a language processor permits one to leverage AI through expert systems (e.g. semantic modelling SPC extension in Poux *et al.*, 2018) such as Neural Networks, Genetics Algorithms or any agent-based technology for Decision Support Systems. Our tests were conducted with an AI pathfinding agent that uses the subspace graph connections to establish possible areas to visit. Based on an initial node and a goal node, the agent determines the nodes succession of the least costly path to the goal. We used a heuristic that works for A* returning the distance between the node and the goal. Then the 3D geometry representing each subspace is tested against "ST3DWithin" SQL statement to establish the occupancy grid of any CEL (Table 9.3). The implementation was made in Python, SQL and JSON.

4.1.9 The GUI

The GUI is conceived so as to answer the 10 usability heuristics for user interface design described in (Nielsen, 1995). Concurrency is also very important, and a platform should be able to scale up to multiple simultaneous connections. As such, a client-side application and RESTful development constitute a good solution for answering efficient interaction and high interoperability. The World Wide Web is a democratized way to share and exchange information. It constitutes a long-term mean to collaborate and is independent of the location, which is very important considering the need to be

Table 9.3 Example of voxel space generalization from 3D Point Cloud for A* pathfinding

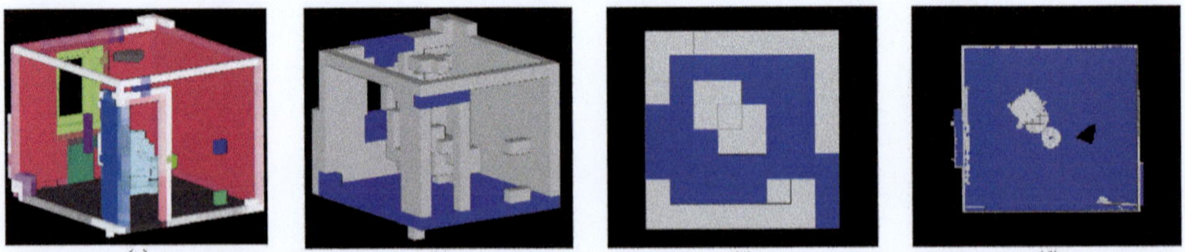

(a) (b) (c) (d)

Source: (a) Raw point cloud from TLS; (b) voxelization at different Octree LoD levels; (c) segmentation: (d) voxel-based topology featuring.

on site to work with digital copies. The web application is implemented in a WebGL framework and is accessible on any HTML5-compatible browser. The server has different roles and one interaction is to allow the user to view different point clouds. Each point cloud is linked to a corresponding database stored on the server. When the user selects a part of a point cloud, the coordinates are sent to the server, which will execute different queries to determine which object is selected by comparing the intersection of these coordinates with the related geometry (e.g. bounding box) of each object in the database. Once the corresponding object is determined, the extension "Pgpointcloud" allows us to retrieve in JSON format the data related to each point constituting the patches of the object. When viewing an object, the client makes AJAX calls to the server to retrieve all this data. Once retrieved, the client can then process them to display each point in the point cloud representing the object. The server is developed in Python using the Django library, psycopg2 and json libraries, which makes it possible to set up this kind of infrastructure quickly and easily.

4.2 *SPC requirements benchmarking*

In order to test the suitability of the SPC infrastructure to efficient decision support systems, the following requirements were evaluated:

- (R1) Allow point cloud data loading, insert and updates; key look-up search, nearest-neighbour search and cluster analysis; outlier identification, histogram and density estimation, random sampling; filtering capabilities based on spatial attribute and semantics (Dobos *et al*., 2014).
- (R2) Visualization and interaction with in-base data;
- (R3) Support KE processes for semantic segmentation/classification;
- (R4) Possibility to attach semantics to point clouds and point cloud subsets;
- (R5) Support for the interpretation and the aggregation of contextual information;
- (R6) Allow spatial, semantic and topology queries;
- (R7) Allow the user to control, manipulate, search, analyze, query and navigate within the data;
- (R8) Support agent-based inference and reasoning;
- (R9) High interoperability with established data models and extensibility through other modules;

The implementation's choice toward a client–server infrastructure (Figure 9.7) was thought to allow point clouds and databases storage on a server. In addition, the application can manage multiple point clouds and databases. All data is therefore centralized with the application, developed and tested on Linux (Ubuntu) and Windows 10, using the described frameworks, languages and libraries in the previous section. The application hierarchy is easy to set up, and administrator sessions for editing, adding or deleting application data are automatically implemented. For example, each object in a point cloud is automatically stored in the application database when a user selects it.

To test the suitability of the infrastructure to point clouds with varying characteristics, the following datasets (Table 9.4) from different sensors/methodologies were integrated:

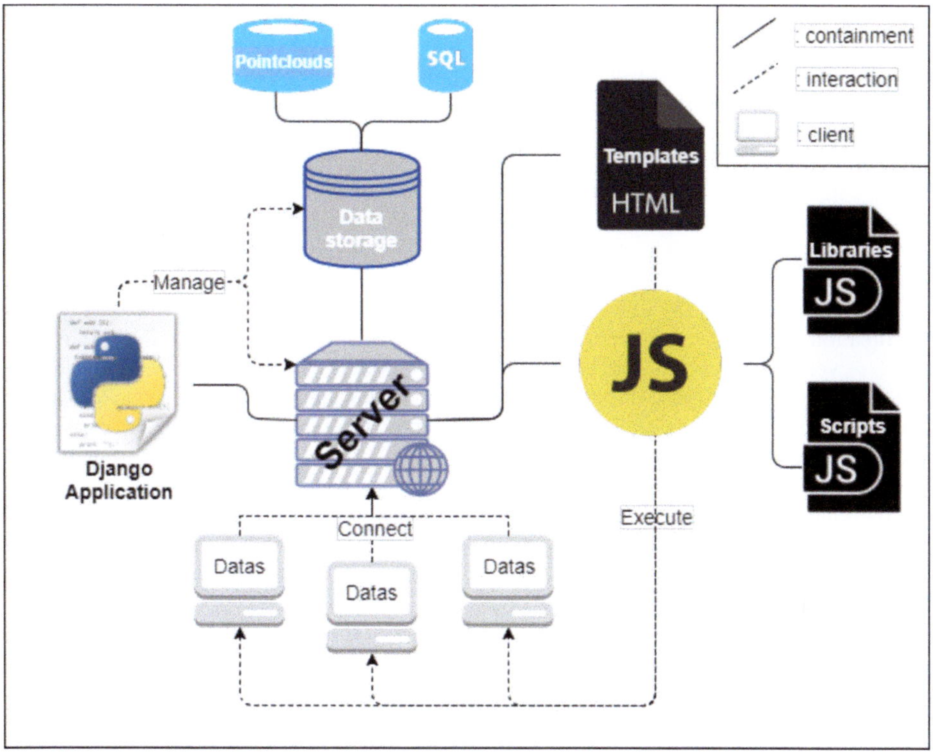

Figure 9.8. SPC client/server infrastructure implementation

Table 9.4 Datasets for benchmarking the SPC infrastructure

PCID	*PCType*	*Sensor/Instrument*	*Point Number (Million)*	*Attributes Number*	*Classification Status*
1	TLS	Leica P30	200	7	Unclassified
2	TLS	Trimble TX5 (eq. Faro Focus 3D)	150	8	Unclassified
3	PS CMOS	Canon 5D Mark III + 24-104 mm	200	6	Unclassified
4	PS CMOS	Canon 70D + Fisheye Sigma 15 mm	450	6	Unclassified
5	IS	Camera Matterport Pro 3D	50	3	Classified
6	2DS+S	Zeb Revo	25	3	Unclassified
7	Lidar	Riegl Litemapper 6800i	70	15	Classified
8	2DS	Velodyne	1	3	Classified
9	2DS+S	NavVis	8	7	Unclassified

Note: TLS = terrestrial laser scanner, PS CMOS = passive sensor CMOS, IS = infrared scanner; 2DS+S = 2D scanner + SLAM

First, we notice that the SPC can directly integrate all tested point cloud with different characteristics while addressing (R1). Mainly, the characteristics of the point cloud data influence the initial connected element detection for non-classified datasets. We also noticed that the quality/representativity of the point cloud data can impact the results. The most influential factors are the irregular point distribution, the point accuracy and the return signal dependence on the physical characteristics of the surface. We observed that the datasets with high noise or with complex structures (indoor and outdoor data) can become problematic for the parsing module. However, device knowledge-based filters and characterization correct this phenomenon, and when the sensor-related device knowledge is formalized in the knowledge layer, it has a limited impact.

Requirements (R2), (R3), (R4) are also validated, with a manual interaction step when the classification module doesn't permit automatic classification. Through the GUI, a Python script on the server is executed. It first determines the selected ConnectedElement object according to the x, y and z coordinates of the ray-casting. To do this, an SQL query is carried out on the database linked to the point cloud and allows us to return the name of the object containing the point (e.g. Table 9.5). (R7) is answered, but the implementation could be extended by providing higher interactivity.

Requirement (R5) for contextual aggregation is limited to established EndPoints such as DBPedia. The integration demands that the concept definitions are the same as the one in the different accessible ontologies. However, as the WWW standards evolve and new semantic web resources become available, it can be extended.

Using PostGIS, Python, SQL, SPARQL statements, information extraction is possible. This gives the SPC infrastructure the ability to fully answer requirement (R6), independently from the initial tested datasets. A problem arises when the generalization geometry is not representative of the described objects (e.g. when noise is too prominent with an incorrect 3D Concave Hull/BB extracted).

Requirement (R8) depends on the depth and completeness of the semantic definition of objects. Indeed, the classification granularity may limit the available operations to a fraction of what is

Table 9.5 Visualization of four point clouds from the benchmarking datasets and visual impact of the voxel-based featuring and grouping before topology

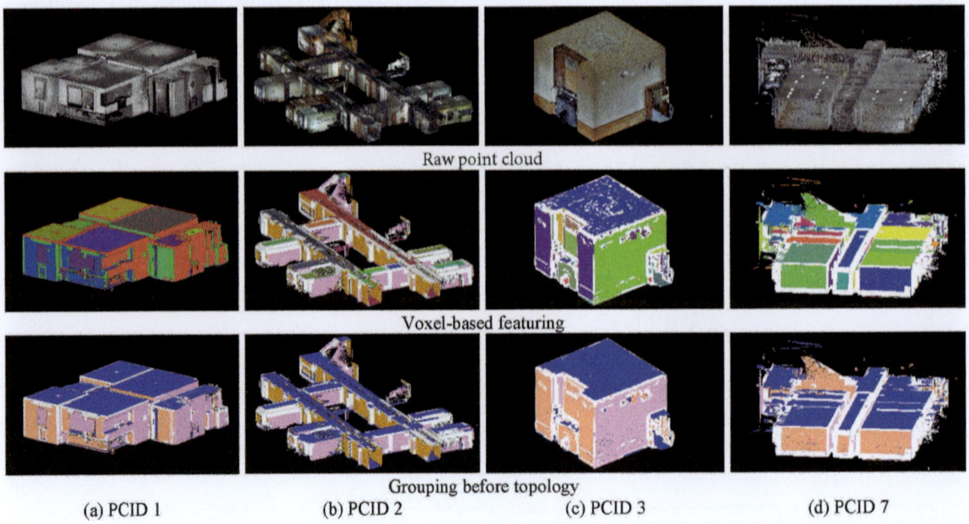

possible due to a lack of specialization. For example, A* pathfinding is possible in all datasets, but for the PCID5, the use of SubSpace and objects definition information gives an additional edge to the deepness of the AI pathfinding. The reasoning module was tested over unclassified datasets, as a classification reasoner, and allowed to automatically detect shapes regarding available knowledge over geometrical and radiometric properties of the point cloud (Poux *et al.*, 2017b). Extending the module by providing domain ontologies and inference rules would prove very interesting for adapted domain-versed classification.

Therefore the (R9) condition is important and was tested. The addition of knowledge "pointers" to the in-base data permits us to use this flexibility concerning every point cloud. The possibility to connect to EndPoints also permits a higher interoperability by gathering established knowledge from a recognized knowledge repository (e.g. DBPedia). As such, information regarding classified elements can directly be linked to extend instance information. This can also be used by the agent support system for inference and better guidance toward decision-making and simulation.

By combining the point cloud database module with the knowledge layer, the SPC infrastructure allows the use of operators either spatial, semantic, topology-based or any combination of them. This was tested over the different datasets, and the reasoner engine Pellet was used to infer new knowledge. Purely spatial operators possess spatial semantics that can be classified as 3D directional operators such as above, under, NorthOf, SouthOf, EastOf, WestOf; metric operators such as distance analysis; topological operators such as touch, contain, equal, inside; Boolean operators such as union, intersection. The ability to play with the generalization possibility provide a very high flexibility regarding possible data analysis and cognitive decision making.

The integration of semantics, e.g. *"The connected element CC0065 is a desk chair named 'comfychair' made in 2018-10-08 for working in front of a PC"* as SQL statements (*INSERT INTO moveableelement_connectedelement_id, type, title, date_prod, kind) VALUES ('65', 'chair', 'comfychair', '2018-10-08', 'working');*) from KE-routines and the knowledge layer permits one to achieve very interesting analysis. For example, the natural-language *"I want to locate the highest table within the room A of the Building B and calculate the free space between its surface and the ceiling for determining the possible extension"* leveraging the linked domain concepts from a level 2 meta-model mirror our real-world information gathering, thus heavily extend possibilities.

Table 9.6 Example of basic SQL statements over the SPC infrastructure

Goal	*SQL Statement*
I want to select the 'semanticpatches' which intersects a defined 2D polygon	SELECT pa FROM semanticpatch WHERE ST_INTERSECT(pa::geometry) = TRUE
I want to select all 'semanticpatches' that have been classified	SELECT pa FROM semanticpatch WHERE spclassifstatus = 1
I want to select the connected element CEL0065	SELECT pa FROM semanticpatch WHERE connectedelement_id = 0065
I want to extract the name of the connected element(s) that include the (x,y,z) position (used through ray-casting)	SELECT name FROM connected_elements WHERE((ST_Z(x, y, z) > ST_ZMin(geom::box3d)) AND (ST_Z(x, y, z) < ST_ZMax(geom::box3d)) AND (ST_Y(x, y, z) > ST_YMin(geom::box3d)) AND (ST_Y(x, y, z) < ST_YMax(geom::box3d))AND (ST_X(x, y, z) > ST_XMin(geom::box3d)) AND (ST_X(x, y, z) < ST_XMax(geom::box3d)));

5 LIMITATIONS, PERSPECTIVES AND POSSIBILITIES

While the SPC infrastructure can play the role of intermediary between real-time acquisition and inference for decision-making without the need to denature point cloud data, the infrastructure presents some limitations that suggest new research directions.

Any modelling choice is arbitrary and depends on the conscious or unconscious aspirations of the designer. Although our work responds to a concern for generalization at a spatio-semantic level, it nevertheless remains that it is not totally independent of a certain context. It is for this reason that we wanted to clearly illustrate a privileged domain of application: indoor environments (for BIM, emergency response, inventory management, UAV collision detection etc.). This choice permits us to explore different scales and configurations for deeply and entirely testing our developments. It is also ideal for the definition of new virtual spaces, and the GIS demand associated with such environment is ever increasing. Therefore, as the formalization of domain constantly evolve, modelling and direct integrations of level 2 domain meta-models remain important.

While three-dimensional spaces are strongly inferred in the SPC model, four-dimensional spaces integrating time or by extension *n*-dimensional spaces are possible characterizations for greater interoperability. Tests were conducted with static point data only, but varying positions in space and time present additional problematics that could be addressed through new modules or an extended SPC data model. While the developed framework constitutes the groundwork of a modular infrastructure that provides direct integration of hard-coded or inferred domain knowledge, future work including the extensibility of the proposed model to other data types, as well as a better integration of learning routines and ontologies as knowledge sources, is very interesting.

The SPC infrastructure permits us to link efficiently AI through agent decision support systems and reasoning. While developments were carried in a 3D voxel space (passable or non-passable at each point) with gravity constraints, using CEL bounding-boxes or an advanced Navmesh could enhance results. This illustrate that finding an efficient AI-based agent can become complicated for one application, so generalizing software agents can become complicated. To have an intelligent agent that performs reactively and/or proactively, interactive tasks need to be tailored to a user's needs without humans or other agents telling it what to do. To accomplish these tasks, it should possess the following general characteristics in regard to Bui and Lee (1999): (1) independence, (2) learning, (3) cooperation, (4) reasoning, (5) intelligence. This relates to the exploratory research field of artificial general intelligence[7] (Goertzel and Pennachin, 2007), which explicitly justifies a need of virtual environments that incentivize the emergence of a cognitive toolkit, such as the SPC infrastructure. As such, SPC-based multi-agent environments provide an opening thanks to their variety (the optimal strategy must be derived optimally) and natural curriculum (the difficulty of the environment is determined by the skill of other agents). This direction, while being extremely exciting to avoid purpose-specific algorithms, is still a research question that need to be further explored, in which deep learning may provide a suitable answer.

6 CONCLUSION

The smart point cloud data model permits us to structure the information (3D geometry and semantics) to leverage knowledge for accessing decision-making support tools and reasoning capabilities. At the frontier between a point cloud GIS system and a spatial infrastructure for agent-based decision support systems, its flexibility allows it to evolve with new developments mainly in artificial intelligence and machine learning. The proposed modular infrastructure includes knowledge discovery processes with knowledge integration and knowledge representation as ontologies, proving efficient context-specific adaptation. Nine point cloud datasets were used for testing the infrastructure, successfully answering identified needs and providing new research directions as modular extensions.

7 ACKNOWLEDGEMENTS

We would like to thank Stuart Cadge and GeoSlam, Philipp Först and NavVis, Christophe Schenke and Région Wallonne for providing point cloud samples respectively concerning the ZEB REVO (PCID6), NavVis (PCID9) and RIEGL (PCID7) datasets.

NOTES

1 The purpose of data mining is to extract knowledge from large amounts of data using automatic or semi-automatic methods.
2 supervised/semi-supervised/unsupervised
3 Attaching meaning to language "objects" is a conceptual phenomenon. As such, this object concept refers to a symbol and conceptualizes a "real-world" entity that shapes the symbol and is attached to a social agreement in an information community (Ogden et al., 1923).
4 The LAS file format is the most common file format for the interchange of three-dimensional point cloud data.
5 Reasoning engines are charged of determining what sorts of computational mechanisms might allow its accessible knowledge to be made available to an agent. What allows humans to behave intelligently is that they can apply their knowledge and adapt/transform it to a new environment to achieve their goals.
6 these were KB determined regarding available product-sheets specifications for each tested point cloud presented in Table 9.4.
7 Artificial general intelligence is the intelligence of a machine that could successfully perform any intellectual task that a human can do, including: reasoning; judgement calls under uncertainty; KR; planning; learning; natural-language communication. Other important capabilities include the ability to sense (e.g. see) and the ability to act (e.g. move and manipulate objects) in the world where intelligent behaviour is to be observed.

REFERENCES

Ben Hmida, H., Cruz, C., Boochs, F. & Nicolle, C. (2012) From unstructured 3D point clouds to structured knowledge – A semantics approach. In: Afzal, M.T. (ed) *Semantics – Advances in Theories and Mathematical Models*. InTech, Rijeka, Croatia, p.284. https://doi.org/10.5772/37633

Biljecki, F., Stoter, J., Ledoux, H., Zlatanova, S. & Çöltekin, A. (2015) Applications of 3D city models: State of the art review. *ISPRS International Journal of Geo-Information*, 4, 2842–2889. https://doi.org/10.3390/ijgi4042842.

Billen, R., Zaki, C., Servières, M., Moreau, G. & Hallot, P. (2012) Developing an ontology of space: Application to 3D city modeling. In: Leduc, T., Moreau, G. & Billen, R. (eds) *Usage, Usability, and Utility of 3D City Models – European COST Action TU0801*. EDP Sciences, Les Ulis, France. p.02007. https://doi.org/10.1051/3u3d/201202007

Boulch, A., Guerry, J., Le Saux, B. & Audebert, N. (2018) SnapNet: 3D point cloud semantic labeling with 2D deep segmentation networks. *Computers & Graphics*, 71, 189–198. https://doi.org/10.1016/J.CAG.2017.11.010.

Bui, T. & Lee, J. (1999) An agent-based framework for building decision support systems. *Decision Support Systems*, 25, 225–237. https://doi.org/10.1016/S0167-9236(99)00008-1.

Clementini, E. & Di Felice, P. (1997) Approximate topological relations. *International Journal of Approximate Reasoning*, 16, 173–204. https://doi.org/10.1016/S0888-613X(96)00127-2.

Cura, R., Perret, J. & Paparoditis, N. (2015) Point Cloud Server (PCS): Point clouds in-base management and processing. *The International Archives of the Photogrammetry, Remote Sensing and Spatial Information Sciences*, II-3/W5, 531–539. https://doi.org/10.5194/isprsannals-II-3-W5-531-2015.

Cura, R., Perret, J. & Paparoditis, N. (2016) *Implicit LOD for Processing, Visualisation and Classification in Point Cloud Servers (Computational Geometry; Computer Vision and Pattern Recognition; Software Engineering)*. Saint Mande. https://doi.org/10.13140/RG.2.1.1457.6400.

Cura, R., Perret, J. & Paparoditis, N. (2017) A scalable and multi-purpose Point Cloud Server (PCS) for easier and faster point cloud data management and processing. *ISPRS Journal of Photogrammetry and Remote Sensing*, 127, 39–56. https://doi.org/10.1016/j.isprsjprs.2016.06.012.

Dobos, L., Csabai, I., Szalai-Gindl, J.M., Budavári, T. & Szalay, A.S. (2014) Point cloud databases. *Proceedings of the 26th International Conference on Scientific and Statistical Database Management (SSDBM '14)*. ACM Press, New York, NY, USA. https://doi.org/10.1145/2618243.2618275

Fayyad, U., Wierse, A. & Grinstein, G. (2002) Information visualization in data mining and knowledge discovery. In: Illustrée (ed). Morgan Kaufmann Publishers Inc, San Francisco.

Goertzel, B. & Pennachin, C. (2007) *Artificial General Intelligence*. Springer, Berlin. https://doi.org/10.1007/978-3-540-68677-4

Gong, J. & Ke, S. (2011) 3D spatial query implementation method based on R-tree. *2011 International Conference on Remote Sensing, Environment and Transportation Engineering*. IEEE, Nanjing, China. pp.2828–2831. https://doi.org/10.1109/RSETE.2011.5964903.

Gong, J., Zhu, Q., Zhong, R., Zhang, Y. & Xie, X. (2012) An efficient point cloud management method based on a 3D R-Tree. *Photogrammetric Engineering and Remote Sensing*, 78, 373–381. https://doi.org/10.14358/PERS.78.4.373.

Gorte, B. & Pfeifer, N. (2004) Structuring laser-scanned trees using 3D mathematical morphology. *20th ISPRS Congress, Commission V. ISPRS*, Istanbul, Turkey.

Guerrero, P., Kleiman, Y., Ovsjanikov, M. & Mitra, N.J. (2018) PCPNet learning local shape properties from raw point clouds. *Computer Graphics Forum*, 37, 75–85. https://doi.org/10.1111/cgf.13343.

Harvey, F., Kuhn, W., Pundt, H., Bishr, Y. & Riedemann, C. (1999) Semantic interoperability: A central issue for sharing geographic information. *The Annals of Regional Science*, 33, 213–232. https://doi.org/10.1007/s001680050102.

Janowicz, K., Schade, S., Bröring, A., Keßler, C., Maué, P. & Stasch, C. (2010) Semantic enablement for spatial data infrastructures. *Trans. GIS*, 14, 111–129. https://doi.org/10.1111/j.1467-9671.2010.01186.x.

Kakde, H. (2005) *Range Searching using Kd Tree*. http://www.cs.utah.edu/~lifeifei/cis5930/kdtree.pdf

Karim, H., Abdul Rahman, A., Boguslawski, P., Meijers, M. & van Oosterom, P. (2017) The potential of the 3D Dual Half-Edge (DHE) Data structure for integrated 2D-Space and scale modelling: A review. In: Abdul-Rahman, A. (ed) *Advances in 3D Geoinformation*. Springer International Publishing, Basel, Switzerland, pp.477–493. https://doi.org/10.1007/978-3-319-25691-7_27

Kolbe, T.H., Gröger, G. & Plümer, L. (2005) CityGML: Interoperable access to 3D city models. In: Van Oosterom, P., Zlatanova, S. & Fendel, E.M. (eds) *Geo-Information for Disaster Management*. Springer International Publishing, Berlin, Germany and Heidelberg, Germany. pp.883–899. https://doi.org/10.1007/3-540-27468-5_63

Kuhn, W. (2005) Geospatial semantics: Why, of what, and how? In: Spaccapietra, S. & Zimányi, E. (eds) *Lecture Notes in Computer Science*. Springer International Publishing, Berlin, Germany and Heidelberg, Germany. pp.1–24. https://doi.org/10.1007/11496168_1

Kurgan, L.A. & Musilek, P. (2006) A survey of knowledge discovery and data mining process models. *The Knowledge Engineering Review*, 21, 1–24. https://doi.org/10.1017/S0269888906000737.

Liu, X., Wang, X., Wright, G., Cheng, J., Li, X. & Liu, R. (2017) A State-of-the-art review on the integration of Building Information Modeling (BIM) and Geographic Information System (GIS). *ISPRS International Journal of Geo-Information*, 6, 53. https://doi.org/10.3390/ijgi6020053.

Maturana, D. & Scherer, S. (2015) VoxNet: A 3D Convolutional Neural Network for real-time object recognition. *2015 IEEE/RSJ International Conference on Intelligent Robots and Systems (IROS)*. IEEE. pp.922–928. https://doi.org/10.1109/IROS.2015.7353481.

Nielsen, J. (1995) 10 Usability Heuristics for User Interface Design. *Conference Companion on Human Factors in Computing Systems CHI 94*, Boston, MA, USA, ACM New York. https://doi.org/10.1145/191666.191729.

Nonaka, L., Takeuchi, H. & Umemoto, K. (1996) A theory of organizational knowledge creation. *International Journal of Technology Management*, 11. https://doi.org/10.1504/IJTM.1996.025472.

Novak, M. (1997) *Intelligent Environments: Spatial Aspects of the Information Revolution* (P. Droege, Ed.). Elsevier, Amsterdam, Netherlands. https://doi.org/978-0-444-82332-8

Ogden, C., Richards, I., Ranulf, S. & Cassirer, E. (1923) *The Meaning of Meaning: A Study of the Influence of Language Upon Thought and of the Science of Symbolism* (A Harvest, Ed.). Harcourt, Brace & World Inc., New York, NY, USA.

Otepka, J., Ghuffar, S., Waldhauser, C., Hochreiter, R. & Pfeifer, N. (2013) Georeferenced point clouds: A survey of features and point cloud management. *ISPRS International Journal of Geo-Information*, 2, 1038–1065. https://doi.org/10.3390/ijgi2041038.

Poux, F. & Billen, R. (2019). Voxel-based 3d point cloud semantic segmentation: Unsupervised geometric and relationship featuring vs deep learning methods. *ISPRS International Journal of Geo-Information*, 8, 213. https://doi.org/10.3390/ijgi8050213

Poux, F., Hallot, P., Neuville, R. & Billen, R. (2016) Smart Point Cloud: Definition and remaining challenges. *The ISPRS Annals of the Photogrammetry, Remote Sensing and Spatial Information Sciences*, IV-2/W1, 119–127. https://doi.org/10.5194/isprs-annals-IV-2-W1-119-2016.

Poux, F., Neuville, R., Hallot, P. & Billen, R. (2017a) Model for reasoning from semantically rich point cloud data. *The ISPRS Annals of the Photogrammetry, Remote Sensing and Spatial Information Sciences*, IV-4/W5, 107–115. https://doi.org/10.5194/isprs-annals-IV-4-W5-107-2017.

Poux, F., Neuville, R., Nys, G.-A. & Billen, R. (2018) 3D Point Cloud semantic modelling: Integrated framework for indoor spaces and furniture. *Remote Sensing*, 10, 1412. https://doi.org/10.3390/rs10091412.

Poux, F., Neuville, R., Van Wersch, L., Nys, G.-A., Billen, R., Wersch, L. . . . Billen, R. (2017b) 3D point clouds in archaeology: Advances in acquisition, processing and knowledge integration applied to quasi-planar objects. *Geosciences*, 7, 96. https://doi.org/10.3390/GEOSCIENCES7040096.

Qi, C.R., Su, H., Mo, K. & Guibas, L.J. (2016) *PointNet: Deep Learning on Point Sets for 3D Classification and Segmentation*. CVPR 2017, in press. https://doi.org/10.1109/CVPR.2017.16.

Qi, C.R., Yi, L., Su, H. & Guibas, L.J. (2017) PointNet++: Deep Hierarchical Feature Learning on Point Sets in a Metric Space. *Conference on Neural Information Processing Systems (NIPS)*. Long Beach, USA.

Richter, R. & Döllner, J. (2013) Concepts and techniques for integration, analysis and visualization of massive 3D point clouds. *Computers, Environment and Urban Systems*, 45, 114–124. https://doi.org/10.1016/j.compenvurbsys.2013.07.004.

Riegler, G., Ulusoy, A.O. & Geiger, A. (2016) *OctNet: Learning Deep 3D Representations at High Resolutions*. https://arxiv.org/pdf/1611.05009.pdf, https://doi.org/10.1109/CVPR.2017.701.

Ross, L. (2010) *Virtual 3D City Models in Urban Land Management – Technologies and Applications*. Technischen Universität Berlin.

Smith, B. & Varzi, A. (2000) Fiat and bona fide boundaries. *Philosophy and Phenomenological Research*, 60, 401–420. https://doi.org/10.2307/2653492.

Tangelder, J.W.H. & Veltkamp, R.C. (2007) A survey of content based 3D shape retrieval methods. *Multimedia Tools and Applications*, 39, 441–471. https://doi.org/10.1007/s11042-007-0181-0

van Oosterom, P., Martinez-Rubi, O., Ivanova, M., Horhammer, M., Geringer, D., Ravada, S., . . . Gonçalves, R. (2015) Massive point cloud data management: Design, implementation and execution of a point cloud benchmark. *Computer Graphics*, 49, 92–125. https://doi.org/10.1016/j.cag.2015.01.007.

Wang, J., Lindenbergh, R. & Menenti, M. (2017) SigVox – A 3D feature matching algorithm for automatic street object recognition in mobile laser scanning point clouds. *ISPRS Journal of Photogrammetry and Remote Sensing*, 128, 111–129. https://doi.org/10.1016/j.isprsjprs.2017.03.012.

Weinmann, M., Hinz, S. & Weinmann, M. (2017) A hybrid semantic point cloud classification-segmentation framework based on geometric features and semantic rules. *PFG – Journal of Photogrammetry, Remote Sensing and Geoinformation Science*, 85, 183–194. https://doi.org/10.1007/s41064-017-0020-5.

Weinmann, M., Jutzi, B., Hinz, S. & Mallet, C. (2015) Semantic point cloud interpretation based on optimal neighborhoods, relevant features and efficient classifiers. *ISPRS Journal of Photogrammetry, Remote Sensing*, 105, 286–304. https://doi.org/10.1016/j.isprsjprs.2015.01.016.

Xiong, X., Adan, A., Akinci, B. & Huber, D. (2013) Automatic creation of semantically rich 3D building models from laser scanner data. *Automation in Construction*, 31, 325–337. https://doi.org/10.1016/j.autcon.2012.10.006.

Zhou, Y. & Tuzel, O. (2017) VoxelNet: End-to-End Learning for Point Cloud Based 3D Object Detection. *ArXiv*. https://doi.org/1711.06396v1.

Zlatanova, S. & Rahman, A. (2002) Trends in 3D GIS development. *Journal of Geospatial Engineering*, 4, 71–80.

Zlatanova, S., Rahman, A.A. & Shi, W. (2004) Topological models and frameworks for 3D spatial objects. *Computers & Geosciences*, 30, 419–428. https://doi.org/10.1016/j.cageo.2003.06.004.

Laser Scanning – Riveiro and Lindenbergh
© 2020 ISPRS, ISBN 978-1-138-49604-0

Chapter 10

Integration of TLS and sonar for the modelling of semi-immersed structures

Emmanuel Moisan, Philippe Foucher, Pierre Charbonnier, Samuel Guillemin, Mathieu Koehl and Pierre Grussenmeyer

ABSTRACT: In this chapter, we address the topic of 3D reconstruction of semi-immersed structures from laser and sonar data. We take advantage of recent sonar technologies whose static implementation is close to that of terrestrial laser scanners and which also provide 3D point clouds. After introducing the acquisition devices and their implementation on site, we focus on the characteristics of 3D sonar data and their preprocessing. Then we introduce an original method for combining the heterogeneous data (laser and sonar), captured in different environments. The resulting model was used as a reference to evaluate a mobile mapping system developed later on. Finally, we propose a summary of the contributions and some possible improvements.

1 INTRODUCTION

Recent years have seen the development of 3D reconstruction applications for semi-immersed structures such as ports, dams, or bridge piers. In this chapter, we focus on the example of canal tunnels. Although they are relatively few in number,[1] these infrastructures represent a security issue, an economic interest, and a heritage preservation matter. Keeping them in good condition requires regular inspection. Moreover, since most of them were built in the 19th century, they are often poorly documented. In this context, having 3D models as accurate as possible of the "whole tube" of these tunnels (i.e. their underwater parts as well as those out of water) represents a real interest for managers and experts.

Several technologies are available for surveying semi-immersed structures. Photogrammetry (Stylianidis *et al.*, 2016) may be used for both underwater and out-of-water parts of the structure as e.g. in Menna *et al.* (2015). The turbidity of water is however an obstacle to the use of image-based methods, and sonar technologies are therefore often preferred for underwater recording. Several recent systems combine dynamic laser and sonar acquisition methods from a ship whose trajectory is determined by a global navigation satellite system (GNSS)/inertial navigation system (INS) combination, see e.g. (Mitchell *et al.*, 2011; Rondeau *et al.*, 2012). When GNSS signals are not accessible, positioning data can be obtained from omnidirectional laser scans (Papadopoulos *et al.*, 2014) or, as we have recently proposed in Moisan (2017), thanks to camera positions provided by photogrammetry. The uncertainties on the positioning of these mobile mapping systems lead to a loss of precision in the 3D models obtained, and it may be interesting to compare them with reference models obtained by more precise techniques.

In this chapter, the focus is on building 3D models as accurate as possible. For the aerial part, out of water, it is classic to use laser scanning in a static acquisition framework, which makes it possible to reach accuracies better than the centimetre. Today, there are recent sonar technologies which, thanks to a mechanical scanning system, allow the use of multibeam echosounders with an implementation close to that of static laser scanners, even if the precision obtained in comparison with the laser is less (Lesnikowski and Rush, 2012; Moisan *et al.*, 2016). The method described in this chapter, whose scope extends beyond canal tunnels alone, mixes mechanical scanning sonar (MSS; Lesnikowski and Rush, 2012) and terrestrial laser scanner (TLS; Grussenmeyer *et al.*, 2016) technologies to construct a reference model of the infrastructure. This one, directly usable by the manager for the documentation or the inspection of the infrastructure, can also serve to evaluate the results obtained by other surveying methods, in particular mobile mapping systems.

The data obtained by the two technologies are of different natures. In particular, as we illustrate in this chapter, sonar is less directional than laser, which should be taken into account in the experimental setup. Moreover, sonar data is very noisy and prone to artifacts that require adequate pre-processing. Robust estimation techniques should systematically be used as soon as sonar data are concerned.

The sonar and laser data are acquired in different environments, so the question arises of their consolidation to form a complete 3D model of the tunnel tube. Indeed, except in the particular case in which one can take advantage of the variations in water height due to the tide, the point clouds obtained by the two techniques have no overlap zone. There is therefore a need to develop specialized registration procedures. The one we propose in this chapter makes use of the continuity of the infrastructure elements inspected, of target objects visible in the two environments, and of the shape of the interface between them (i.e. the water line).

The outline of the chapter is as follows. We begin (section 2) by describing the laser and sonar sensors used in our experiments and how they are implemented, with a focus on the specificities of the sonar data and their preprocessing. The proposed registration procedure of sonar and laser data is then described in section 3. We show in section 4 the resulting 3D model of a tunnel entrance and how it was used later on to evaluate the results of the mobile mapping system introduced in (Moisan, 2017). Finally, an analysis of the method and possible improvements are proposed in section 5.

Note that part of the material used in this chapter has already been published in Moisan *et al.* (2015a) under open access license. A preliminary version was also published by ISPRS in Moisan *et al.* (2015b).

2 LASER AND SONAR DATA RECORDING

In this section, we introduce the scanning devices we use, the constraints, and the experimental setup for the survey, as well as the necessary pre-processing of the data. We focus on sonar data, which have several specificities compared to laser data.[2]

2.1 *Overview of the scanning devices*

The above-water acquisitions were performed using one of the latest Faro® terrestrial laser scanners (TLS; Grussenmeyer *et al.*, 2016), namely the Focus 3D X330; see Figure 10.1. The technology used to measure the distance between the device and objects is phase difference (Vosselman and Maas, 2010). Sweeping the laser beam through 360° horizontally and almost 360° vertically scans its visible environment over a range from 0.6 m to 330 m.

The device chosen for the 3D digitization of the underwater parts is a mechanical scanning sonar (MSS; Lesnikowski and Rush, 2012), the BV5000 by Blueview® (Teledyne), whose main characteristics are provided in Figure 10.2. It is based on a multibeam echosounder (MBES), the MB 1350, whose very high frequency (1.35 MHz) allows a high resolution in distance estimation, since the associated wavelength is of the order of 1 mm, but at the price of a reduced range (30 m). The MB 1350 device is mounted with a vertical swath direction on a pan-tilt rotation system that enables scanning the environment 360° around; see Figure 10.3. Since its swath aperture is limited, tilting the sensing head is necessary to increase the vertical scan range.

Placed on a tripod at the bottom of the canal (provided the floor is strong and flat enough), the BV5000 operates in a rather similar way to a TLS. However, the sonar technology used makes a significant difference in terms of resolution. Indeed, the sonar beam divergence is important (1°), compared to that of a TLS (0.011°). As illustrated on Figure 10.4, the footprint of the sonar signal on the scanned surface may vary drastically with the distance and incidence angle.

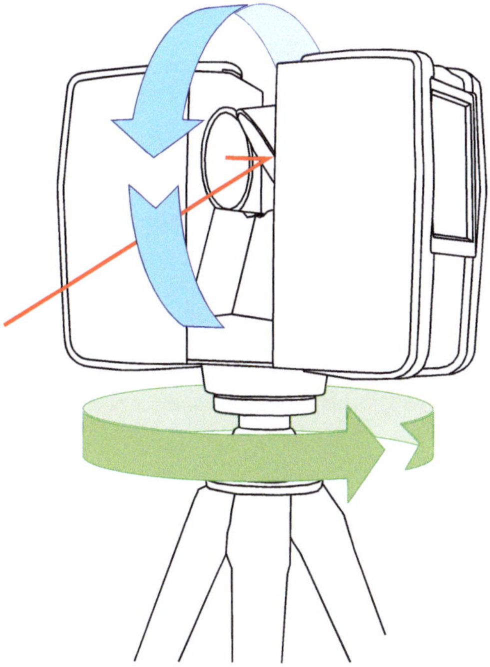

	Faro Focus 3D X330
beam width	2.25 mm + 2 × 0.011°
ranging error	±2 mm (10–25 m)
maximum range	330 m
field-of-view (vertic./horiz.)	300°/360°
horizontal resolution	0.035° (6 mm@10 m)
vertical resolution	0.035° (6 mm@10 m)

Figure 10.1. Schematic representation and characteristics of the Faro® Focus 3D X330 TLS.

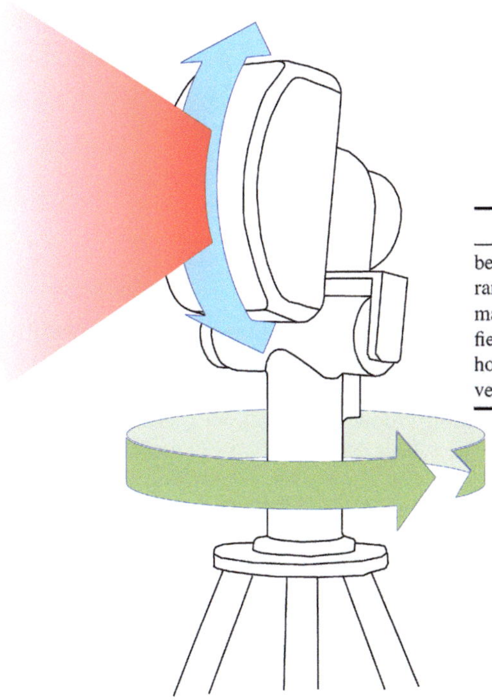

	Blueview BV5000
beam width	1° × 1°
ranging error	15 mm
maximum range	30 m
field-of-view (vertic./horiz.)	45°/360° (320°/360°)
horizontal resolution	~ 0.09° (16 mm@10 m)
vertical resolution	0.18° (30 mm@10 m)

Figure 10.2. Schematic representation and characteristics of the Blueview® BV5000 MSS.

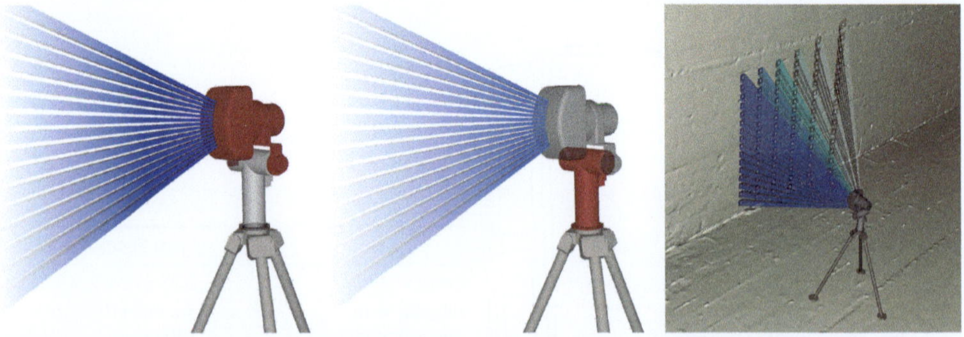

Figure 10.3. Illustration of the two components of the BV5000 MSS, i.e. the MBES MB 1350 (left) and its rotation mechanism (center); illustration of the operation of the MSS placed on a tripod at the bottom of the canal (right).

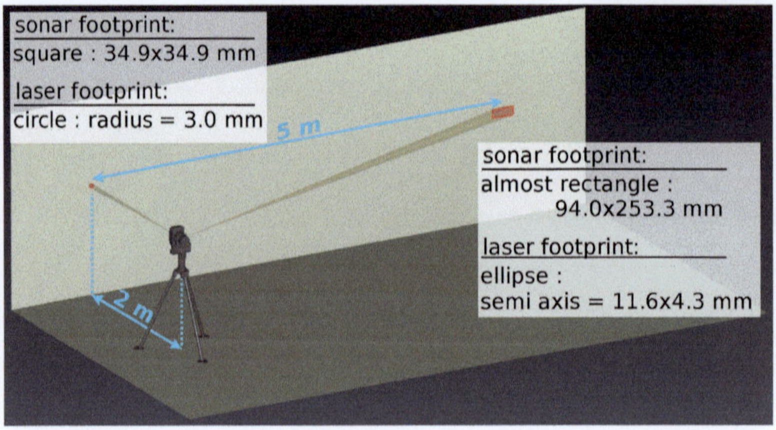

Figure 10.4. Illustration of the sonar footprint on the canal wall, at two different distances from the sensor, compared to what the equivalent laser footprint would be.

2.2 *Experimental setup*

Our test structure, the Niderviller tunnel, is located near Strasbourg (France), on the Marne-Rhine canal. Drilled between 1839 and 1845, it is straight, 475 m long, in masonry and bordered by a pedestrian path. The radius of the vault is 4 m, the width of the canal is 6 m, and the water depth is about 2.5 m. In order to avoid hindering fluvial traffic (about 7000 boats per year), the acquisitions were concentrated on half a day and limited to the tunnel entrances. The 3D model whose reconstruction is described below is that of the south entrance of the tunnel (see Figure 10.5).

At each entrance, two TLS scans were performed, from each bank of the canal. To register point clouds, spherical targets were placed in the scanning area. In order to geo-reference the model, the coordinates of sphere centers were established with traditional surveying methods based on a set of reference points implemented on site, in the French geodetic system, RGF 93, and the French vertical reference height system, NGF-IGN 69.

Two static sonar acquisitions, 10 m apart, were performed with the MSS, one inside the tunnel and the other outside. The MSS on its tripod was immersed in the canal from a boat, as the image in Figure 10.6 shows. We used three different tilt angles (0°, −9°, and −30°) to survey the canal floor and walls. Each static acquisition took about 30 min, which is longer than for TLS scans.

Figure 10.5. General view of the south entrance to the Niderviller tunnel, located in northeastern France.

Figure 10.6. Laser acquisition (top). Top view of the acquisition layout, showing the positions of the TLS, MSS, and ladders – LD – at the tunnel entrance (bottom left). Illustration of sonar and tripod immersion (bottom right), where the underwater elements (sonar, ladder, canal wall, and bottom) have been recreated.

The sonar scans can be consolidated, as in the case of TLS, using specially designed targets (Lesnikowski and Rush, 2012), but we did not use any for this experiment. In addition, the positions of the sonar acquisitions are not visible from the surface, and the registration of sonar data with laser data is therefore problematic. To facilitate this operation, it is advisable to use target objects visible simultaneously in both environments. This approach was applied in Menna *et al.* (2015), to consolidate under-water and above-water photogram-metric acquisitions, for example. In our experiment, two wooden ladders (3.6 m high and 0.32 m wide), duly weighted, were partly immersed on the side of the canal and scanned both by the TLS and the MSS. Wood was preferred to metal as the ladder material so as to obtain clearer sonar echoes. The precise dimensions and geometric characteristics of the ladders (parallelism, orthogonality, and inter-rung distances) had previously been measured using a TLS.

2.3 *Sonar data pre-processing*

As shown in Figure 10.7, the sonar signal can be perturbed by various effects that may be related to the technology used, the experimental configuration or the properties of the observed objects (Lesnikowski and Rush, 2012).

The most visible artifacts are due to signal reflections on the water surface. They can be easily removed, provided that the water level is known, which may require external measurements. In this experiment, the ladders were used for this purpose. The reflection phenomenon is also observed for other surfaces, such as walls or the floor. Other artifacts are attributed to acoustic phenomena occurring in water, especially in the presence of cables, fishes, or suspended particles. This results in "ghost" objects or systematic shapes occurring when the signal is backscattered by a surface. In our case, all these outliers were manually filtered out before the sonar point clouds were consolidated.

Some artifacts appear to be related to the scanning mechanism. For example, we noticed that some profiles, which correspond to successive acquisition angles, were very similar. One possible explanation is that the rotating mechanism temporarily jammed while the system continued to increment profiles. This error has the consequence of altering the geometry of the point cloud and is difficult to correct. Moreover, geometric distortions may occur if the tilt angle is incorrect (Lesnikowski and Rush, 2012). We have shown (Moisan, 2017) that it is possible to re-calibrate this angle provided the geometry of the structure is known (for example, a lock can be used for this purpose). Finally, as can be seen in Figure 10.7, acquisitions are not only absent in a circular region under the sensor (which is normal) but also in a particular angular sector. This effect, whose origin is yet unexplained, was systematic in our experiments and could also be observed in Thomas

Figure 10.7. Rendering of a raw sonar point cloud. In green, artifacts due to the reflection of the signal on the water surface. In yellow, acoustic anomalies in water. In red, acoustic phenomena due to signal backscattering. Acquisition anomalies (on the wall) probably caused by a temporary stop of the system rotation, as well as a lack of measurement (on the floor), are also visible.

(2011, p. 76). Note that it is possible that these defects have been corrected in the recently launched new version of the BV 5000 rotary system.

The noise inherent in sonar technology, as well as the unequal resolution associated with the variable signal footprint on surfaces, give the raw MSS data a granular appearance that can be seen on Figure 10.7. In our work, we chose to perform at the same time the sonar data denoising and their meshing in order to obtain a 3D model. This approach uses visual control of the result to reveal measurement errors and guide denoising. These errors are visible in Figure 10.8 (center), where a first mesh was made from the raw data: the surface obtained under these conditions is irregular and shows discontinuities.

To remove noise and reduce artifacts, two techniques are available. The first consists of resampling the point cloud, i.e. selecting points in the cloud so that they are uniformly spaced according to a minimum distance criterion. These points are then used as mesh vertices. The price to pay with this method is that some details may be lost. The second way is to estimate the surface closest to the points using robust estimators. For this purpose, new points can be interpolated. However, the risk is to obtain a too-smoothed model.

In practice, the point cloud meshing can be done using specific software, some of which are presented in Remondino (2003). In our case, we used 3DReshaper, which combines the two meshing methods mentioned above in order to successively refine the model by re-using the point cloud. In this way, the underwater model results from a *coarse-to-fine technique*. The process starts with a large-scale mesh made by selecting points according to a minimum distance criterion. Then points are picked again in the cloud or computed by interpolation to progressively increase the mesh resolution. Point selection involves either a distance-to-mesh or a maximum surface deviation criterion. The parameters are empirically tuned by an operator, and the process requires a trade-off between detail preservation and noise suppression.

The denoising step requires many manual operations, such as removing anomalies or correcting errors in meshes. Hence, it involves a great deal of interpretation. However, it provides visually correct results, as shown by the manual superposition of the resulting mesh on an archive image (that was not used in the process); see Figure 10.9.

Figure 10.8. Partial view of the canal bottom and side wall near a cofferdam groove: raw sonar point cloud (left), result of a direct meshing, with meshing errors in blue (center), final result (right).

Figure 10.9. Photograph on the northern entrance to the tunnel during its emptying for maintenance in 2009 (left). Visualization of the MSS denoised model (in grey) superimposed on the image after manual orientation based on the masonry block and the cofferdam grooves.

3 LASER AND SONAR DATA ALIGNMENT AND GEO-REFERENCING

In this section, we present our proposed method for registering sonar data on laser data to form the complete 3D reference model. We recall that this model will be used to evaluate the accuracy of models provided by a mobile mapping system. 3D model comparison can be done in any arbitrary coordinate system. However, our test site is equipped with accurate geodetic reference points, so the model can be georeferenced without additional complexity.

The usual approach consists in aligning the models by using elements common to the scans. For example, in this work the consolidation of laser point clouds is carried out using spheres measured by at least two acquisitions. Since the coordinates of the sphere centers are known, the geo-referencing of TLS point clouds can be deduced in a straightforward way. However, the MSS does not measure out of water and conversely, the TLS does not acquire points under water, so the scans of the two devices do not have any overlap area. Thus, the alignment of the laser and sonar models raises a problem.

The development of the solution is based on the fact that, despite the lack of overlap, the laser and sonar models share some information. First, elements of the environment can be modeled by geometric entities, and some of these are measured by both TLS and MSS. Second, the only contact element between the two models is the projection of the water line on the structure. In addition, this silhouette presents salient elements that can be used to determine horizontal translation. Finally, targets can be used to link the two environments. In our case, the targets are wooden ladders that are partly immersed during the acquisitions.

Based on these observations, a process was implemented. This is represented in Figure 10.10. It breaks down the consolidation into three steps that we will now describe.

3.1 *Orientation correction*

First, the orientation of the underwater 3D model is corrected. For this purpose, geometric entities common to both environments (water/air) are modeled manually. For example, the canal wall forms a plane that is acquired by both MSS and TLS devices. The elements estimated in the sonar and laser model are planes or straight lines (for example the edges of the cofferdam grooves). Thus, correcting the orientation of the sonar model is tantamount to aligning the normals of the planes and the guiding vectors of the homologous lines (see Figure 10.11). The rotation matrix is

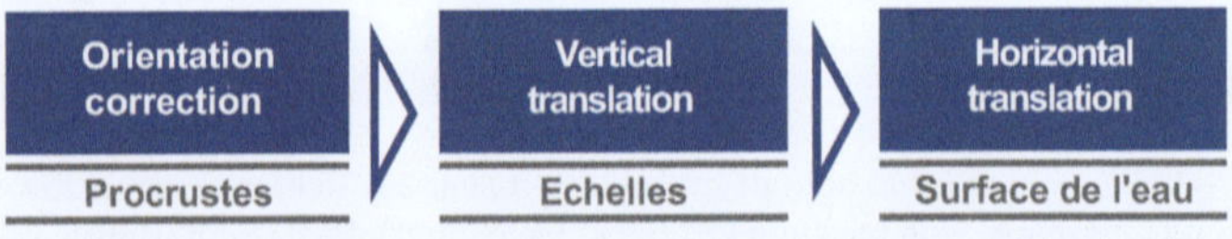

Figure 10.10. Flowchart of the alignment process of the 3D sonar model on the 3D laser model.

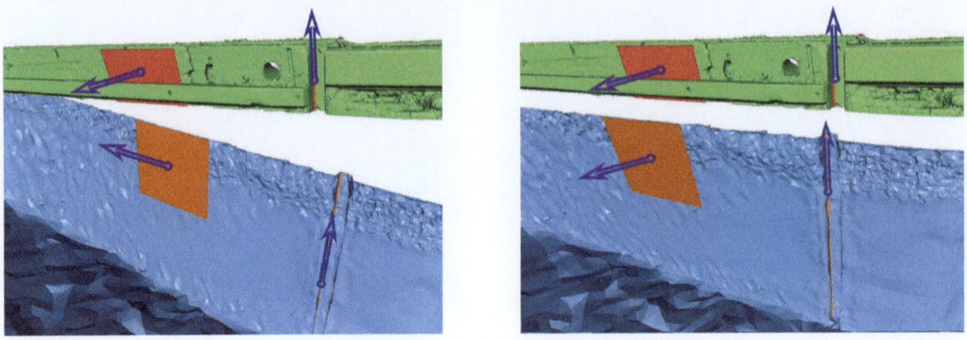

Figure 10.11. Illustration of the orientation of the 3D laser model (green) and the 3D sonar model (blue) before (left) and after (right) orientation correction by the Procrustes method.

estimated from the two sets of vectors by the Procrustes method. In our case, four geometric primitives are used (two lines and two planes) to orient the underwater model. More specifically, we use the solution described in Golub and Van Loan (2012) to perform the alignment.

3.2 *Vertical translation*

Once the orientation of the sonar model has been corrected, the second step consists in determining the vertical translation to be applied to the sonar model in order to place it at the correct altitude. This correction can be derived from the models of the ladders extracted from the sonar and laser data, provided the geometry of the ladders (especially their inter-rung distances) is known accurately. While ladders can easily be modelled in the laser point cloud, things are more difficult for sonar data, which are heavily perturbed and contain many *outliers*. For this purpose, a robust automatic algorithm has been developed. The starting data is the ladder point cloud, which is first segmented manually in the global cloud. Next, the ladder modeling process follows three steps described in the diagram in Figure 10.12. Two of these steps, which will be described later on, use M-estimation (Huber, 1981), whose basis ideas will be first recalled.

M-estimators. The adjustment by M-estimators replaces the minimization of the quadratic sum of the residues (of the least squares) by an energy of the form:

$$J(\theta) = \sum_i \rho(r_i) \tag{10.1}$$

where θ is the vector of the model parameters, ρ is a non-square potential function (or cost function), as shown in Figure 10.13, and r_i are the residuals, i.e. the differences between the observations and their predictions by the model. In the half-quadratic framework, see e.g. (Charbonnier *et al.*, 1997); it was shown that minimizing J is equivalent to minimizing:

$$J^*(\theta, b) = \sum_i b_i r_i^2 + \Psi(b_i) \tag{10.2}$$

where Ψ is a convex potential whose expression derives from that of ρ and b_i is an auxiliary variable. Its role is both to reduce the weight of outliers and to linearize the estimation problem. Indeed, J^* is quadratic with respect to r (therefore, w.r.t. θ in a linear regression) when b is fixed

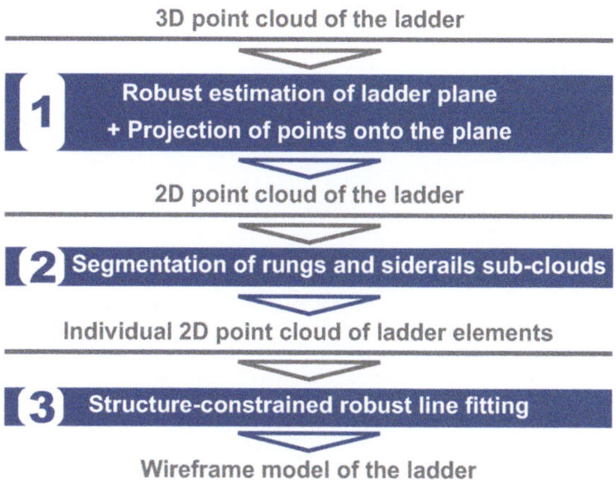

Figure 10.12. Diagram of the ladder modelling process.

and convex w.r.t. b when r is fixed, with a minimum obtained for $b = \rho'(r)/2r$. The exploitation of these mathematical properties leads to algorithms called iterative reweighted least squares (IRLS), which alternate until convergence the calculation of b and the minimization of the quadratic criterion J^*.

The b weights are adjusted for each iteration according to the value of the residues. In addition, $\rho'(r)/2r$ is a decreasing function, such that $b \simeq 1$ for small residues and $b \to 0$ for large ones. In this way, random errors (which correspond to small residuals) are treated in the ordinary least squares manner, while small weights are attributed to large errors, thereby reducing their influence on the estimate. Our experience is that the convergence properties of this algorithm, deterministic and fast, are even better when implemented using the Smooth Exponential Family potentials (Figure 10.13) in continuation, i.e. by decreasing α during iterations.

Step 1: robust estimation of the ladder plane. The first step in the process is estimating the ladder plane, defined by an origin and two orthonormal basis vectors. The least squares

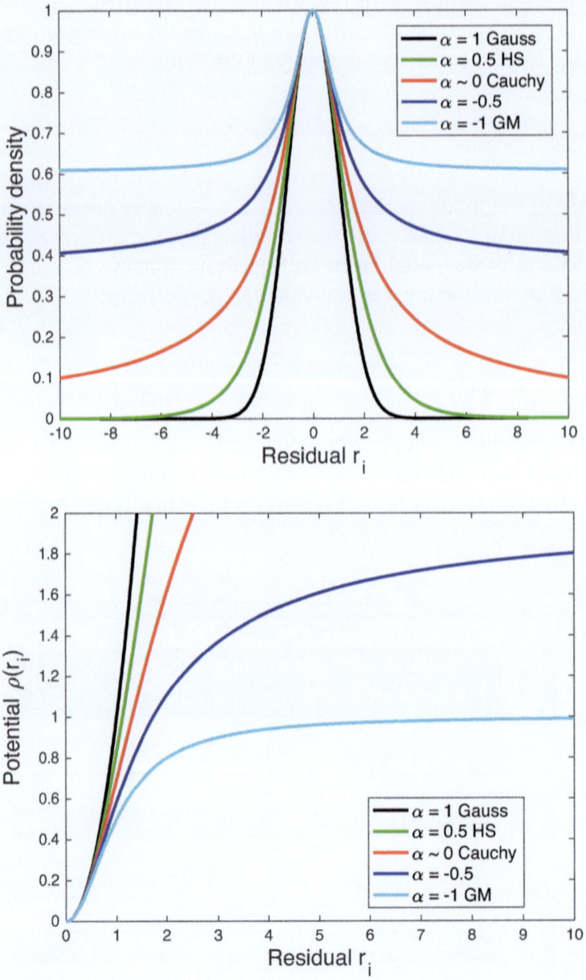

Figure 10.13. Representation of probability density functions (top), and corresponding potential functions (bottom), for the Smooth Exponential Family (Tarel *et al.*, 2002): $\rho(r_i) = \frac{1}{\alpha}\left[\left(1+r_i^2\right)^{\alpha} - 1\right]$

solution for this type of regression (akin to principal component analysis) is given by the mean of the 3D points for the origin and by the eigenvectors related to the two largest eigenvalues of their covariance matrix, for the basis vectors. The solution associated with minimizing J^* is similar, except that the mean and covariance matrix are weighted by the b weights. More details on the algorithm can be found in Moisan *et al.* (2015a). Once the origin and basis vectors are estimated, all points on the ladder are projected onto the plane, and the rest of the process is executed in two dimensions.

Step 2: segmentation of ladder elements. The robust estimation of the ladder plane provides axes that follow fairly well the directions of the uprights and bars. Hence, the distribution of 2D point coordinates along both axes contains peaks that correspond to the ladder elements. These distributions are approximated using Parzen-window density estimation with Gaussian kernels at two different resolutions. The rough locations of peaks are determined using a kernel resolution (bandwidth) of 1 cm. Then the intervals between peaks are analyzed using a bandwidth of 1 mm, using a method which is similar to the *fine-to-coarse* histogram analysis technique proposed by Delon *et al.* (2007). This method is successively applied to the horizontal and vertical coordinates, so the uprights are first separated from the ladders, then the rungs are extracted individually, see Figure 10.14.

Step 3: structure-constrained robust fitting. The last step consists in modelling the ladder by a set of lines, taking into account its geometric characteristics. The structure of the ladder is such that the rungs are parallel, as well as its uprights, while rungs and uprights are orthogonal. The inter-rung and inter-uprights distances are known with precision thanks to a preliminary TLS survey of the ladder. Orthogonality can be exploited by applying a 90° rotation to the bar clouds. In this way, only one direction must be estimated, simultaneously with one origin for the rungs and one for the uprights. Once again, an M-estimator is used for this purpose. We refer the reader to Moisan *et al.* (2015a) for a complete presentation of this algorithm.

Figure 10.14 illustrates the ladder modelling process and shows a result obtained using a commercial software (that does not take into account any *a priori* information on the ladder structure). Our method correctly models the ladder, despite the very high noise level in the data.

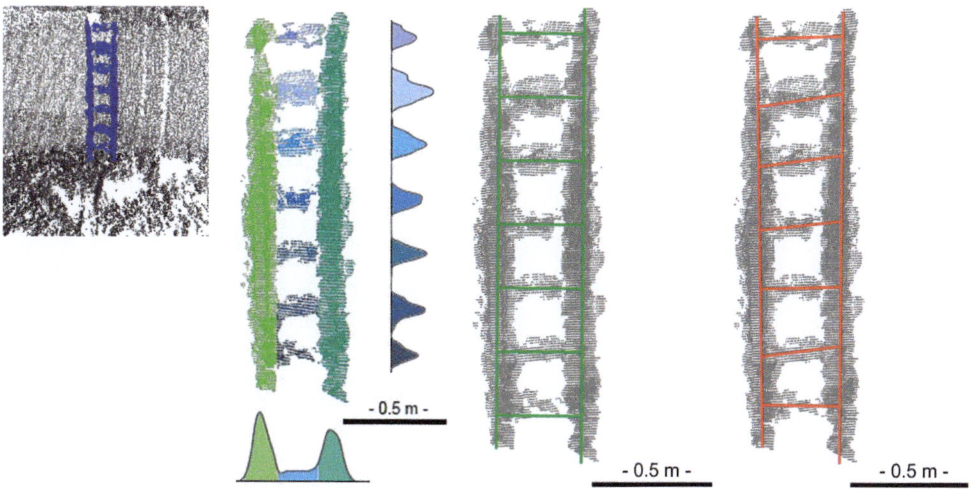

Figure 10.14. Illustration of the ladder estimation process. From left to right: extraction of the points corresponding to the ladder in the raw cloud; segmentation of the ladder elements using histogram analysis; structure-constrained robust estimation of the ladder; non-robust and unconstrained estimation obtained with commercial software.

3.3 *Horizontal translation*

Finally, the silhouettes of the water line on the structures determined in the sonar and laser models are used to estimate the horizontal translation. This operation cannot be performed earlier in the processing chain because the estimation of the position of the water line in the sonar digitization is too imprecise in general. At this stage, however, the sonar model is at the right height and the altitude of the water line is known, so its silhouette can be determined without difficulty. Then the horizontal translation vector is readily estimated using a 2D version of the ICP (Besl and McKay, 1992) algorithm.

In this section, we have presented the original method we developed to align the sonar data to the laser model. Figure 10.15 visually summarizes the three main steps of the method by focusing on one of the ladders used as a target object.

4 RESULT AND APPLICATION

In this section, we show the reference model obtained using the proposed method. Then we illustrate its use to evaluate an image-based model combining a photogrammetric reconstruction of the tunnel vault and sidewalls with dynamically acquired SONAR data.

4.1 *Obtained 3D reference model*

At the end of the three steps of our alignment chain, the sonar model is aligned with the laser model. Thus, one obtains a complete and geo-referenced model, which is presented in Figure 10.16. Two types of complementary rendering of the model are proposed: mesh and point cloud. This last form of representation is complementary to mesh visualization because shadowing effects highlight certain details. The result obtained is visually very satisfactory. However, we observe that the model is not of a very homogeneous aspect, because of the different nature of the acquisitions.

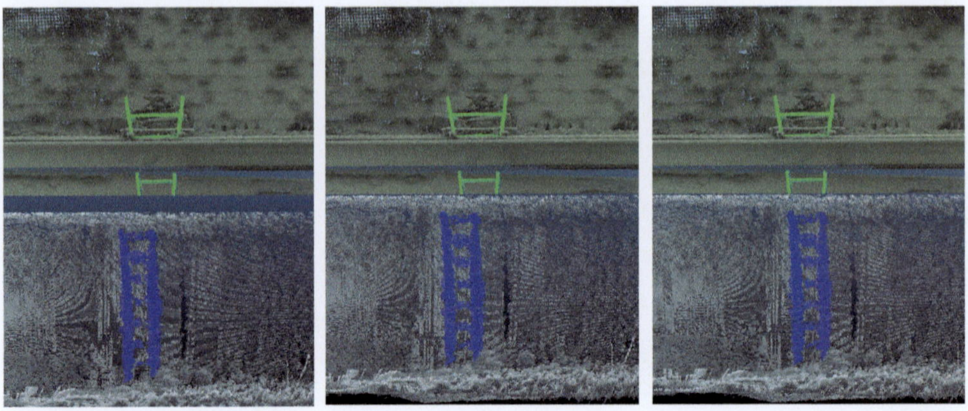

Figure 10.15. TLS and MSS point clouds after orientation correction (left), vertical (center) and horizontal translation correction (right). The upper part of the ladder is partly masked by a boom.

Figure 10.16. Visualization of the 3D reference model of the southern entrance to the Niderviller tunnel in the form of a cloud of points (top) and a mesh (bottom). The red disc in the middle of the canal indicates the position of the MSS acquisition.

4.2 *Evaluation of a dynamic image-based 3D recording method*

The reference model was used to evaluate the 3D reconstructions obtained by an experimental mobile mapping system later developed. This is based on a prototype for dynamic sonar data and image acquisition onboard a ship, shown in Figure 10.17. It consists of 6 industrial cameras, and a multibeam echo sounder (MB 1350) attached to the front of the boat. Due to the limited acquisition angle of the sensors, several round trips in the tunnel, with different sensor orientations, were

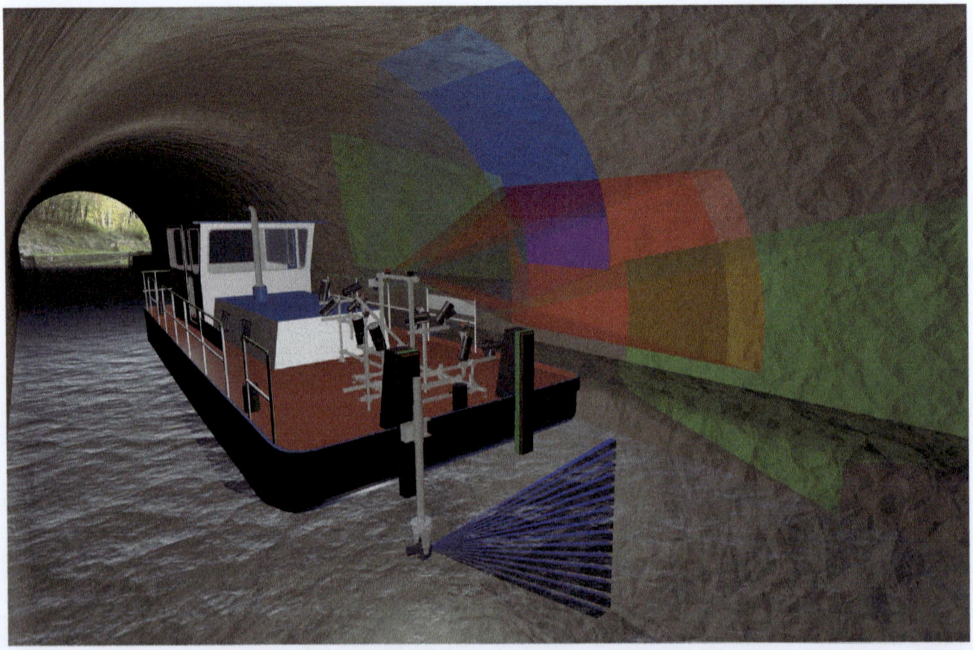

Figure 10.17. Schematic view of the prototype mobile mapping system in the tunnel, featuring the surfaces covered by the cameras and sonar at each acquisition.

necessary to survey the whole structure. The data acquired in the Niderviller tunnel represented 90,000 images and 120,000 sonar profiles.

The above-water part of the 3D model was reconstructed from photogrammetric techniques (Luhmann *et al.*, 2014). Because of the large mass of images to be processed, these were ordered in order to reconstruct the models by sections, which were later assembled. Targets referenced in the French geodetic system (RGF 93 and NGF-IGN 69) visible in the images were used to geo-reference the reconstructed model. The reconstruction of the underwater part consisted in aligning the sonar profiles acquired by the MS 1350, along the acquisition trajectory. Usually, in bathymetry, this trajectory is given by GNSS monitoring, possibly coupled to an inertial navigation system (INS). However, in tunnels, GNSS information is unavailable, and an INS, used alone, may drift quickly. To tackle this problem we took advantage of the properties of photogrammetry, which allows not only to reconstruct the 3D model of the vault, but also to estimate the positions and orientation of the cameras at each acquisition. The relative (fixed) position of the cameras being calibrated beforehand, we were able to obtain the sonar trajectory from the trajectory of the cameras. The underwater model could then be reconstructed in the same coordinate system as the vault model with which it was then directly combined to obtain the final 3D model of the tunnel tube, shown in Figure 10.18.

A visual comparison between the 3D model from the mobile mapping system and the reference model is proposed in Figure 10.19, showing a good qualitative agreement between the models. A quantitative assessment was obtained by calculating the distances between the models. The results showed that for the out-of-water part, the precision of the photogrammetric reconstruction is in the centimetre range. As might be expected, the sonar model is less accurate, since the accuracy is in the decimeter range.

Figure 10.18. 3D model of the southern entrance of the Niderviller tunnel obtained from dynamic acquisitions. The color strips in the underwater model correspond to the 6 passages of the sonar.

Figure 10.19. Visual comparison by superimposing the dynamic model (in colour) and the reference model (in grey).

5 SUMMARY

In this chapter, we considered the problem of the 3D reconstruction of a particular semi-immersed structure, namely a canal tunnel. The proposed method is based on the joint use, in a static acquisition framework, of two complementary technologies: terrestrial laser scanning and mechanical sonar scanning. Both TLS and MSS devices directly produce point clouds. However, the underlying technologies are different, and therefore the obtained data are heterogeneous.

As we have illustrated in this chapter, sonar is less directive than laser and, in a narrow environment like a canal, the size of the MBES footprint grows quickly with distance, which necessarily impacts the quality of the data. In this work, we used a distance of 10 m between acquisition locations, but our experience shows that closer MSS locations should be recommended.

Moreover, sonar data are very noisy, and it is necessary to pre-process the acquisitions in order to obtain an exploitable model. In addition, it is strongly recommended to use robust methods to tackle high noise levels and artifacts. The solution we propose, which combines denoising and meshing, involves a large part of human interpretation. We believe that acoustic and image processing techniques (such as moving least squares, bilateral filtering; Tomasi and Manduchi, 1998) or non-local means filtering (Buades *et al.*, 2005) should be explored to devise more automatic and data-driven denoising methods.

In order to obtain a complete model of the surveyed canal tunnel entrance, we had to align the 3D sonar and laser models. This operation is problematic since there is no overlap between the sonar and laser measurements. The proposed method takes advantage of the continuity of semi-submerged elements (belonging to the structure or attached to it, such as the ladders), which are digitized on each side of the water surface. In particular, a robust estimation technique was implemented to model the ladders in the sonar point cloud. More appropriate targets, easier to detect in the sonar cloud, could be devised. In the spirit of Menna *et al.* (2015), one could use a target made up of two spheres installed at each end of a pole, one solid for the laser and the other in wire for the sonar.

The laser and sonar data integration method described in this chapter is original and could be applied to many other semi-immersed structures. The resulting 3D reference model can be used by infrastructure managers and experts (for inspection or documentation of structures) and by engineers (for the development of high-performance mobile mapping systems). Our semi-immersed structure digitization technique involves up-to-date sonar and laser technologies. However, one may foresee that the progress of technology will further improve the performances of the acquisition devices so that more accurate and useful models can be expected using the proposed methodology. For example, features under development such as sonar depth and imaging data fusion will provide textured underwater models, making them more interpretable for visual inspection tasks.

NOTES

1 There are 33 canal tunnels in France, representing 42 km of the 8500 km of waterways managed by Voies Navigables de France.

2 The preprocessing and alignment of laser data is done by classical techniques, see e.g. Grussenmeyer *et al.* (2016).

REFERENCES

Besl, R.J. & McKay, N. (1992) A method for registration of 3-D shapes. *IEEE Transactions on Pattern Analysis and Machine Intelligence*, 14(2), 239–256.

Buades, A., Coll, B. & Morel, J.M. (2005) A non-local algorithm for image denoising. *Proceedings of Conference on Computer Vision and Pattern Recognition (CVPR)*. Vol. 2. IEEE, San Diego, USA. pp.60–65.

Charbonnier, R., Blanc-Féraud, L., Aubert, G. & Barlaud, M. (February 1997). Deterministic edge-preserving regularization in computed imaging. *IEEE Transactions on Image Processing*, 6(2), 298–311.

Delon, J., Desolneux, A., Lisani, J.-L. & Petro, A.B. (January 2007). A nonparametric approach for histogram segmentation. *Transactions on Image Processing*, 1, 253–261.

Golub, G. & Van Loan, C. (2012) *Matrix Computations, Volume 3 of Johns Hopkins Studies in the Mathematical Sciences*. 4th ed. Johns Hopkins University Press, Baltimore and London.

Grussenmeyer, P., Landes, T., Doneus, M. & Lerma, J.-L. (July 2016) *Basics of Range-Based Modelling Techniques in Cultural Heritage 3D Recording, Chapter 6*. Whittles Publishing, Dunbeath, Caithness, Scotland. pp.305–368.

Huber, P.J. (1981) *Robust Statistics*. John Wiley and Sons, New York, NY, USA.

Lesnikowski, N. & Rush, B. (January 2012) *Spool Piece Metrology Applications Utilizing BV5000 3D Scanning Sonar: High Resolution Acoustic Technology for Underwater Measurement*. Technical Report, Blue View Technologies, Seattle, USA.

Luhmann, T., Robson, S., Kyle, S. & Boehm, J. (2014) *Close-Range Photo Grammetry and 3D Imaging*. 2th ed. Walter de Gruyter, Berlin and Boston.

Menna, F., Nocerino, E., Troisi, S. & Remondino, F. (April 2015) Joint alignment of underwater and above-the-water photogrammetric 3D models by independent models adjustment. *ISPRS – International Archives of the Photo grammetry, Remote Sensing and Spatial Information Sciences*. Vol. XL-5/W5, Piano di Sorrento, Italy. pp.143–151. ISPRS/CIPA Workshop "Underwater 3D recording and Modeling".

Mitchell, T.J., Miller, C.A. & Lee, T.R. (June 2011) Multibeam surveys extended above the waterline. *Proceedings of WEDA Technical Conference and Texas A&M Dredging Seminar*, Nashville, USA.

Moisan, E. (2017) *Imagerie 3D du "tube entier" des tunnels navigables*. PhD Dissertation, University of Strasbourg, France. In French.

Moisan, E., Charbonnier, P., Foucher, P., Grussenmeyer, P., Guillemin, S. & Koehl, M. (December 2015a). Adjustment of sonar and laser acquisition data for building the 3D reference model of a canal tunnel. *Sensors*, 15(12), 31180–31204. Special Issue Sensors and Techniques for 3D Object Modeling in Underwater Environments.

Moisan, E., Charbonnier, P., Foucher, P., Grussenmeyer, P., Guillemin, S. & Koehl, M. (April 2015b) Building a 3D reference model for canal tunnel surveying using sonar and laser scanning. *ISPRS – International Archives of the Photogram-metry, Remote Sensing and Spatial Information Sciences*. Vol. XL-5/W5, Piano di Sorrento, Italy. pp.153–159. ISPRS/CIPA Workshop "Underwater 3D recording and Modeling".

Moisan, E., Charbonnier, P., Foucher, P., Grussenmeyer, P., Guillemin, S., Samat, O. & Pagès, C. (July 2016) Assessment of a static multibeam sonar scanner for 3d surveying in confined subaquatic environments. *ISPRS – International Archives of the Photogrammetry, Remote Sensing and Spatial Information Sciences*. Vol. XLI-B5. Praha, Czech Republic. pp.541–548.

Papadopoulos, G., Kurniawati, H., Shafeeq Bin Mohd Shariff, A., Wong, L.J. & Patrikalakis, N.M. (2014) Experiments on surface reconstruction for partially submerged marine structures. *Journal of Field Robotics*, 31(2), 225–244.

Remondino, F. (February 2003) From point cloud to surface: The modeling and visualization problem. *ISPRS – International Archives of the Photo gramme try, Remote Sensing and Spatial Information Sciences*. Vol. XXXIV-5/W10. Tarasp-Vulpera, Switzerland. International Workshop on Visualization and Animation of Reality-based 3D Models.

Rondeau, M., Leblanc, E. & Garant, L. (September 2012) Dam infrastructure first inspection supported by an integrated multibeam echosounder (MBES)/LiDAR system. *Proceedings of the Canadian Dam Association Annual Conference*. Saskatoon, Canada.

Stylianidis, E., Georgopoulos, A. & Remondino, F. (July 2016) *Basics of Image-Based Modelling Techniques in Cultural Heritage 3D Recording, Chapter* 5. Whittles Publishing, Dunbeath, Caithness, Scotland. pp.253–304.

Tarel, J.P., Ieng, S.S. & Charbonnier, P. (May 2002) Using robust estimation algorithms for tracking explicit curves. In: Heyden, A., Sparr, G., Nielsen, M. & Johansen, P. (eds) *Computer Vision — ECCV 2002*. Lecture Notes in Computer Science, Vol. 2350. Springer, Berlin, Heidelberg. pp.492–507.

Thomas, L.H. (2011) *The A.J. Goddard: Reconstruction and Material Culture of a Klondike Gold Rush Sternwheeler*. Master's Thesis, Texas A&M University, USA.

Tomasi, C. & Manduchi, R. (1998) Bilateral Filtering for Gray and Color Images. *International Conference on Computer Vision (ICCV)*. IEEE, Bombay, India. pp.839–846.

Vosselman, G. & Maas, H.G. (2010) *Airborne and Terrestrial Laser Scanning*. Whittles, Dunbeath, Caithness, Scotland.

Laser Scanning – Riveiro and Lindenbergh
© 2020 ISPRS, ISBN 978-1-138-49604-0

Chapter 11

Integral diagnosis and structural analysis of historical constructions by terrestrial laser scanning

Luis J. Sánchez-Aparicio, Susana Del Pozo, Pablo Rodríguez-Gonzálvez, Ángel L. Muñoz-Nieto, Diego González-Aguilera and Luís F. Ramos

ABSTRACT: In addition to drawing and mapping plants, cross sections and ortophotos, 3D models from terrestrial laser scanning technology can offer relevant information for the integral diagnosis and structural analysis of historical constructions. Both the metrics offered by the terrestrial laser scanner point cloud and the spectral backscattered reflectance data acquired by this technology, can be used to map and evaluate structural and chemical damages of historical constructions in a semiautomatic and robust way. Under this basis, the present chapter attempts to offer an overview of most suitable methodologies in this regard. Specifically, a multisensory approach is proposed and validated through which the diagnosis of the Master Gate of Saint Francisco, inside the fortress of Almeida (Portugal), has been analyzed in terms of loss of material suffered over time. The terrestrial laser scanner point cloud was also used to extract the current resistance sections of the most damaged parts of the barrel vault, allowing the evaluation of its structural stability under different postulated casuistic: (i) in case of presenting greater material losses and (ii) in case of a settlement of the construction.

1 INTRODUCTION

Since the development of the first terrestrial laser scanner in 1999 (MacDonald, 2006), the use of terrestrial laser scanning (TLS) has largely spread as a suitable measurement technique in the field of geomatics. TLS has become more popular due to its cheapening and its evolution in terms of its adaptation to the demand, equipment miniaturization and greater performance of the current TLS models. Two fields where this popularization was noteworthy have been architecture (Sánchez-Aparicio *et al.*, 2014; Korumaz *et al.*, 2017) and civil engineering (Yang *et al.*, 2018; Bautista-De Castro *et al.*, 2018). Reverse engineering is commonly used to document historical constructions for rehabilitation and conservation works or simply to put them in value or provide support to disseminate them. The generation of orthophotos or images from uncommon points of view (Hori and Ogawa, 2017) as well as their integration to virtual-reality environments (Fernandez-Palacios *et al.*, 2017) are some of the multiple applications offered by 3D point cloud models.

Thereby point clouds have become a common input in graphic engineering processes. Floor plans, cross-sections, 3D models, orthophotos, etc. derived from them bring new opportunities to civil engineering and architecture, which are adapting its work procedures to improve the management of this type of spatial information.

From the combination of radiometric and geometric information coming from point clouds and the possibility to enrich point clouds with thematic information coming from external sensors (e.g. digital or thermographic cameras; Costanzo *et al.*, 2014), TLS surveys have overcome the first approach of reverse engineering, aiming its use in integral diagnostics of historical constructions.

On the one hand, geometrical strategies such as principal component analysis (PCA), random sample consensus (RANSAC; Schnabel *et al.*, 2007) or the characterization of point clouds

(CANUPO; Brodu and Lague, 2012) allow taking advantage of the geometrical features presented in point clouds, allowing the automatic segmentation of point clouds, offering new possibilities for the diagnosis of conservation state of historical constructions.

On the other hand, different approaches for the spectral analysis of laser intensity data, based on the TLS radiometric calibration (Del Pozo *et al.*, 2016a) and classification (Sánchez-Aparicio *et al.*, 2016), have proved to be suitable for the detection of certain types of chemical and biological damages in historical constructions.

In order to perform an exhaustive analysis and diagnosis of the structural, chemical and biological damages of constructions, the use of additional measurement techniques is recommended in order to enrich the result obtained with the TLS approach. As examples, mechanical tests to evaluate the properties of materials, Raman spectroscopy to evaluate the chemical composition of the indicators of damage or boroscopic tests can be used among others.

This chapter deals with the full exploitation (i.e. geometric and radiometric) of terrestrial laser scanning together with other complementary techniques: (i) sonic tests to evaluate the physical properties of the stone; (ii) impact-eco methods to evaluate the thickness of the barrel vault supports; (iii) double punch tests to analyze the mechanical properties of the mortar; and (iv) boroscopic inspections to evaluate the inner layers of the construction. The main goal is to accurately and semi-automatically map and quantify damages of constructions by using this approach. A wide diversity of damage indicators such as structural deformations, biological colonization, salt crusts, moisture and black crusts, among others, are the main goal of the proposed integral diagnosis.

Under this basis, the present chapter attempts to demonstrate the applicability of the multisensory strategy based on TLS technology in the Master Gate of San Francisco (Figure 11.1). This historical construction is one of the most representative elements of the Restoration fortress of Almeida in Portugal. Having an Italian trace, it was conceived to defend the Portuguese border against the Spanish and French invasions during the 17th century (Figure 11.1a). For more information about its history and architecture see Arce *et al.* (2018) and Sánchez-Aparicio *et al.* (2018b).

This chapter is organized as follows: section 2 presents TLS technology as a suitable support tool to analyze the state of conservation of constructions describing the data acquisition and

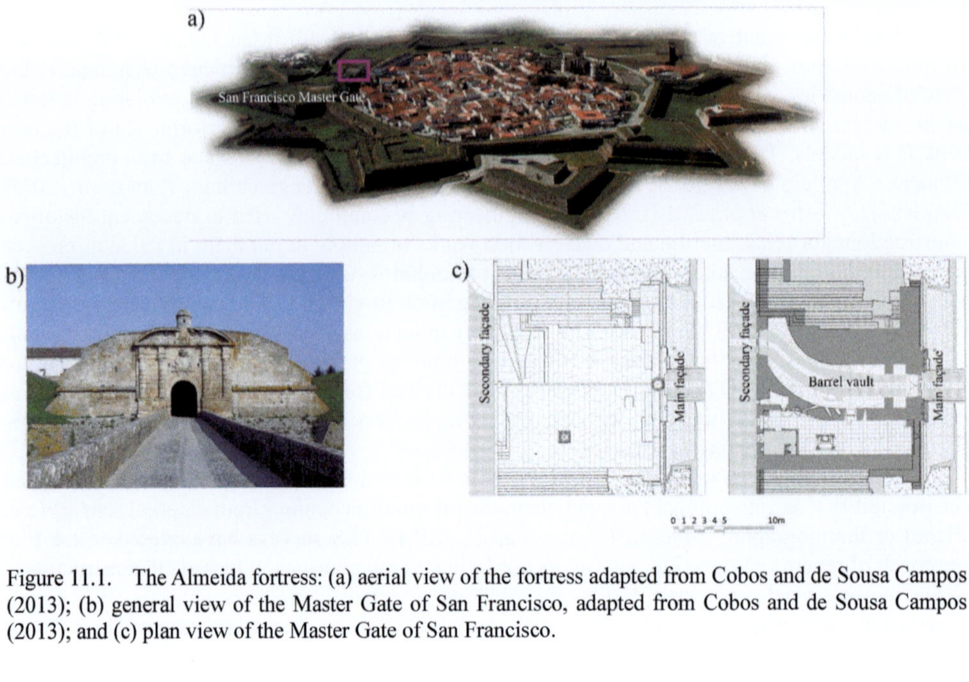

Figure 11.1. The Almeida fortress: (a) aerial view of the fortress adapted from Cobos and de Sousa Campos (2013); (b) general view of the Master Gate of San Francisco, adapted from Cobos and de Sousa Campos (2013); and (c) plan view of the Master Gate of San Francisco.

preprocessing steps required to take full advantage of this technology; section 3 exposes the results of the construction diagnosis as well as the structural analysis of its current state of conservation complemented by predictive evaluations; and finally, section 4 includes the conclusions arising from the use of this technology to succeed documentation, reconstruction and conservation state analysis of most common damages of historical constructions.

2 TLS AS A SUPPORT TOOL TO DIAGNOSE THE CONSERVATION STATE OF HISTORICAL CONSTRUCTIONS

The current section will show the methodology proposed and applied to the study case. As Figure 11.2 illustrates, a first visual inspection is necessary and key to determine which methods and instrumentation will be used. Assuming TLS the best technology to acquire in a non-invasive way large amounts of geometric and radiometric information about constructions, we explain this approach first. After TLS data acquisition, the processing of geometric and radiometric data as well as its integration is explained. In this process, different processing algorithms are suggested in order to obtain suitable results. Finally other complementary approaches (sonic, impact-eco, double punch and boroscopic tests) are described and analyzed with the aim of improving the final structural analysis and diagnosis of constructions.

2.1 *Visual inspection*

Previous to the multisensor campaign, a visual inspection of the construction was required to select the most appropriated methodology to study the current state of conservation of the construction as well as to plan the data acquisition. Since some indicators of damage were observed (white and black crusts, lichens as well as material losses; Figure 11.3), it was decided to use a combination of the following sensors: (i) a lightweight terrestrial laser scanner to digitalize the construction as well as to map the different damage presented on it; (ii) some sonic and double punch tests to characterize the mechanical properties of the masonry; (iii) some impact-eco tests to evaluate the thickness of the supports; and (iv) some boroscopic tests to evaluate the inner composition of the historical construction.

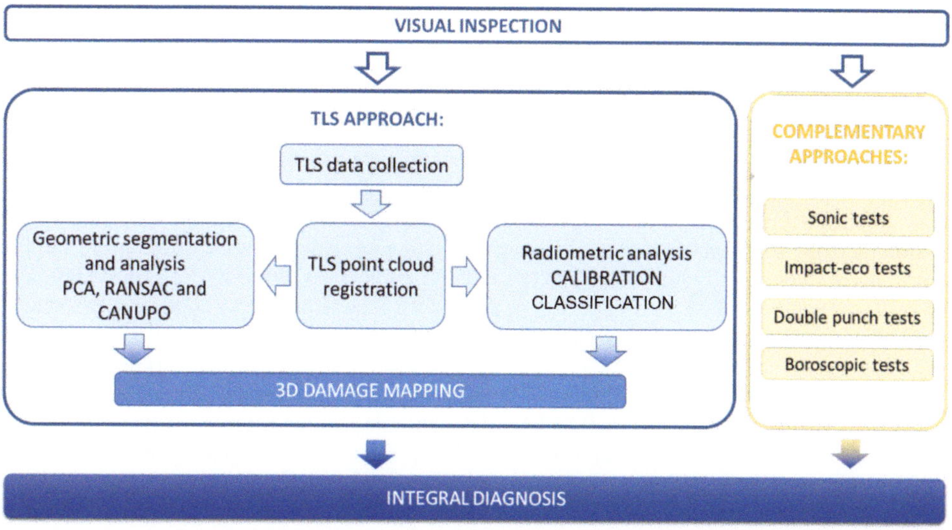

Figure 11.2. Proposed workflow to perform integral diagnosis of constructions on the basis of TLS technology.

a)

b)

c)

Figure 11.3. Current state of conservation of the Master Gate of San Francisco (Almeida, Portugal): (a) main façade on which is possible to observe the strong biological colonization (orange lichens); (b) detail of the current state of the barrel vault with presence of black and white crusts; and (c) detail of the material losses in the lower part of the supports.

It should be mentioned that during the visual inspection a close relationship between the material losses of the barrel vault and the presence of some biological and chemical pathologies was observed. Black crusts indicate a polluted atmosphere that promotes the disaggregation of the masonry blocks. This damage process is also linked with the emergence of white crusts due to the hidratation and deshidratation cycles of the melting salts.

2.2 *TLS point cloud: data collection and processing*

Given the great potential of TLS technology to carry out 3D surveys of any object in a non-invasive way, this technology is rated as one of the most suitable to reconstruct and document historical constructions as well as to study possible deteriorations suffered through time (Sánchez-Aparicio *et al*., 2016). Besides offering a great flexibility in data capture and a good level of precision in 3D measurements, this active-contact-free technology can also record a set of complementary spectral information, the so-called intensity values usually belonging to the infrared spectrum. In addition, most TLS devices are equipped with an external digital camera, so additional RGB information can be used to compose a photorealistic model of the historical construction. A complete review of the most common TLS systems applied to historical constructions is provided in Riveiro *et al*. (2016).

To analyze the state of deterioration of the Master Gate of San Francisco, it was decided to use a phase-shift TLS, specifically the FaroFocus3D 120 (Figure 11.4). This TLS device operates at the wavelength of 905 nm (near infrared) and measures distances in the range of 0.60 to 120 m with a point measurement rate of 976,000 p/s and a nominal precision of ±2 mm at 25 m.

Given the complexity of the case study chosen, especially because of its magnitude (around 689 m^2 in plant) and its complex geometry, a large number of scan stations were required to properly document the entire structure. In this regard, the lightness of the FaroFocus3D sensor allowed us to easily perform the data acquisition since 26 scan stations were required to document the entire

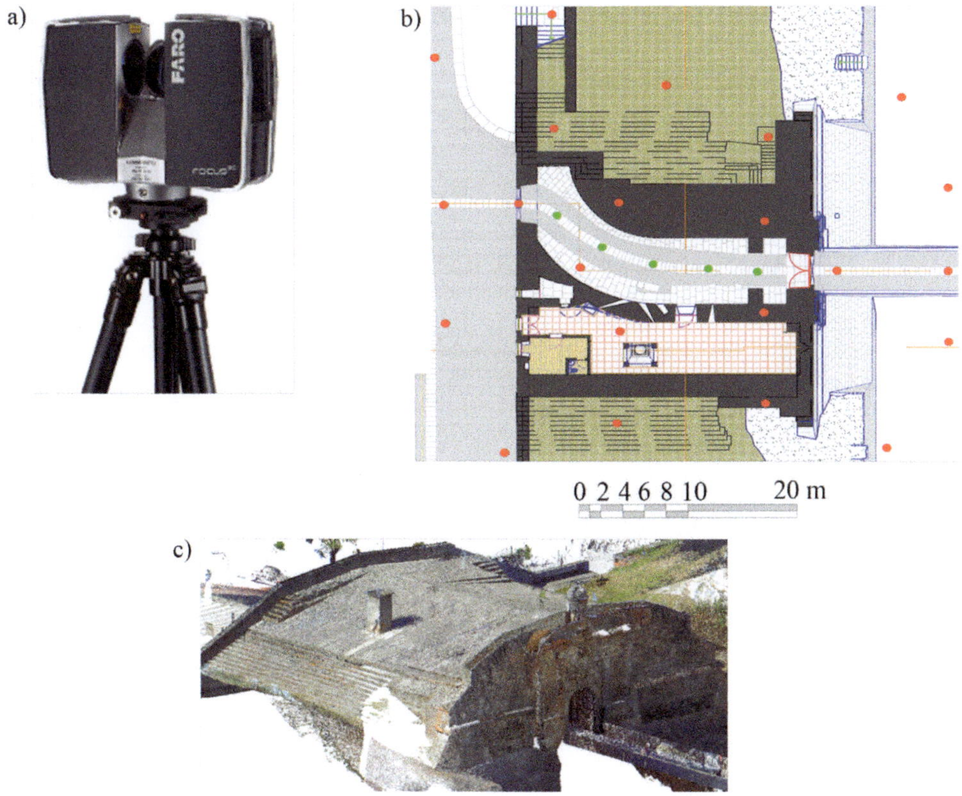

Figure 11.4. TLS campaign: (a) Faro Focus3D 120 scanner; (b) scheme of the laser scanning network of data collection; and (c) final 3D photo-realistic point cloud of the Master Gate of San Francisco. Yellow points represent scan stations carried out to capture the external envelop, whereas the green points represent scan stations to acquire the barrel vault.

construction: 11 for the two façades, 5 for the barrel vault and 10 for the roof of the structure. The registration of the different point clouds was carried out by means of the iterative closest point algorithm (Besl and McKay, 1992). As a result, a point cloud of 712 million points was obtained after the alignment process using the FaroFocus3D processing software, SCENE®, with a registration error of 0.0005 ± 0.003 m. The aligned point cloud had to be simplified, guaranteeing a homogenous final 3D model in order to reduce computation time and display the subsequent diagnosis analysis. This procedure was performed applying a decimation filter based on the distance among points. At this stage CloudCompare software was used, setting a threshold of 5 mm (best compromise between resolution and computational costs). As a result, a final subsample 3D point cloud of 105 million points was obtained.

2.3 *Geometric analysis of the TLS point cloud*

The 3D point cloud, as a cluster of millions of points defined in a Cartesian coordinate system, contains great potential information for the diagnosis of historical constructions. Under this basis, the following section will describe the results obtained after the application of the RANSAC and CANUPO algorithms over the decimated point cloud of the Master Gate of San Francisco.

Previously, the point cloud of the structure was divided into its different constructive parts, specifically: (i) the main façade; (ii) the inner façade; (iii) the roof; and (iv) the barrel vault. This process was performed through the open-source software CloudCompare.

2.3.1 RANSAC algorithm

Usually, the deformation analysis of constructions is mainly carried out by means of two approaches (Sánchez-Aparicio *et al.*, 2018b): (i) point-to-point distance comparison between point clouds of different epochs or (ii) distance comparison between points and the geometric primitives (basically planes but also cylinders, semispheres etc.) that represent the ideal geometry of the constructive element. For the diagnosis of the current state of conservation of the Master Gate of San Francisco, the second approach was applied.

This procedure consists of determining the hypothetical initial state of the constructive element to evaluate deviations suffered over time. For that end, the methodology represented in Figure 11.5 is proposed. The first stage (Figure 11.5a) consists of carrying out a RANSAC classification (Schnabel *et al.*, 2007) in which the TLS point cloud is divided into different groups as representative of each plane. This algorithm is based on the evaluation of the normal of each point and their neighbors. Given that RANSAC provides an excessive number of categorical planes due to spatial discontinuity between some parts of the point cloud, the second strategy (Figure 11.5b) tries to unify groups belonging to similar categorical planes. In order to correctly define the reference planes that belong to each categorical group, it is necessary to carry out a third process. Once the final reference planes are determined, the last process consists of mapping the differences between the obtained reference model of the constructive element and the real one. This process is performed by calculating distances between points of the real model and the reference planes (Figure 11.5c). This procedure was performed by using Matlab® and the open source software CloudCompare.

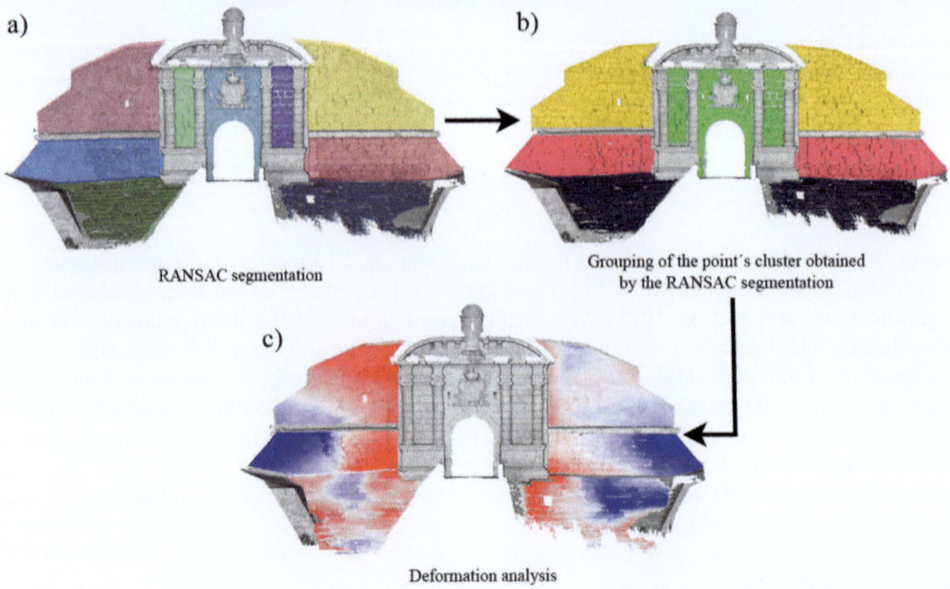

Figure 11.5. Graphical representation of the approach used to evaluate the possible out-of-plane deformations suffered by the construction. P1 is the centroid of the cluster of points inside the blue rectangle and P2 is the centroid shifted in the z axis. (a) Shows, in different colors, the results of the RANSAC shape detection for planes; (b) represents the results of clustering similar planes; and (c) shows the results of the deformation analysis.

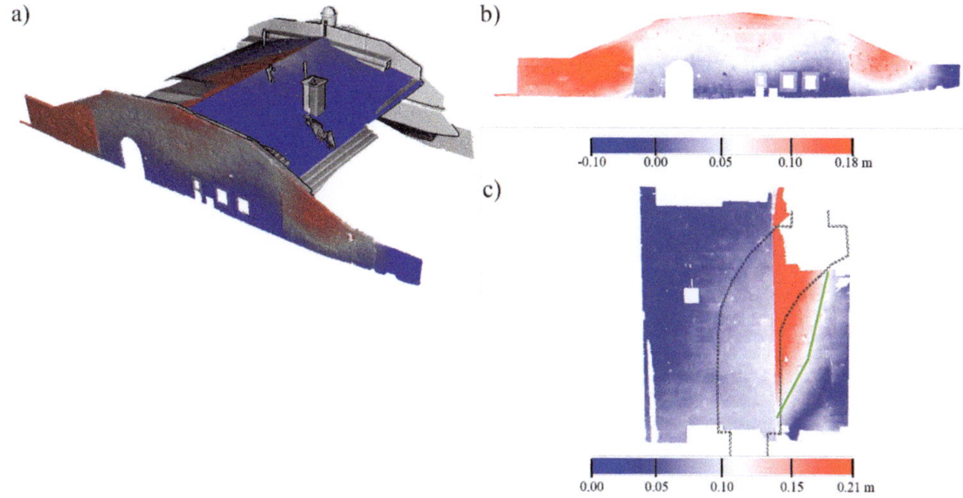

Figure 11.6. Deformation analysis based on the approach defined by Sánchez-Aparicio *et al.* (2018b); (a) secondary façade and (b) the roof. The pseudo-color map was created in the open-source software Cloud-Compare®. In black dash line is the position of the barrel vault, and in green is the position of a major crack detected on the roof during the visual inspection.

As a result, it was possible to obtain the deviations between the parametric shape considered as reference (initial state) and the current state captured by the TLS. Being especially useful were the data provided by this strategy in the secondary façade and in the roof (Figure 11.6).

For further information, see Sánchez-Aparicio *et al.* (2018b).

2.3.2 CANUPO classifier

Initially designed for the segmentation of natural spaces, the CANUPO algorithm is a multiscale classification method able to segment point cloud according with the geometrical features contained in the points. To this end, the CANUPO classifier evaluates the local dimensionality of each point (Brodu and Lague, 2012).

The concept of local dimensionality represents what the point cloud looks like from a geometrical point of view and at a given location and in a specific scale. To this end, a PCA (eigenvectors and eigenvalues of the covariance matrix) between each point and its neighborhood is carried out, obtaining three eigenvalues ($\lambda_1 > \lambda_2 > \lambda_3$). Then the variance in dimension is computed following equation 1.

$$p_i = \frac{\lambda_i}{\lambda_1 + \lambda_2 + \lambda_3} \qquad (1)$$

where p_i is the variance; and λ_i is the eigenvalue evaluated on which i represents the dimension of the eigenvalue evaluated {1,2,3}. The proportion between eigenvalues (p_1, p_2, p_3) define a measure of how much a point can be considered as a 1D/2D or 3D entity. These possible proportions can be drawn as a triangle whose vertexes represent the ideal entities (line, plane and sphere; Figure 11.7a). All the possible cases are included inside the triangle´s area. The closer is the calculated proportion to a vertex the higher is the similarity to an entity.

It is worth mentioning that CANUPO classifier evaluates the local dimensionally among points at different scales and thus, different neighborhoods' sizes. This multiscale criteria aim

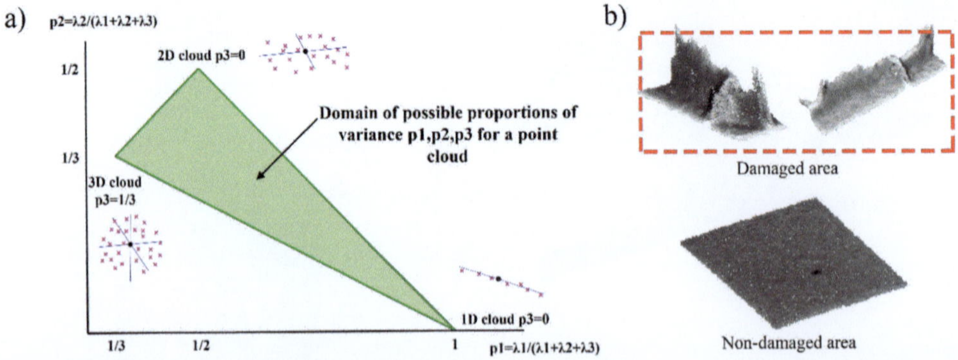

Figure 11.7. CANUPO classifier: (a) classification of the points according with the variance, adapted from (Brodu and Lague, 2012) and (b) samples extracted from the Master Gate point cloud and used to train the CANUPO classifier. In figure a, magenta dots represent the neighbors of the point considered (black circular dot), and the blue lines represent the eigenvectors obtained during the PCA analysis of the point and its neighbors.

to achieve a high degree of classification success in comparison with the use of a unique neighborhood's size (Brodu and Lague, 2012). This criterion requires the use of three parameters: (i) minimum scale, (ii) maximum scale and (iii) size of the step to increment the scale between evaluations.

Under the basis previously defined, the workflow used for CANUPO classifier can be summarized in the following stages (Brodu and Lague, 2012): (i) the training phase and (ii) the segmentation stage. During the first stage, the classifier uses the local dimensionality of each point at a different scales (different neighborhoods' sizes) to perform a linear discriminant analysis (Fisher, 1936). This analysis allows obtaining the hyperplane of maximal separability (Fisher, 1936) between informational classes and, thus, the segmentation rule required to classify the point cloud. As it was exposed, CANUPO classifier evaluates, at different scales, the dimensionality of each point. To this end, it is required to define a work-scale, with a maximum and minimum value and an interval of variation. For the present evaluation, a work scale that ranges from a minimum value of 0.05 m to a maximum value of 0.90 m with intervals of 0.05 m was used. These parameters were derived from the experimental studies carried out by Sánchez-Aparicio *et al.* (2018b), giving the best classifier in terms of accuracy (balance accuracy of 0.93 and fisher discriminant ratio of 10.13) and visual appearance. This classifier was obtained with samples of nondamaged areas and areas characterized by strong material losses coming from the point cloud of the Master Gate (Figure 11.7b).

Once the segmentation rule was obtained (hyperplane of maximal separability), the barrel vault was segmented into the two informational classes (damaged and nondamaged areas) previously defined through the use of the open-source software CloudCompare (Figure 11.8).

2.4 *Radiometric analysis of the TLS point cloud*

The backscattered laser intensity data is an added value to the metrical information provided by TLS technology. Since the proportion of radiation reflected by each material and for each specific wavelength is different, this information can be used to analyze materials and pathologies. In this way, not only physical but also chemical damages presented in historical constructions could be quantified and monitored over time (Del Pozo *et al.*, 2016b). Next, the two main processes required to exploit the radiometric data captured by TLS devices are described.

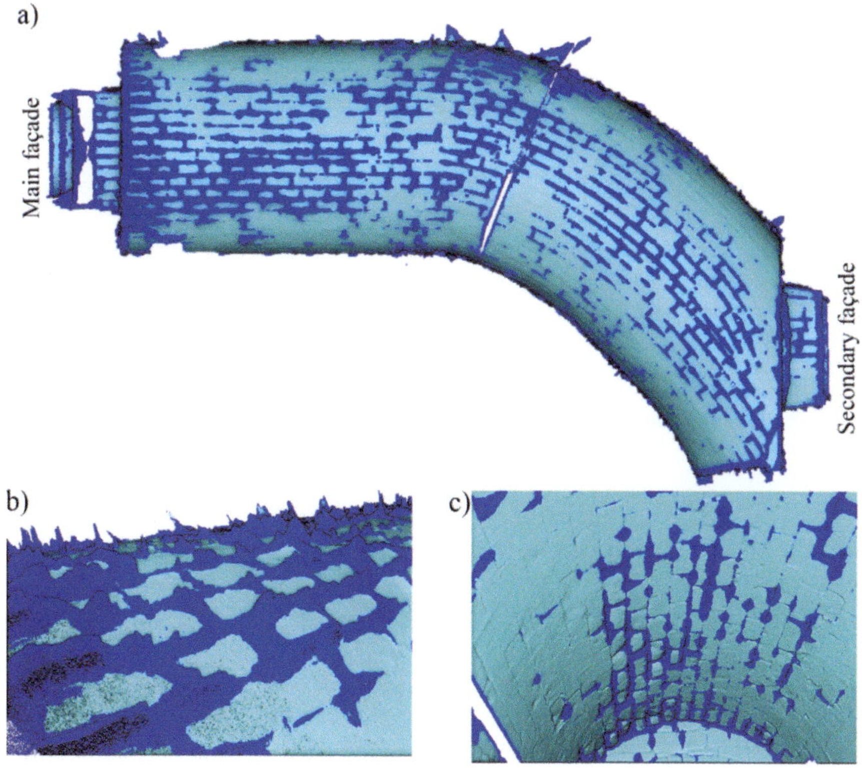

Figure 11.8. Results obtained with the CANUPO classifier: (a) general view of the barrel vault; (b) detail of the damages detected by the classifier; and (c) inner view of the results obtained. In black-blue the points classified as damaged areas and in light-blue the nondamaged areas.

2.4.1 Radiometric calibration of the TLS

TLS detectors receive the backscattered beam laser reflected by each material, and then it is converted into an electronic signal proportional to the incoming radiance (Figure 11.9). So prior to the radiometric analysis of constructions, a radiometric calibration of the sensor used in data collection is recommended. In this way, reflectance physical values are analyzed instead of digital levels conditioned by the internal alterations of the TLS used in each case. When using TLS technology, this process consists of analyzing the TLS intensity and studying the transformations that take place inside the device and carry out the reversal of such process.

The backscattered TLS signal depends on the distance between the scanner and the measured element and the incoming beam incident angle (Kaasalainen *et al*., 2011). The distance effect consists of intensity attenuation proportional to the square of the distance considered. On the other hand, the incidence angle effect is related not only to the scanning geometry but also to the scattering properties of the surface of the material under study. It affects the backscattered intensity according to Lambert's Law so that the higher the incidence angle, the smaller the amount of light coming back to the TLS.

The transformations occur inside the FaroFocus3D 120® were studied for the range of distance between 2 and 36 m. For this range, a logarithmic behavior of the backscattered intensity in reflectance values, *r;* was obtained and adjusted by the next empirical equation:

$$\rho_{(dis\tan ce-range)} = e^{a\cdot\times b} \times b \times d^2 \times e^{c1\times I} \quad (2)$$

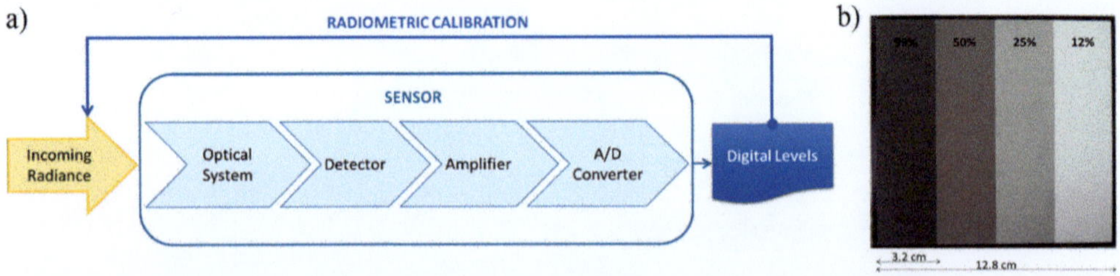

Figure 11.9. Radiometric calibration of the TLS: (a) transformations of incoming radiance inside TLS sensor and (b) the reference panel (Spectralon®) used for the calibration of the FaroFocus3D 120 with four different reflectance panels.

Table 11.1 Radiometric calibration coefficients of the FaroFocus3D 120® device

Range of distances	*a*	*b*	c_1	R^2
3–5.25 m	−1.0928	3.0295 10^{-5}	0.006397	0.9868
5.25–9 m	−0.1134	4.9446 10^{-7}	0.005911	0.9932
9–36 m	0.0214	3.9072 10^{-7}	0.005415	0.9966

where *a* and *b* were two coefficients related to the signal attenuation, *d* is the distance, *c1* is the gain of the FaroFocus3D 120® and *I* is the raw intensity value in digital levels (11 bits).

The coefficients related to the signal attenuation have the values shown in Table 11.1 for the different ranges of distances. As mentioned in Sánchez-Aparicio *et al.* (2018b), for this range of distances, a logarithmic behavior of the backscattered intensity of the FaroFocus3D 120® is observed. The distances for which the TLS was radiometrically calibrated are the most commonly used to digitalize historical constructions and for which there is an optimization between the number of laser stations and time spent in data acquisition.

2.4.2 Radiometric classification

In order to accurately classify the materials of a historical construction as well as its possible chemical damages, the TLS used should be calibrated, and so the point cloud intensity values should be reflectances instead of digital levels. Then a classification process is performed in which each point is classified based on its reflectance value. The two main generic classification approaches are the supervised and unsupervised methods. Table 11.2 shows the most used classification algorithms.

Below, Figure 11.10 shows the classification result of a historical construction applying the fuzzy k-means unsupervised approach, one of the most widely used and the one being used in this work. This classification was carried out in the software Matlab®.

2.5 *Additional sensors to complement the diagnosis*

Thanks to the application of the approaches shown in previous sections (section 2.3 and section 2.4) it was possible to characterize great part of damages presented on the construction. These areas need to be analyzed carefully in order to understand the current stability of the barrel vault. To this end, it the use of additional strategies was required with the aim of complementing the information obtained by the TLS, namely (Figure 11.11): (i) two indirect sonic tests for the mechanical characterization of the granite used in the construction, obtaining an average P wave of 1475 m/s and an average R wave of 758 m/s; (ii) six impact-eco tests for the evaluation of thickness support,

Table 11.2 Most common classification techniques. Adapted from (Li *et al.*, 2014)

Classification technique	*Characteristics*	*Examples of classifiers*
Pixel-based techniques	Each pixel is assumed pure and typically labelled as a single class	*Unsupervised*: K-means, ISODATA, SOM, hierarchical clustering *Supervised*: Maximum likelihood, Minimum distance-to-means, Mahalanobis distance, Parallelepiped, K-nearest neighbors. *Machine learning*: artificial neural network, classification tree, random forest, support vector machine, genetic algorithms.
Sub-pixel-based techniques	Each pixel is considered mixed, and the areal portion of each class is estimated	Fuzzy classification, neutral networks, regression modelling, regression tree analysis, spectral mixture analysis.
Object-based techniques	Geographical objects, instead of individual pixels, are considered the basic unit	Image segmentation and object-based image analysis techniques: E-cognition, ArcGis Feature Analyst.

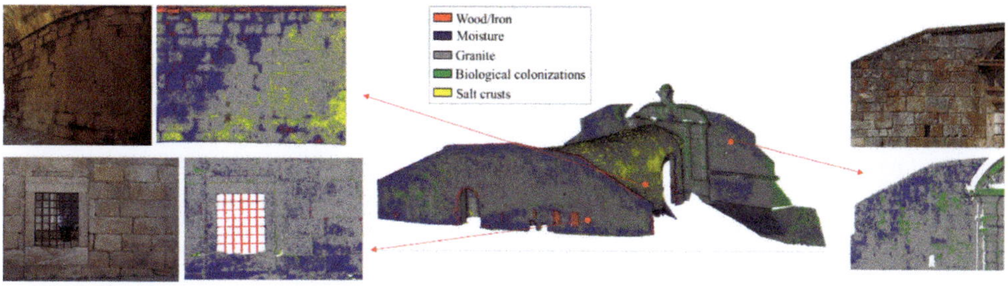

Figure 11.10. Results obtained after the application of the unsupervised fuzzy k-means classification algorithm.

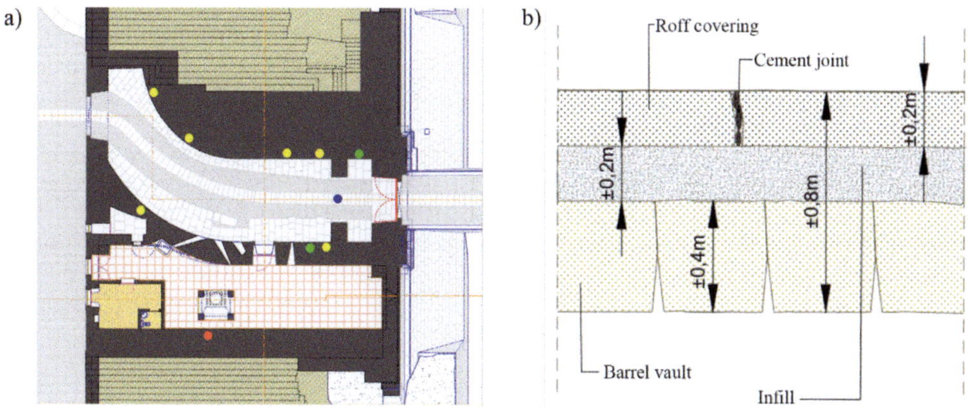

Figure 11.11. Additional sensors used to complement the diagnosis: (a) location of the complementary tests carried out in the Master Gate and (b) results obtained from the boroscopic test carried out. Yellow points indicate the impact-eco tests; green points correspond to the places on which the sonic tests were carried out; red points indicate the location on which the mortar specimens were extracted for the double punch tests and blue points the position on which the boroscopic tests was performed.

throwing an average value of 1.77 m; (iii) 10 double punch tests to evaluate the compressive strength of the mortar used in the masonry, obtaining a compressive strength of 5.80 MPa; and (iv) one boroscopic test to evaluate the inner composition of the construction.

For more details about the physical and operational fundaments of these, the reader is referred to specialized bibliography (Miranda *et al.*, 2012; Matysek *et al.*, 2017).

3 DIAGNOSIS OF THE CONSTRUCTION

According to the results provided by the visual inspection, a diagnosis of the construction is required in order to evaluate the current conservation state trough the geometrical and radiometrical analysis of the point cloud and the information provided by the additional sensors. This diagnosis was carried out at two levels: (i) analysis of the indicator of damage and (ii) evaluation of the current and future structural stability of the barrel vault.

3.1 *Analysis of the indicators of damage*

Starting from the results obtained during the application of the different algorithms proposed in the present chapter (RANSAC shape detector for the analysis of the façade´s deformations, CANUPO for the evaluation of the material losses and fuzzy k-means for the analysis of crusts, moistures and the biological activity; Figure 11.8; Figure 11.10), a robust 3D damage mapping was carried out using the open-source software CloudCompare. During this evaluation it was possible to analyze all the damages presented in the construction into a 3D context, allowing the evaluation of the material losses presented in the construction.

Considering the result of this 3D damage mapping, it was possible to conclude that the most preoccupant indicator of damage is the material disaggregation, appearing in 8.45% of the total area of the barrel vault (Figure 11.8) and whose origin can be attributed to the presence of melting salts and black crusts (Figure 11.10). With respect to the melting salts, which represents 4.14% of the total surface of the barrel vault, these compounds can be attributed to possible salts infiltrations coming from the soil as well as the roof covering (made with cement mortar joints; Figure 11.11). Regarding the black crusts/moisture, it was possible to observe a presence of this damage in 14.17% of the barrel vault. These damages are linked with the large amount of biological colonization (orange nitrophylic lichens that colonizes 4.14% of the façades; Figure 11.10). This type of lichens has developed an extra resistance to those environments rich in nitrogen oxide ($NO_{x)}$, a chemical compound that can be attributed to polluted environments. These alterations are promoting intense material losses and thus a reduction of the bearing capacity of the construction.

Regarding the deformation of the façades, we did not observe the presence of any out-of-plane collapse mechanism (Figure 11.6). Attributing the deviation to construction defects or derived from the intensive history of the construction, especially for the works carried out in 1986 (Sánchez-Aparicio *et al.*, 2018a). Regarding the roof's deformation (Figure 11.6b), the discrepancies observed, with a maximum value of 0.21 m, can be attributed to a possible accommodation of the infill, explaining the presence of the large longitudinal crack observed during the visual inspection (Sánchez-Aparicio *et al.*, 2018b).

3.2 *Evaluation of the current structural stability*

In contrast with other structural systems such as concrete or steel-based structures, traditional masonry structures were conceived to support vertical loads (mainly gravity loads). The stability of these type of structures, characterized by a low tensile strength and an excellent compressive behavior, is usually studied by means of the limit analysis theory (Heyman, 1966). Taking this into account, the present section will evaluate the structural stability of the barrel vault (construction element with more conservation problems) under the basis of the discrete limit analysis for masonry construction initially defined by Livesley (1978). This approach, conceived for the estimation of the stability of the arches, can be extended for the evaluation of masonry barrel vaults. In this case, the barrel vault is considered as a finite succession of arches along a specific path and thus, the stability of this constructive element depends on the stability of each arch (Heyman, 1966).

Taking this into consideration, the first step required for the evaluation of the barrel vault passes through the choice of the most critical arches. Understanding it, as those arches with less resistance section and thus with more probability of failure by equilibrium (the truss line cannot be contained into the geometry of the arch) or resistance (local stresses higher than the compression strength of the material). Under this basis, and considering the disposition of the different constructive elements, as well as the data provided by the CANUPO classifier, two sections corresponding to the most damaged part of the structure were evaluated (Figure 11.12): (i) one section in the vicinity of the intersection with the main façade – this section corresponds with the most damaged section of all the barrel vault – and (ii) one section at the mid-spam of the barrel vault. These sections were extracted with the open-source software CloudCompare and vectorized in the software AutoCAD®.

Once the different sections have been extracted, the next step required for the evaluation of its stability is a pass through the limit analysis of each arch. To this end, the commercial software limit state RING® was used. This software allows the evaluation of the stability of masonry

a)

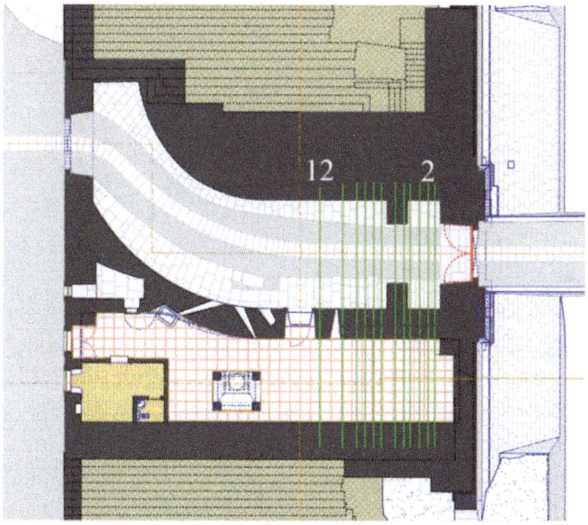

b)

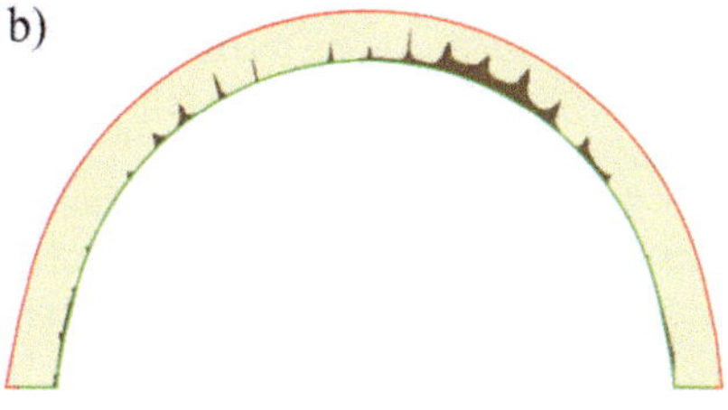

c)

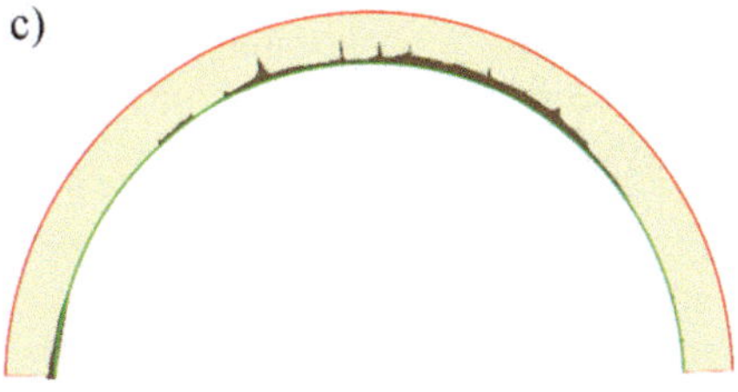

Figure 11.12. Sections used to evaluate the stability of the barrel vault: (a) plant view, (b) detail of section 2; and (c) detail of section 12. In red the intrados of the arches and supports, in green the extrados of the initial geometry, in light brown the current section and in dark brown the estimated material losses.

Table 11.3 Effective sections (thickness) used to simulate the generalized material loss presented on the arches

Section	*Initial thickness (mm)*	*Effective thickness (mm)*	*Reduction (%)*
2	400	343	14.3
12	400	366	8.5

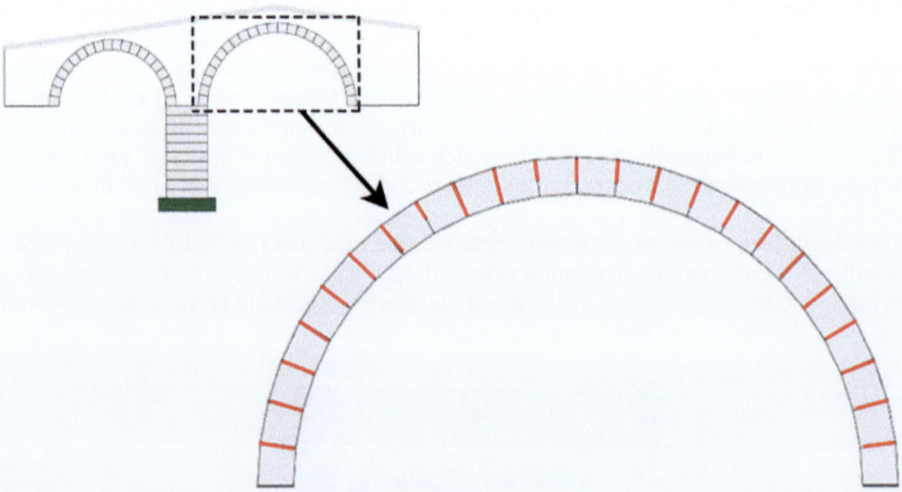

Figure 11.13. Structural model used to simulate the stability in section 12. In red it is possible to observe the real damage of the barrel vault.

arches through the use of an improved version of the discrete limit analysis proposed by Livesley (1978) on which is possible to consider additional effects such as sliding between blocks (with or without dilatancy) or failure by crushing (Gilbert and Melbourne, 1994). Complementary to the approach previously defined, the following assumptions were considered in order to simulate the stability of the structure (Table 11.3; Figure 11.13): (i) each arch was simulated according to their effective thickness; (ii) the lack of material in each joint was simulated in order to take into account the reduction of the contact area; (iii) the compressive strength of the model considered was the strength obtained by means of the double punch test carried out on the mortar, assuming a value of 5.80 MPa; (iv) a frictional model was taken into account in order to simulate possible sliding problems, and this model considered a frictional coefficient of 0.6; (v) a Mohr-Coulomb failure criterion was taken into account for the simulation of the infill, assuming a cohesion value of 30 KPa and a frictional angle of 30°; (vi) a Boussinesq load distribution model with a limitation angle of 30° was considered in order to simulate the dispersion of the loads inside the infill of the construction; and (vi) a live load of 5 kN/m was applied on the structure in order to simulate the possibility of having a dense group of people reunited and locate on the top of the structure. This load was considered since each year, in Almeida, a representation of the battle between France and Portugal is carried out (Arce *et al.*, 2018).

In general terms, and in contrast with the strong material losses presented in the structure, the barrel vault shows large values of stability (Table 11.4), showing a minimum safety factor of 7.42 in the union with the main façade (area most affected by the environmental agents; Figure 11.14).

Table 11.4 Results of the structural analysis carried out in section 2 and section 12

Section	*Collapse mechanism*	*Safety factor (nondamaged state)*	*Safety factor (damaged state)*	*Safety factor reduction (%)*
2	4 plastic hinges	16.90	7.42	43.90
12	4 plastic hinges	16.90	11.50	29.59

a)

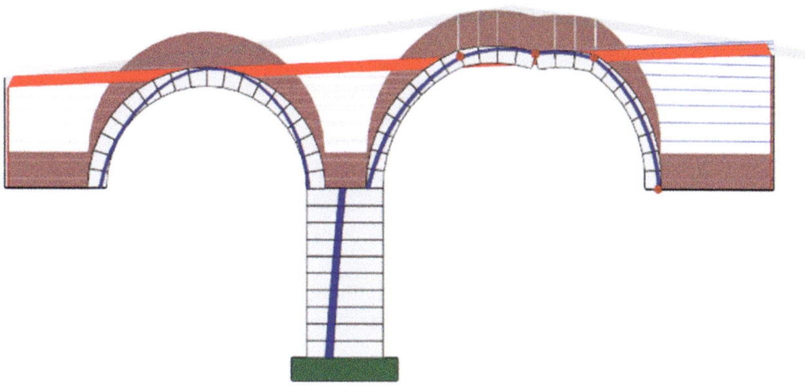

b)

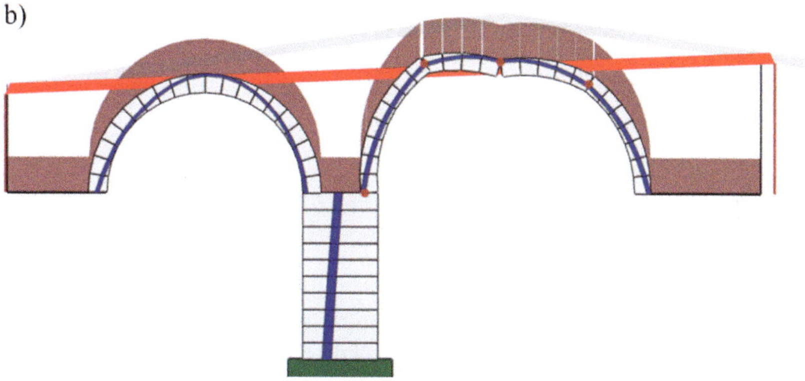

Figure 11.14. Collapse mechanism of the structure: (a) collapse mechanism carried out by the creation of four hinges at section 2 and (b) collapse promoted by the formation of four hinges at section 12.

3.2.1 Stability of the construction in case of continuing the material losses

Taking into account the two possible causes of collapse of the barrel vault, (i) failure by equilibrium and (ii) failure by crushing, seems to be logical that if the material losses continue (as a consequence of the disaggregation of the masonry promoted by the melting salts and the contamination) the structural stability of this construction element will decrease. Under this basis, several predictive evaluations were carried out with the aim of analyzing the influence of disaggregation of the material in the stability of the barrel vault. To this end, several numerical evaluations were carried out, increasing the level of damage in each one (effective section and material losses at joint level; Table 11.5).

Table 11.5 Evolution of the structural stability of the barrel vault with respect to the material losses. Between brackets the structural stability without considering the compressive failure.

Section	*Evolution of the safety factors according with the material losses*					
	Current	*+20%*	*+40%*	*+60%*	*+80%*	*+100%*
2	7.42 (11.7)	4.07 (6.29)	2.69 (5.71)	1.30 (3.02)	–	–
12	11.5 (23.9)	9.2 (15.7)	6.6 (9.9)	5.4 (7.7)	3.5 (4.9)	1.09 (3.21)

As expected, if the disaggregation of the masonry blocks continues, the safety factor of the barrel vault decreases as a consequence of the reduction of its resistance section and a reduction of the contact area between elements (equilibrium and resistance failure mechanisms; Table 11.5). This value promotes the creation of a collapse mechanism with four hinges and with a similar topology to those obtained previously (Figure 11.14). According to the results of these numerical evaluations, it is possible to conclude that the most critical part, in terms of stability, is the entrance of the main façade. In this part, if the damage increases in 65% to 70% with respect to the current one, the collapse can be produced in presence of an agglomeration of people on the roof (Table 11.5). With regards to the medium section of the barrel vault, it is possible to observe higher safety factors at the same increment of the damage (Table 11.5). The collapse of this part will be produced in a situation on which the level of disaggregation was 200% of the actual damage (Table 11.5).

It is worth mentioning that for lower levels of disaggregation, the contribution of the compressive strength to the stability of the model is higher than in large levels (Table 11.5). At medium-high levels of damage, the truss line has more problems to be contained inside the arch's limits. However, at high levels of damage (around 180%), the crushing effect (failure by resistance) starts to be more important due to the presence of small contact areas, generating high compressive stresses (Table 11.5).

3.2.2 Stability of the construction in case of support's failure

Apart from the possible reduction of the resistance section of the barrel vault as a consequence of the action of melting salts and traffic contamination, another equilibrium failure can appear as a consequence of a movement of the supports. This type of failure can be presented in two ways: (i) as a consequence of loss of the interaction between the infill (passive pressure) and the supports of the barrel vault or; (ii) as a result of the settlement of the supports. Under this basis, additional structural evaluations were carried out.

If the interaction between the infill and the support disappear, the collapse of the arches considered will be promoted by the presence of four plastic hinges, three in the main barrel vault and one in the lower part of the main barrel's support (Figure 11.15). In this situation the safety factor decreases to a value of 2.29 for section 2 and value of 6.6 for section 12. According with this results, and assuming that a recommended safety factor for masonry supports should be 3.0 (Heyman, 1966), it is possible to conclude that if the interaction between infill and support disappears, the supports of the entrance, in the presence of an agglomeration of people, can suffer some cracks (safety factor lower that 3.0) preserving its stability (safety factor higher than 1.0; Heyman, 1966).

In case of a failure by the settlement of the central support, the structure can collapse as a consequence of a multi-span mechanism. This mechanism is composed by a total of eight plastic hinges (four in the main barrel vault and four in the secondary barrel vault; Figure 11.15b).

a)

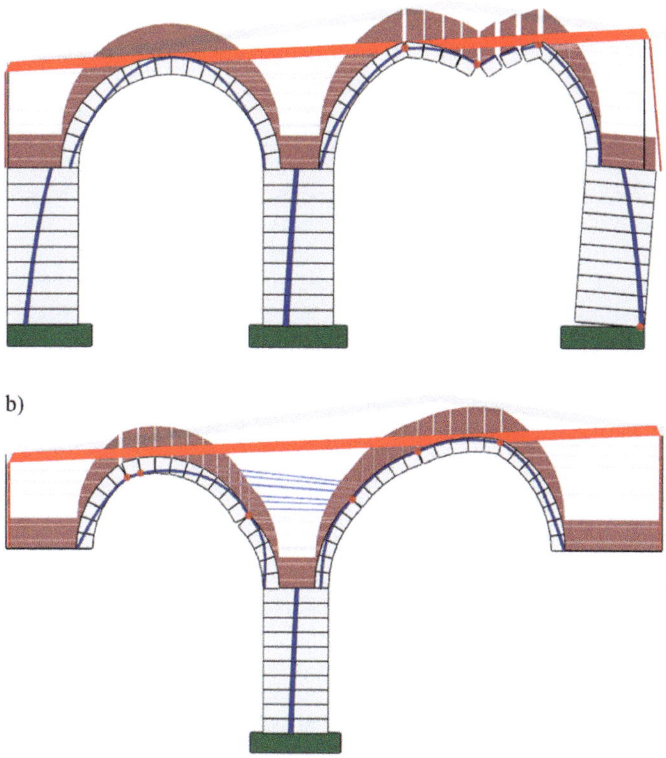

b)

Figure 11.15. Collapse mechanism as a consequence of the failure of the supports: (a) failure of section 2 in case of which the interaction between infill and support disappears and failure of section 2 in case of settlement of the central support.

4 CONCLUSIONS

Through a selected study case, this chapter demonstrates the suitability of TLS point clouds as a support in structural analysis and diagnosis of constructions. Geometric and radiometric procedures have been analyzed and integrated in order to assess their properties to guarantee an accurate and meaningful 3D mapping of chemical damages and structural analysis.

Radiometric and geometric TLS data are finally integrated into a multisensory framework that allows an integral diagnosis, making it possible to explain the main conservation problem of the study case: the strong material losses. A TLS point cloud was used to extract the effective sections of the most damaged arches of the barrel vault, allowing the evaluation of the current and future structural stability under different casuistic: (i) in case of having more material losses and (ii) in case of settlement.

By means of the radiometric information provided by TLS, biological and chemical damages were also accurately mapped, making possible a complete pathological diagnosis. For this purpose, we also have pointed out that the radiometric calibration of the TLS is required in order to transform the point cloud digital levels into reflectance values. Then a fuzzy k-means classification algorithm was applied in which each point is classified based on its reflectance value. These pathological processes are linked with the material losses. The black and the white crusts are promoting the material disaggregation (reduction of the resistance section of the barrel vault). The strong biological colonization (nitrophylic lichens) is suggesting the presence of pollutants in the atmosphere, which explains the existence of black crusts and thus the material disaggregation.

With respect to the TLS data acquisition, it is required to take into account the following aspects: (i) the complexity and size of the construction and (ii) the average density of the point clouds. For the first one, it is recommended to use a light-weight laser scanner due to the high amount of scan stations needed. Also it would be taken into account the physical principle of the laser, phase shift scanners being the most suitable scanner for those constructions in which is not possible to station near the construction (e.g. historical bridges). Concerning the average density of the data acquired, it is recommended to use of a maximum density of 5 mm in order to have a good compromise between resolution and computational costs.

Additional information coming from indirect sonic tests, impact eco tests, double punch tests and boroscopy inspections was also used to complement TLS data. The main aim was the mechanical characterization of the granite and mortar used as well as the inner composition between the roof and the barrel vault.

The workflow provided has proven useful for the evaluation of the current conservation state and the current structural stability of historical constructions.

With regards to the study case, and in contrast with the strong material losses, it is possible to conclude that the construction showed enough bearing capacity (safety factor of 7.42). However, if the material losses continue, the structure can collapse as a consequence of the reduction of the effective section of the barrel vault. The reduction of its effective section depends of the hidratation/deshidration cycles of the melting salts (white crusts) for which would be required the use of additional tests such as Raman spectroscopy or accelerated degradation tests.

In order to avoid a possible collapse of the barrel vault, it is recommended to carry out restoration actions focused on the waterproofing of the roof, the cleaning of the masonry blocks by means of mechanical or laser procedures and the retrofitting of the masonry joints with a compatible mortar, as well as traffic restrictions.

REFERENCES

Arce, A., Ramos, L.F., Fernandes, F.M., Sánchez-Aparicio, L.J. & Lourenço, P.B. (2018) Integrated structural safety analysis of San Francisco Master Gate in the Fortress of Almeida. *International Journal of Architectural Heritage*, 1–18.

Bautista-De Castro, Á., Sánchez-Aparicio, L.J., Ramos, L.F., Sena-Cruz, J. & González-Aguilera, D. (2018) Integrating geomatic approaches, Operational Modal Analysis, advanced numerical and updating methods to evaluate the current safety conditions of the historical Bôco Bridge. *Construction and Building Materials*, 158, 961–984. https://doi.org/10.1016/j.conbuildmat.2017.10.084.

Besl, P.J. & McKay, N.D. (1992) A method for registration of 3-D shapes. *IEEE Transactions on Pattern Analysis and Machine Intelligence*, 14(2), 239–256.

Brodu, N. & Lague, D. (2012) 3D terrestrial lidar data classification of complex natural scenes using a multiscale dimensionality criterion: Applications in geomorphology. *ISPRS Journal of Photogrammetry and Remote Sensing*, 68, 121–134.

Cobos, F. & João dos Santos de Sousa Campos. (2013) *Almeida, Ciudad Rodrigo: la fortificación de la Raya Central*. Consorcio Transfronterizo de Ciudades Amuralladas.

Costanzo, A., Minasi, M., Casula, G., Musacchio, M. & Fabrizia Buongiorno, M. (2014) Combined use of terrestrial laser scanning and IR thermography applied to a historical building. *Sensors*, 15(1), 194–213.

Del Pozo, S., Herrero-Pascual, J., Felipe-García, B., Hernández-López, D., Rodríguez-Gonzálvez, P. & González-Aguilera, D. (2016a) Multispectral radiometric analysis of façades to detect pathologies from active and passive remote sensing. *Remote Sensing*, 8(1), 80.

Del Pozo, S., Sánchez-Aparicio, L.J., Rodríguez-Gonzálvez, P., Herrero-Pascual, J., Muñoz-Nieto, A., González-Aguilera, D. & Hernández-López, D. (2016b) Multispectral imaging: Fundamentals, principles and methods of damage assessment in constructions. *Non-Destructive Techniques for the Evaluation of Structures and Infrastructure*, 11, 139.

Fernandez-Palacios, B.J., Morabito, D. & Remondino, F. (2017) Access to complex reality-based 3D models using virtual reality solutions. *Journal of Cultural Heritage*, 23, 40–48.

Fisher, R. (1936) Linear discriminant analysis. *Annals Eugenics*, 7, 179.

Gilbert, M. & Melbourne, C. (1994) Rigid-block analysis of masonry structures. *Structural Engineer*, 72(21).

Heyman, J. (1966) The stone skeleton. *International Journal of Solids and Structures*, 2(2), 249–279.

Hori, Y. & Ogawa, T. (2017) Visualization of the construction of ancient roman buildings in ostia using point cloud data. *The International Archives of Photogrammetry, Remote Sensing and Spatial Information Sciences*, 42, 345.

Kaasalainen, S., Jaakkola, A., Kaasalainen, M., Krooks, A. & Kukko, A. (2011) Analysis of incidence angle and distance effects on terrestrial laser scanner intensity: Search for correction methods. *Remote Sensing*, 3(10), 2207.

Korumaz, M., Betti, M., Conti, A., Tucci, G., Bartoli, G., Bonora, V., . . . Fiorini, L. (2017) An integrated Terrestrial Laser Scanner (TLS), Deviation Analysis (DA) and Finite Element (FE) approach for health assessment of historical structures. A minaret case study. *Engineering Structures*, 153, 224–238. https://doi.org/10.1016/j.engstruct.2017.10.026.

Li, M., Zang, S., Zhang, B., Li, S. & Wu, C. (2014) A review of remote sensing image classification techniques: The role of spatio-contextual information. *European Journal of Remote Sensing*, 47(1), 389–411.

Livesley, R.K. (1978) Limit analysis of structures formed from rigid blocks. *International Journal for Numerical Methods in Engineering*, 12(12), 1853–1871.

MacDonald, L. (2006) *Digital Heritage*. 1st ed. CRC Press, Boca Raton, FL, USA.

Matysek, P., Seręga, S. & Kańka, S. (2017) Determination of the mortar strength using double punch testing. *Procedia Engineering*, 193, 104–111.

Miranda, L.F., Rio, J., Miranda Guedes, J. & Costa, A. (2012) Sonic impact method–A new technique for characterization of stone masonry walls. *Construction and Building Materials*, 36, 27–35.

Riveiro, B., Conde-Carnero, B. & Arias-Sánchez, P. (2016) Laser scanning for the evaluation of historic structures. *Civil and Environmental Engineering: Concepts, Methodologies, Tools, and Applications*. IGI Global. pp.807–835.

Sánchez-Aparicio, L.J., Del Pozo, S., Ramos, L.F., Arce, A. & Fernandes, F.M. (2018a) Heritage site preservation with combined radiometric and geometric analysis of TLS data. *Automation in Construction*, 85, 24–39.

Sánchez-Aparicio, L.J., Del Pozo, S., Ramos, L.F., Arce, A. & Fernandes, F.M. (2018b) Heritage site preservation with combined radiometric and geometric analysis of TLS data. *Automation in Construction*, 85(Supplement C), 24–39. https://doi.org/10.1016/j.autcon.2017.09.023.

Sánchez-Aparicio, L.J., Del Pozo, S., Rodríguez-Gonzálvez, P., Herrero-Pascual, J., Muñoz-Nieto, A., González-Aguilera, D., . . . Solla, M. (2016) Practical use of multispectral techniques for the detection of pathologies in constructions. *Non-Destructive Techniques for the Evaluation of Structures and Infrastructure*, 11, 253.

Sánchez-Aparicio, L.J., Riveiro, B., Gonzalez-Aguilera, D. & Ramos, L.F. (2014) The combination of geomatic approaches and operational modal analysis to improve calibration of finite element models: A case of study in Saint Torcato Church (Guimarães, Portugal). *Construction and Building Materials*, 70, 118–129.

Schnabel, R., Wahl, R. & Klein, R. (2007) Efficient RANSAC for point-cloud shape detection. *Paper Read at Computer Graphics Forum*. Available from: https://diglib.eg.org/handle/10.2312/CGF.v26i2pp214-226 [accessed July 19, 2019].

Yang, H., Xu, X. & Neumann, I. (2018) Optimal finite element model with response surface methodology for concrete structures based on terrestrial laser scanning technology. *Composite Structures*, 183, 2–6. https://doi.org/10.1016/j.compstruct.2016.11.012.

Laser Scanning – Riveiro and Lindenbergh
© 2020 ISPRS, ISBN 978-1-138-49604-0

Chapter 12

Laser scanning for bridge inspection

Linh Truong-Hong and Debra F. Laefer

ABSTRACT: A large portion of bridges was built after a second World War, mostly over a designed 50-year lifespan. As such, the bridges' performance is significantly reduced because of excessive usage, overloading, material aging, and environmental impacts. Accurate assessment of a bridge's condition is needed for maintaining a safe, functional, and reliable structure. In general inspection, visual inspection by on-site inspectors is the predominant method, which is to detect structure changes, identify structural defects, and determine the overall condition. The visual inspection process is subjective and highly dependent upon an inspector's experience, especially when working in adverse conditions (poor weather and access difficulties). A vantage point of laser scanning, particularly terrestrial laser scanning (TLS), is to flexibly capture the bridge structure with high details of geometry and texture of the structures' surface. That can be an alternative method to fully or partially replace the visual inspection. Thus, this chapter presents the state of the art of TLS for bridge inspection. This investigation spans from geometric modelling to measurement of structural deformation to detection of structural surface deficiencies.

1 INTRODUCTION

The most recent American Society of Civil Engineers' (ASCE) Report Card on infrastructure reported more than 50% of 614,387 of the nation's bridges are more than 40 years old, with about 9.1% of them subjected to structural deficiencies (ASCE 2017). Similar quantities were also found in other countries. For example, in a recent survey of bridges from six European nations, most of the bridges were built in the period 1945–1965, undergone significant deterioration (Pakrashi *et al.*, 2011). The same trend was also found in Japan, where the majority of bridges were constructed from 1960 to 1980 without the high quality (Fujino and Abe, 2003). With a designed 50-year lifespan, the bridges' current performance was significantly reduced because of excessive usage, overloading, material aging, and environmental impacts. As such, accurate assessment of a bridge's condition is needed for maintaining a safe, functional, and reliable structure.

A bridge is often constructed from various materials including concrete, steel, masonry, and even timber, and bridge inspection tasks are needed to detection changes, identify structural deficiencies, and determine the overall condition. Structural deterioration can be determined through identifying cross-sectional and longitudinal geometric changes, which may be caused by scaling, spalling, chemical attack, bleeding, corrosion, or cracking (Figure 12.1; New York State Department of Transportation, 2014). Structural changes may manifest as alterations in the condition of the connections, deformations, distortions, or embedment loss, all of which can impact structural integrity. Additionally, inspection is needed to update asset inventories. Important information may involve geometric characteristics such as vertical clearance and structural defects, as well as loading capabilities. In such endeavors, laser scanning can capture the current geometry and some of the structural deficiencies, as will be the topic of this chapter.

Although many types of bridge inspection exist, a general inspection with a frequency of 24 months is commonly required for highway bridges (New York State Department of Transportation, 2014), which involves a general recommendation and a condition rating for further actions.

Figure 12.1. Common surface defects in a bridge structure: (a) spalling; (b) water bleed; (c) crack; (d) corrosion

In this type of inspection, visual inspection by on-site inspectors is the predominant method. This persists despite significant effort having been devoted to developing and applying nondestructive methods over the last two decades. The visual inspection process is subjective and highly dependent upon an inspector's experience, especially when working in adverse conditions (poor weather and access difficulties; Phares *et al.*, 2004). Visual inspection may also require rather significant logistics and bridge closure (Metni and Hamel, 2007).

In contrast, laser scanning is a noncontact measurement method acquiring a massive amount of three-dimensional (3D) topographic data. Representations of visible surfaces are captured highly accurately and quickly. The laser sensors can be integrated into various platforms to capture objects from air (airborne and unmanned aerial vehicle, UAV), mobile (e.g. car, boat and train), or ground and can offer various flexible data capture systems to maximize data coverage of a bridge structure (Truong-Hong Linh and Laefer, 2014; Riveiro *et al.*, 2013; Laefer, 2013). The vantage point of the data capture platform must be considered carefully. For example, aerial laser scanning (ALS) often captures vertical side surfaces and a bridge's deck with low density (Laefer, 2013), while mobile laser scanning (MLS) only collects a point cloud of the structure's surfaces along a routine of a vehicle or boat (Williams *et al.*, 2013). Missing data and/or low density may provide insufficient for surface damage detection. Thus, this chapter focuses on terrestrial laser scanning (TLS) for bridge inspection, as the scanning unit has the flexibility to capture the bridge structure from different locations with a million data points per second with millimeter accuracy. This investigation spans from geometric modelling to measurement of structural deformation to detection of structural surface deficiencies.

2 DATA ACQUISITION

A bridge poses greater challenges for data capture than most buildings due to the shape of the structural components, their orientation, and self-shadowing issues (Hinks *et al.*, 2009). The superstructure (i.e. deck, slab, diaphragms, and girders) is more complex than the substructure. Moreover, there is also great variation in bridge superstructures (from simple box girders to more complex beam girders to much more complicated trusses). Therefore, to provide sufficient laser scanning data points of a bridge structure for inspection, the scanning strategy must be planned with care and extensive consideration of the structure and the local environment as a unique scenario with the goal of maximizing data coverage, minimizing the number scan stations, and providing a sufficient quantity and quality of data points for each subsequent postprocessing step.

This usually requires a prescan site visit to identify suitable scan station locations and to consider the elevation of each with respect to minimizing occlusions (including terrain and self-shadowing). Minimizing the number of scan stations reduces time. On site, this relates to moving the scanner and scanning the targets to aid in co-registration. Multiple scans also extend the processing time, as demonstrated in recent work by Ruodan and Brilakis (Lu and Brilakis, 2017) in the scanning of 10 bridges with a medium size. In that work, the scanning of a bridge on average required 2.82 hours across 17.4 scan stations. As the scanning time at each station averaged only 6.31 minutes with a Faro Focus3D X330, this implies that the logistics time was about 36% of the total time on site. Importantly, each scan requires significant postprocessing time for registering and data cleaning. Lu and Brilakis (Lu and Brilakis, 2017) reported this to be on average 12.12 hours for each bridge. Moreover, error propagation from registering data sets of multiple scan stations can cause negative effects on further applications despite several methods (Kang *et al.*, 2009; Yan *et al.*, 2017; Bae and Lichti, 2008), offering automatic data registration with relatively high accuracy.

Data quantity is mostly related to the selected sampling step (the distance between two consecutive points), which should be selected by the scanner operator based on the smallest object that the user wants to detect. This is highly application specific. For geometric modelling, the smallest size of a concrete structure is in centimeter order, while it is a millimeter order in steel structures. Thus, identifying or reconstructing 3D models of the steel structure is always a big challenge because of insufficient data points describing edges of the structures. Anil *et al.* (2012) measured dimensions of steel's section by a point-to-point method, which is the distance between two points on the edge of the section. This implies there are at least three data points on the edge of the section to support users in identifying the edge. This can translate the sampling step should be smaller than a half of edges's length of the cross-section. Additionally, TLS captures surfaces of structural components as discrete points. This results in points not necessarily landing on the exact outer edge of a member or even overestimating due to small registration errors. As an example, Tang *et al.* (2009) reported that the structure size manually measured from data points can be smaller than a real dimension of about 56 mm.

The sampling step needed for surface damage detection (i.e. spalling or scaling) is much smaller than that required for geometric modelling, because damage tends to be (at least initially) small in size. For example, AASHTO (American Association of State Highway and Transportation Officials, 2011) designates a diameter of 152 mm of spalling as the documentation threshold of damage to rate a bridge for a condition of fair. Laefer *et al.* (2014) showed experimentally that the minimum crack detectable is 2 times the sampling step. To detect moderate cracking in reinforce concrete structure is 0.3 mm to 1.27 mm (American Association of State Highway and Transportation Officials, 2011), a sampling step of a point cloud must be no more than 0.15 mm. Figure 12.2 clearly illustrates the region containing surface spalling scanned with different sampling steps. As such, the scanner must locate at the short distance away from the structure's surfaces to capture sufficient data points, which may cause difficulty in logistics and requires a long scanning time and large storage memory.

A point cloud's quality, herein measured as the noise level of the data points considered, may cause an adverse impact on data processing for both geometric modelling and surface damage detection. Four main factors impact the quality of the point cloud: (i) the physical properties of

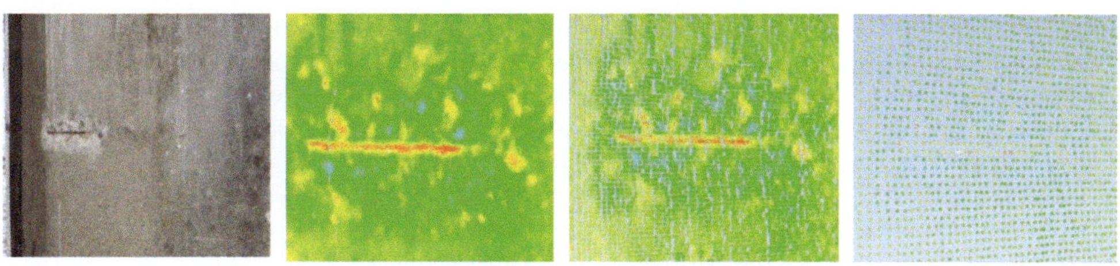

a) b) c) d)

Figure 12.2. Impact of a sampling step on surface damage recognition: (a) photograph of a surface damage; (b) a sampling step of 0.8 mm @10 m; (c) a sampling step of 1.6 mm @10 m; (d) a sampling step of 3.1 mm @10 m.

the scanner, (ii) environmental conditions, (iii) surface properties, and (iv) scanning geometry (Soudarissanane *et al.*, 2011). The first two relate to the scanner unit and atmospheric conditions (e.g. humidity, temperature and wind) where the laser pulses traverse from the sensor to a structure's surface and back, while the third factor relates to the reflectivity and roughness of the encountered surface. The last is mainly relevant to range measurement and incidence angle, which affect the footprint of the laser beam on the structure's surface and a return signal. The footprint increased and the return signal decreased when the range measurement and incidence angle increased. When investigating influence of the range measurement and incidence angle on the noise level, Soudarissanane *et al.* (2011) show that the noise level is less than 2 mm if the scanner is located less than 30 m from the surface with an incidence angle of no more than 40° when the Leica HDS6000 was used to capture data. Moreover, when measuring the crack width with 15 mm wide, Laefer *et al.* (2009) state the minimum errors of the crack width detection can achieve when the structure's surface is scanned from 12.0 m to 15.0 m with the scan angle no more than 45°. Similarly, Laefer *et al.* (2014) report that the minimum crack width of 1 mm is detectable when the crack face is captured from an orthogonal distance no more than 7.5 m with the incidence angle less than 15°. Thus, expanding the scan angle increases the positional error. The scanner should be placed at locations that minimize the scan angle. Finally, the range measurement in conjunction with the scanning angle also impacts intensity values of individual points (Soudarissanane *et al.*, 2011), which affects the accuracy in surface damage detection based attributes of a point cloud.

Finally, while direct access on the bridge may be beneficial in terms of access and minimizing occlusion, traffic or even natural frequency excitation due to wind may induce vibrations in the scanner that will negatively impact the positional accuracy of individual points. Thus, inspectors or surveyors should select an appropriate TLS scanner and a tripod and prepare a data collection plan to reduce the noise level due to these factors.

3 GEOMETRIC MODELLING

A goal of geometric modelling is to create 3D models of a bridge by fitting point clouds of the bridge components to known generic surfaces or volumes. Since the 3D geometric model is often used for either bridge information modelling (BrIM), bridge management, or assessment based numerical modelling (i.e. finite element analysis), a solid model is the preferred embodiment instead of a surface representation. To achieve those common approaches, involve either constructive solid geometry (CSG) or boundary representation (B-rep) (Laefer *et al.*, 2009). CSG constructs components by Boolean operations based a series of simple primitive solids (e.g. box, sphere, cylinder, or torus) in a library, stores a data construction in a CSG tree, and computes boundary representation. The representation can produce continuous and watertight models, but there is no unique CSG tree describing to the specific object (Goldman, 2009). In contrast, B-rep stores objects' boundaries regardless of construction method or spatial subdivision, in which a solid is decomposed into a cell with a simple topological and geometric structure. In B-rep, the solid models are described as a blanket of faces, edges and vertices, and the relationships between entities are described in terms of topology recorded (Laefer *et al.*, 2009). The geometry consists of vertices, curves, and surfaces along with numerical data describing position, size, and orientation, while the topology describes adjacency and connectivity of the geometry (Laefer *et al.*, 2009).

By using CSG, the structural component is divided into a series of simple components, which can be modelled by simple volumetric primitives such as cuboids, cylinders, spheres, so on. These primitives can fit to a point cloud of the component/subcomponents by determining optimal parameters representing the primitives to satisfy the smallest deviation, which is often defined through the orthogonal distance from the point cloud to the surface of the model. A prerequisite step is to extract the point cloud of the component and subcomponents from an entire data set of the bridge, which can be done by implementing existing segmentation methods such as a region growing (Vo *et al.*, 2015; Rabbani *et al.*, 2006), random sample consensus (RANSAC) (Schnabel *et al.*, 2007), or Hough transform (Tarsha-Kurdi *et al.*, 2007). Irrespective of the segmentation methods implemented, users must select or tune parameters to extract the desired sections of the point clouds, thereby segregating the surface/components. For example, with a region growing, the users must choose optimal parameters to control a growing process, which can be residual and angle thresholds (Rabbani *et al.*, 2006). However, although the segmentation method can extract the point cloud on the same surface (planar or curve or free-form surface), developing automatic processes to recognize those segments of a specific component or subcomponents remains a major challenge. In practice, this task is often done manually.

When using B-rep to represent the solid model, there are two main approaches to model geometric surfaces of a structural component: (i) fit the geometric primitives directly from their point clouds or (ii) extrude cross-sections along a sweeping profile. In the first method, commercial software has tools allowing users to segment a point cloud of the structural component's primitives (e.g. surfaces), which can be done through a region growing or RANSAC method, for example. The surface is then fitted by built-in simple geometric primitives (e.g. planes, cylinders, and spheres or even irregular shapes). Boundaries can also be obtained from the intersection of multiple geometric primitives. In practice, the modelling procedure is often done within a commercial software for object reconstruction from laser scanning data (Minehane *et al.*, 2014; Mailhot and Busuioc, 2006; Lu and Brilakis, 2017). For example, a 3D model of parts of a bridge on M18, Ireland (Figure 12.3), was done by the authors with a Leica cyclone software V9.1.3 (Leica Geosystems, 2017). the Leica cyclone's built-in tools allow to select a point cloud of a component surfaces, and then users can choose the fitting surface to create the surface. Moreover, in an attempt at introducing automatic geometric modelling process, Walsh *et al.* (2013) used a region-growing technique with a smoothness constraint to extract a point cloud belonging to individual surfaces of a concrete pier cap (Figure 12.3). Next, a planar surface was fitted to the point cloud of each individual surface using a least-squares approach. The complete 3D model can be achieved by modifying these fitting surfaces to their intersections based on surface locations, orientations, and sizes (Figure 12.3). However, this work also requires user interaction to extract edge points or groups of points.

The second modelling method creates surfaces of a 3D solid model by extruding cross-sections along a sweeping profile. The method starts by extracting cross-sections within the point clouds, which can be done by selecting the sub-point cloud within the interval thickness, along a longitudinal direction of the component (Laefer and Truong-Hong, 2017). The polygon can be fitted through the point cloud of the cross-section, in which the Euclidean distance between the data points and the fitted polygon is used to determine the best polygon. Next, the algorithm often needs

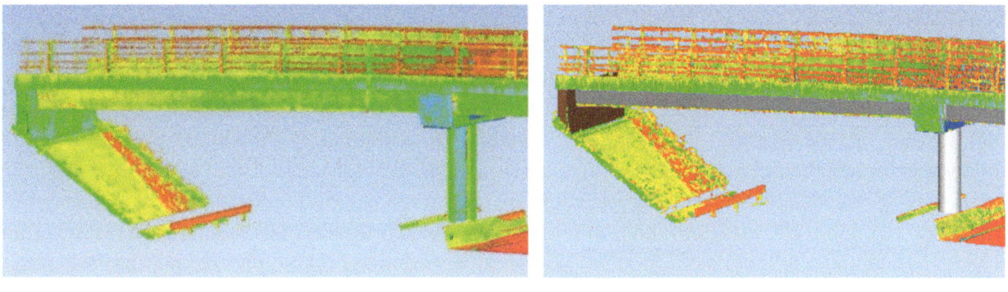

a) b)

Figure 12.3. 3D model of parts of the bridge manually created within a Leica cyclone V9.1.3: (a) a point cloud; (b) fitting objects created by using built-tools within Leica cyclone V9.1.3.

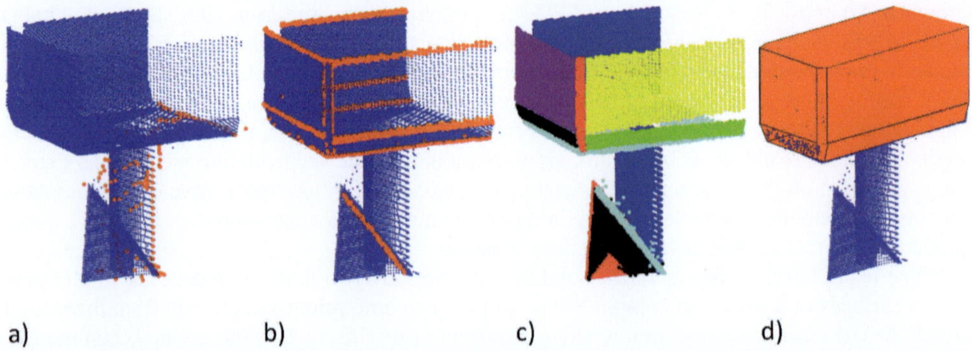

Figure 12.4. Creating a 3D model of a pier cap: (a) a point cloud of a pier cap; (b) edge point extraction by users; (c) point clouds of individual surfaces; (d) a complete 3D model (Walsh *et al.*, 2013).

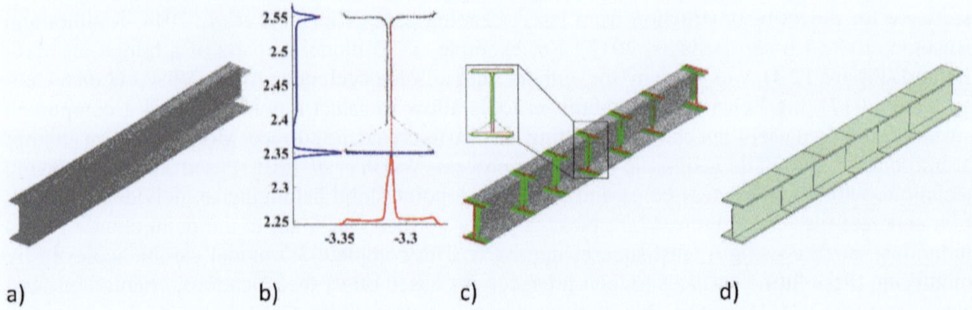

Figure 12.5. Generating a 3D model of a steel beam: (a) point cloud; (b) estimation of section dimensions; (c) cross-sections identification; (d) 3D model generated by sweeping through cross-sections.

to determine the trajectory of the component, thereby allowing the cross-section sweeping. In such an approach, structural deformation may be overlooked, which can lead to incorrect reconstruction. To overcome this issue, Laefer and Truong-Hong (2017) generate multiple cross-sections along the longitudinal direction and then the 3D model of the component can be achieved by extruding these cross-sections (Figure 12.5).

Another approach involves spatial decomposition methods, such as quadtree and octree decomposition techniques for two dimensions (2D) and three dimensions (3D), respectively. These can be used to directly convert a point cloud to a finite element mesh (FEM) for structural analysis (Truong-Hong *et al.*, 2013; Hinks *et al.*, 2012). Both methods share strategies to generate meshes from a point cloud, but the quadtree method is for generating quadrilateral meshes in 2D, while the octree method is for hexahedral meshes in 3D. In a first stage, the bounding box B(Ω) of Ω Î R^2 (2D) or R^3 (3D), which bounds all data points of structure, is recursively subdivided into a set of disjoint cells in 2D or voxels in 3D until terminate criteria are reached. For example, the termination criterion can be a minimal cell/voxel size (Ayala *et al.*, 1985), a predefined maximum tree depth (Pulli *et al.*, 1997), or maximum data points in a cell/voxel (Wang *et al.*, 2005). Since FEM elements must be continuous, this dictates that the element must share common element faces, edges, and nodes to adjacent elements. As such, a pure quadtree/octree with even-sized cells/voxels is often to use in a subdivision strategy, although an adaptive quadtree/octree can provide a more efficient approach in terms of execution time and storage.

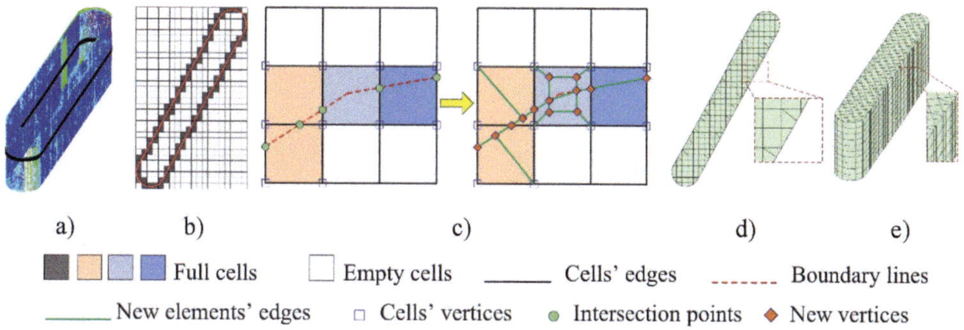

Figure 12.6. Generating the 2D aspect of the cross-section and the 3D mesh of a pier: (a) point cloud of pier rendering by intensity; (b) pure quadtree decomposition (white cells are empty, grey cells are full, and black dots are data points); (c) template meshes to create new elements for boundary cells; (d) quadrilateral meshes of the cross-section; (e) hexahedral meshes of the pier by extruding the quadrilateral meshes.

The cell/voxel can be classified into various categories, based on its relative position to the sensor or object (Pulli *et al*., 1997), on specific types of data points within a cell/voxel (Truong-Hong *et al*., 2013), or the number of data points within a cell/voxel (Truong-Hong and Laefer, 2017). For example, in Truong-Hong *et al*. (2017), the voxels were classified as "full" if the voxel contained an equal or larger number of points compared to a predefined threshold (and here is one point) and "empty" if less.

In a second stage, the initial mesh (quadrilateral or hexahedral meshes) from stage 1 is refined by the structure's boundaries. Boundaries can be 2D lines in the 2D domain (Truong-Hong and Laefer, 2017) or a triangular surface mesh (Gibson, 1995). The refinement process divides the cell/voxels on the structure's boundaries into the number of the smaller cells/voxels with different shapes depending on intersection between the boundary (a 2D line or 3D surface) and the cells/voxels. To provide for the refinement process, various template mesh configurations were developed to find a reasonable transition from the cell/voxel models to the new mesh (Schneiders *et al*., 1996). Finally, only mesh elements representing the structure need to be stored for further computational modelling. The quadrilateral or hexahedral meshes inside the structure's boundary can be identified by adopting a flying voxel method proposed by Truong-Hong *et al*. (2012a), which obviates the need for storage where building openings (e.g. windows) exist.

Adopting the method proposed by Truong-Hong and Laefer (2017) to generate hexahedral meshes of a pier (Figure 12.6a), the pure quadtree method is used to decompose a bounding box of a point cloud of a cross-section into uniform cells (Figure 12.6b). The cells are classified as full and empty based on whether the cell contains the points of the cross-section. Template meshes are created and mapped to the full cells to improve an accurate boundary of the component. The number of new elements are created based on intersection of the full cell and boundary lines generated from the point cloud of the cross-section (Figure 12.6c). The complete 3D mesh can be achieved by sweeping the quadrilateral mesh of a cross-section along the direction of the pier (Figure 12.6d and 12.6e).

By using a similar strategy, Yan *et al*. (2015) used a voxel grid to generate hexahedral finite meshes of a bridge from a point cloud (Figure 12.7a). The method starts by subdividing a bounding box into a uniform distribution of voxels. Next, the voxels are classified as external, boundary or internal. The grid-based mesh generation method proposed by Schneiders (1996) is then employed to modify the boundary voxels to represent the real boundary of the structure. A continuous mesh can be obtained by modifying nodes and edges to ensure adjacent mesh elements share the same nodes, edges, and faces (Figure 12.7b).

a) b)

Figure 12.7. Generating hexahedral meshes from a point cloud: (a) a point cloud of the bridge; (b) hexahedral mesh of the bridge (Yan *et al.*, 2015).

4 STRUCTURAL INSPECTION

4.1 *Deformation and vertical measurements*

Laser scanning can be used to measure overall displacements of a bridge (Lichti *et al.*, 2002; Zogg and Ingensand, 2008; Lovas *et al.*, 2008), vertical clearance (Riveiro *et al.*, 2013; Liu *et al.*, 2012), and deformations/distortions of each member (Truong-Hong and Laefer, 2015a). Those applications are often used in loading tests and structure assessments. As the measurement of a structure's deformation requires millimeter accuracy, the TLS and MLS point clouds are commonly used. However, MLS integrated into a vehicle is limited in acquiring data points with respect to the field of view of the scanner along the vehicle's routine. Thus, MLS data is often restricted for use in establishing in-situ vertical clearance (Gong *et al.*, 2012).

A deformation is defined as the distance from a deformed surface to a non-deformed surface. Note that the deformed surface is the surface of the structural component under loaded conditions or service, while the reference surface is the surface of the structure under unloaded conditions (or only self-weight). The deformation can be obtained through point-to-surface, point-to-cell, or cell-to-cell approaches with the workflows shown in Figure 12.8.

After scanning and registering all point clouds for a single bridge into a single coordinate system, the next step involved extraction of the data points of the reference and sampling surfaces, which were known as the reference and sampling data sets, respectively. This is done either manually, semi-automatically, or automatically. Automatic segmentation can be achieved by using existing methods such as region growing–based methods (Rabbani *et al.*, 2006; Vo *et al.*, 2015). Moreover, in measuring deformation of a structure in the middle of its service lift, documentation of the initial geometry may not be available. Thus, the reference surface is assumed to be planar and rectilinear throughout (Truong-Hong and Laefer, 2015a).

In the point-to-surface approach (M1), the deformation is a distance from the data points to a fitting surface deriving from either the reference data or the sampling data. However, as the shape of the deformed surface (or the sampling data set) is often unknown, the fitting surface derived from the sampling data can result in relatively low accuracy (Tang *et al.*, 2010). As the shape of the reference surface is theoretically known, a close form of the reference surface can be defined and the fitting surface (S_r) through the reference data ($\mathbf{P}_s$) can give high accuracy. Then the distance $d(\mathbf{P}_s, S_r)$ is computed. In practice, as the reference data may be subjected to a large amount of noise or contain irrelevant points, the point-to-surface method may generate large errors (Truong-Hong and Laefer, 2015a). Importantly, the point-to-cell (M2) and cell-to-cell (M3) methods can overcome this drawback.

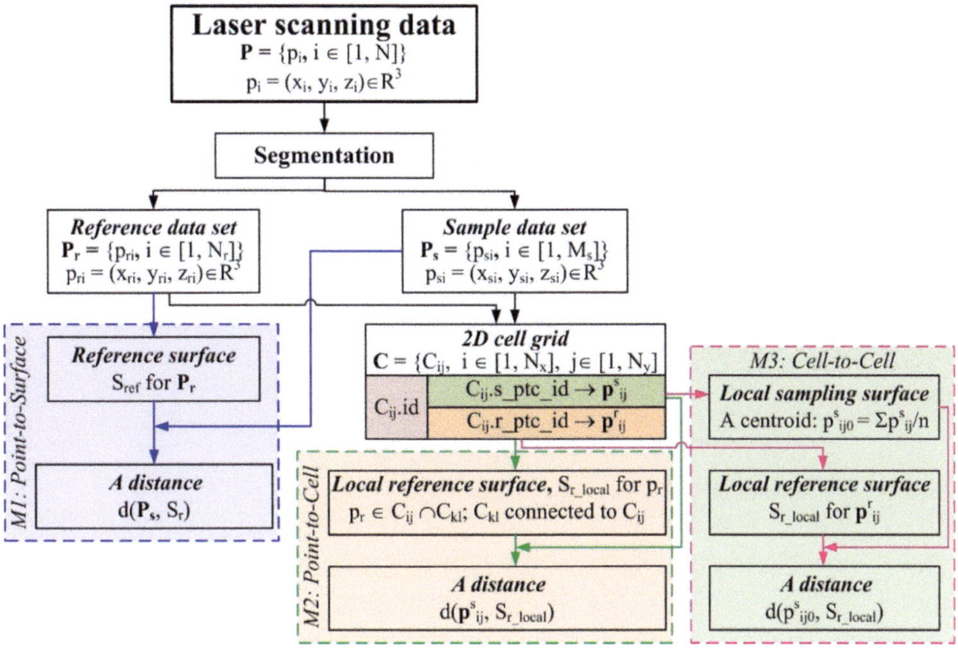

Figure 12.8. Methods for deformation measurement from a point cloud.

In both aforementioned methods, the workaround involves having the reference and sampling data sets initially projected onto a plane of interest (PoI; e.g. an xy plane in the global coordinate system). A 2D cell grid is used to divide a bounded, 2D region of the reference and sampling data sets into a set of uniform cells. Each cell is represented by an index $C_{ij,}$ where i ϵ [1, N_x] and j ϵ [1, N_y] with the dimensions of the cell (Δx and Δy), where the number of the cells (N_x, N_y) along x and y directions can be expressed in Equations 1 and 2.

$$N_x = \left\lceil \frac{(x_{max} - x_{min})}{\Delta x} \right\rceil \quad (1)$$

$$N_y = \left\lceil \frac{(y_{max} - y_{min})}{\Delta y} \right\rceil \quad (2)$$

where x_{min}, x_{max}, y_{min}, and y_{max} are the minimum and maximum coordinates of the reference and sampling data sets, while the notation [ ] means that the value is rounded up to the nearest integer.

Each cell C_{ij} has two lists for indexing the reference and sampling data points, as notated by C_{ij}.r_ptc_id and C_{ij}.s_ptc_id, respectively. This storage strategy allows easy retrieval of the corresponding reference ($\mathbf{p}^r_{ij}$) and sampling data points ($\mathbf{p}^s_{ij}$) within the respective cell, C_{ij}. Note that, only cells (C_{ij}) containing both $\mathbf{p}^r_{ij}$ and $\mathbf{p}^s_{ij}$ are used for determining the extent of deformation.

In the point-to-cell method (M2), the deformation is defined as the distance from a data point p_{si} in the sampling data set to the local surface at the projection of p_{si} on the reference surface (Truong-Hong and Laefer, 2015a). As deformation of the structure is assumed to be small, the projection of the sampling points ($\mathbf{p}^s_{ij}$) in cell C_{ij} can be located on the local reference surface (S_{r_local}) by fitting using the reference points ($\mathbf{p}^r_{ij}$). However, this assumption can be inappropriate when the number of $\mathbf{p}^r_{ij}$ is small. Additional reference points ($\mathbf{p}^r_{lk}$) located in eight adjacent neighbor cells $\mathbf{C}_{kl}$ of C_{ij}

were included to $\mathbf{p}^r_{ij}$ to estimate the S_{r_local}. Thus, the S_{r_local} is estimated from the points $\mathbf{p}^r = \mathbf{p}^r_{ij} \cap \mathbf{p}^r_{lk}$ distributed in a small region (3 × 3 of a cell size), which is assumed to be a planar surface. The centroid (p_0) of S_{r_local} can be expressed in Equation 3, while a principal component analysis (PCA) is employed to determine the surface normal $n = (n_x, n_y, n_z)$, which is the eigenvector corresponding to the smallest eigenvalue determined from the covariance matrix given in Equation 4. The term S_{r_local} can be expressed in Equation 5. Finally, the distance, $d(\mathbf{p}^s_{ij}, S_{r\_local})$ from the sampling points to S_{r_local} can be given as per Equation 6.

$$p_0 = \frac{1}{N}\sum_{i=1}^{N} p_i^r \tag{3}$$

$$C = \sum_{i=1}^{N}\left(p_i^r - p_0\right)\left(p_i^r - p_0\right)^T \tag{4}$$

$$S_{r_{local}} = n_x x + n_y y + n_z z + d \tag{5}$$

where N is the number of data points in $\mathbf{p}^r$ and $d = -n_x x_0 - n_y y_0 - n_z z_0$

$$d\left(p_{ij}^s, S_{r\_local}\right) = \frac{n_x x_{si} + n_y y_{si} + n_z z_{si} + d}{\sqrt{n_x^2 + n_y^2 + n_z^2}} \tag{6}$$

where $\mathbf{p}^s_{ij} = (x_{si}, y_{si}, z_{si}) \in R^3$ is x-, y-, z- coordinates of the sampling data points.

In the cell-to-cell method (M3), it is assumed that for each cell C_{ij}, the local planar surfaces S_{r_local} and S_{s_local} are, respectively, created from the reference ($\mathbf{p}^r_{ij}$) and sampling data points ($\mathbf{p}^s_{ij}$). However, both local surfaces are small in size, and the changes of the surface within the cell is small. Thus, the deformation $d(\mathbf{p}^s_{ij0}, S_{r\_local})$ is herein defined as the distance from the centroid ($\mathbf{p}^s_{ij0}$) of the S_{s_local} to the S_{r_local} by using Equation 6. The centroid $\mathbf{p}^s_{ij0}$ can be determined by using Equation 2, while the PCA is used to compute the normal of the S_{r_local}.

The proposed methods are demonstrated by measuring the vertical clearance of an overpass bridge at the intersection between N25 and Coolballow Rd., Co. Wexford, Ireland (Figure 12.9a). Data for the bridge and road were acquired by a Leica ScanStation P20 TLS unit with a sampling step of 3.1 mm at a range measurement of 10 m (Figure 12.9b). The average distance from the scanner to the bridge deck was around 15.0 m. An octree-based region-growing approach (Vo *et al.*, 2015) was employed to extract the point clouds of the road and the bottom of the bridge's girders (Figure 12.9c).

In this demonstration, the road surface is considered as the reference surface, while the bottom of the bridge's girders is assumed as the sampling surface. When using the point-to-surface method (M1), a fitting surface method is employed to fit the reference surface. In this case, the best-fitting surface was $S_r = 7.94x - 7.96y - 0.002xy - 0.004x^2 + 0.004y^2$ with a root mean square error (RMSE) of 17 mm, where x and y are respectively x- and y- coordinates of the point cloud. The vertical distance $d(P_s, S_r)$ at the points $\mathbf{P}_s$ is the difference between the z-coordinate of $\mathbf{P}_s$ and the value of S_r by substituting x- and y- coordinates of $\mathbf{P}_s$.

When using the point-to-cell (M2) and cell-to-cell methods (M3), a cell size of 0.1 m was selected to decompose both the reference and sampling data sets into uniform cell grids. Normal vectors of the local reference surfaces (S_{r_local}) in the method M2 and M3 can be estimated by using PCA. Next, the orthogonal distance, $d(\mathbf{p}^s_{ij}, S_{r\_local})$ from the sampling data set to the reference surface can be determined by using Equation 6. Thus, the vertical clearance can be given in Equation 7.

$$d_{vert}\left(p_{ij}^s, S_{r_{local}}\right) = d\left(p_{ij}^s, S_{r_{local}}\right) / \cos\alpha \tag{7}$$

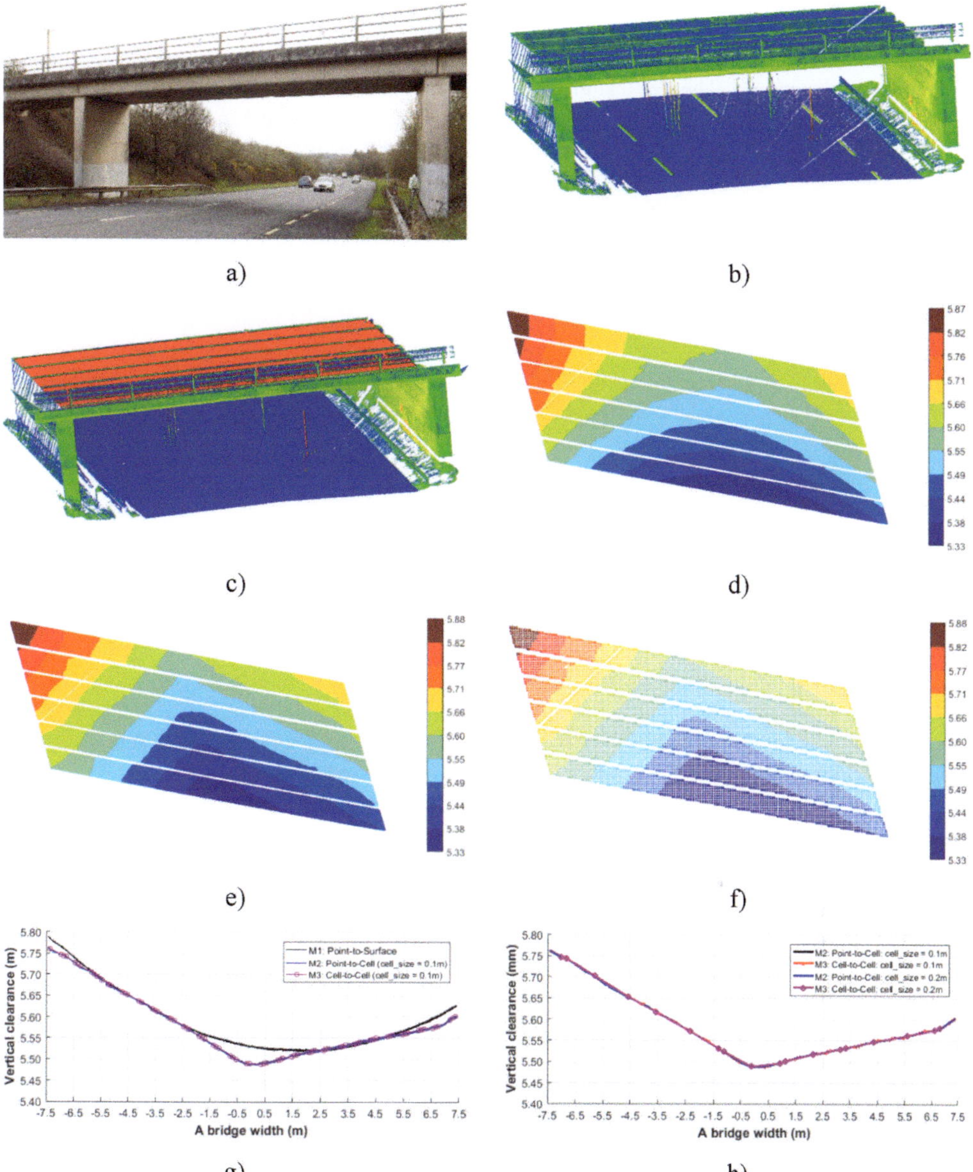

Figure 12.9. Demonstration of the proposed method in vertical clearance measurement: (a) photo of the bridge; (b) a point cloud; (c) a reference and sampling data sets(*); (d) vertical clearance from M1; (e) vertical clearance from M2 (cell size = 0.1 m); (f) vertical clearance from M3 (cell size = 0.1 m); (g) vertical clearance of the cross-section from M1–3; (h) vertical clearance of the cross-section from M2 and M3 with different cell sizes.

Note: (*) Red points are the sampling data and blue points are the reference data.

where α is the angle of the between the normal vector of the S_{r_local} and the unit vector $n_z = (0, 0, 1)$ representing the vertical direction.

Figure 12.9d–12.9e illustrate the vertical clearance of the bridge obtained from each of the different methods (M1–3). The vertical clearance along a cross-section of the road showed that the values derived from the method M1 differed from the method M2 and M3, by more than 40 mm [42.9 mm (M1 vs. M2) and 43.4 mm (M1 vs. M3)]. Moreover, there is the small difference between the vertical clearance derived from M2 and M3, for which the maximum difference was 3.2 mm. That is because the road surface has transverse slope with respect to the horizon, and the implicit fitting surface in M1 does not represent accurately the reference surface. M2 and M3 both generate realistic vertical clearance values with only a small difference between them. Finally, increasing the cell size from 0.1 m to 0.2 m seemly does not cause a significant difference in the calculated vertical clearance. The key factors for the cell size selection should be as follows: (i) a sampling step of the data, (ii) requirement of accuracy, and (iii) computing resource. Notably, Park *et al.* (2017) propose to use double layers of cells to overcome discontinuity of the deformation between adjacent cells due to irregular distribution of the point cloud.

4.2 *Surface deficiency identification*

Surface deficiencies can appear as change in geometry or in surface texture. The first type of damage can be spalling, scaling, disintegration, surface loss, and crack, while the second one is probably water bleeding, chemical attack, corrosion, or crusting. For the former type, the local surface at the damage region is significantly different from the undamaged surface. The difference can be recognized through features of the surface (e.g. a normal vector, curvature, smoothness, or discontinuity), which are determined from a point cloud. For the latter type, the surface texture–based damage, texture information can be obtained from attributes of individual data points, which can be backscatter strength (or intensity values) derived from a laser scanning unit, or red, green, and blue (RGB) colors obtained by mapping images onto a point cloud.

4.2.1 Spalling, scaling, disintegration, and surface loss

At the location of damage, (e.g. spalling), local surfaces at those locations appear significantly different from other surrounding regions. Thus, the local surface features can be indicators for extracting damage. The surface features at a given point $p_i \in \mathbf{P}$ in R^3 can be derived from analysis of its nearest neighbor points, **q**, which is also known as the point features (Weinmann *et al.*, 2014; Pauly *et al.*, 2003).

To identify the location of mass loss of a concrete surface, Teza *et al.* (2009) used a distribution of Gaussian curvature in a subarea. The proposed method first subdivided the point cloud into subareas with a defined size. Next, for each point in a subarea, parabolic fitting was employed to determine the principal curvatures of the surface through its neighbor points, as derived from a range search with a searching radius 2 to 3 times of a sampling step. The authors demonstrated that damaged areas had standard deviations (s) of Gaussian curvature distribution that differed significantly from undamaged areas. For this, both the median and STD of the Gaussian curvature of undamaged areas were considered as thresholds. The method was applied to a concrete bridge on the A16 Napples-Canosa motorway in the Avellino province in Italy. The bridge was scanned with a Riegl LMS Z-420i. Subareas were approximately 10 times the sampling step, resulting in damage detection of 84% and 87% for a column and a girder, respectively. However, the method failed if noise in the data exceeded 0.8 cm to 1.0 cm.

Interestingly, Liu and Chen (2013) reported that the surface curvatures are more sensitive to surface roughness, making it harder to establish a threshold to detect damage. The authors proposed using distances from data points to the flat reference plane and gradients of the points. This is based on assumptions that undamaged surfaces are flat. These reference planes were used to establish an average distance and point gradient against which to identify defects. Next, the selected surface was divided into grids, and if more than 50% of the number of the points in the grid section are defective points, then that portion of the grid is considered damaged. The area and volume of each damaged section are computed based on the four boundary points of the grid and its average depth. Although the method can detect the defective area with an error less than about 1%, the study showed that establishing the distance threshold is difficult. However, the gradient-based threshold is suitable for identifying defects at edges of the structures.

A volume change due to spalled concrete can be determined by comparing a cross-section of damage area and ones of a nondamaged area, as was done by Olsen *et al.* (Olsen *et al.*, 2010) who sub-divided a structural component into multiple 10 mm slices. A polygon loop wrap was used to fit through a point cloud of each section after manually removing mixed data points. The spalled volume was then computed by using an area of the fitting polygon and the interval thickness compared to the undamaged component. The drawback of this approach is that the volume of the undamaged surface must be known. In contrast, Mizoguchi *et al.* (2013) used a region-growing method to extract point clouds of undamaged regions for estimating a scaling depth. In that implementation, the initial seeding points were manually selected, and the points were merging if the distance from the points to the fitting surface of the current region was less than the threshold. A primitive surface of the final region was created by using a least squares method. The scaling depth was the signed distance from data points belonging to the damaged regions to the primitive surface. The proposed method was implemented to determine the total scaling depth of piers of the old bridge scanned by RIEGL VZ-400. The experimental work generated an average absolute error of 2.28 mm (s = 3.60 mm) when comparing to manual measurement.

In all of these methods, success depends on user-defined thresholds. To overcome this limitation, the following example introduces an automatic approach by the authors to identify damage locations of a girder's segment containing spalling (Figure 12.10a). The damaged region was captured by a Leica ScanStation P20 with a sampling step of 6.1 mm at the measurement range of 10 m, from a distance approximately 2.2 m from the damaged region. Generally, in structures, undamaged surfaces are a large subset and mostly smooth planes. As such, identifying the undamaged surfaces is easier than identifying the damaged surfaces. Thus, the proposed method starts by segmenting the point cloud of the undamaged surfaces using existing segmentation methods (e.g. region growing–based methods or RanSAC). Herein, the voxel-based region growing method as proposed by Vo *et al.* (2015) was employed to segment the point cloud of a girder's section (Figure 12.10b), in which the input parameters were as follows: voxel size of 20 mm, angle threshold of 5° and residual threshold of 5 mm. Results of the segmentation are shown in Figure 12.10c, where black points (unsegmented points not belonging to any segment) are possibly the damaged regions, while other colors describe the undamaged surfaces.

The unsegmented points can be the point cloud of the damaged regions and others (e.g. mixed pixels or small undamaged patches of structures). As such, further filtering of the unsegmented points is needed. For this, the density-based clustering techniques DBSCAN (Ester *et al.*, 1996) was employed to split the un-segmentation points into multiple clusters, in which e and minPts were empirically set equal as 5 times the sampling step and 16 points, respectively. The clustering process divides the unsegmented points (Figure 12.10c) into two clusters (rendering in yellow and red in Figure 12.10d). Next, the filtering process classified these clusters as real and unread damage.

A cluster is considered as damage if it satisfies the four conditions as per Equation 8. The first condition is that an equivalent diameter (r_{eqv}) of the cluster should be equal or larger than a predefined threshold (r_0), which is herein chosen r_0 = 152 mm corresponding to the smallest spall considered as a fair condition as per AASHTO (American Association of State Highway and Transportation Officials, 2011). This condition aimed to eliminate small clusters. The equivalent diameter can be determined from an area of the cluster projected onto its planar fitting surface. The second condition is based on there being a homogeneous data point distribution, which implies that the point density of the damaged surface is mostly similar to ones of the undamaged surfaces. This condition is aimed to remove clusters in the form of noise or mixed data points. The third condition

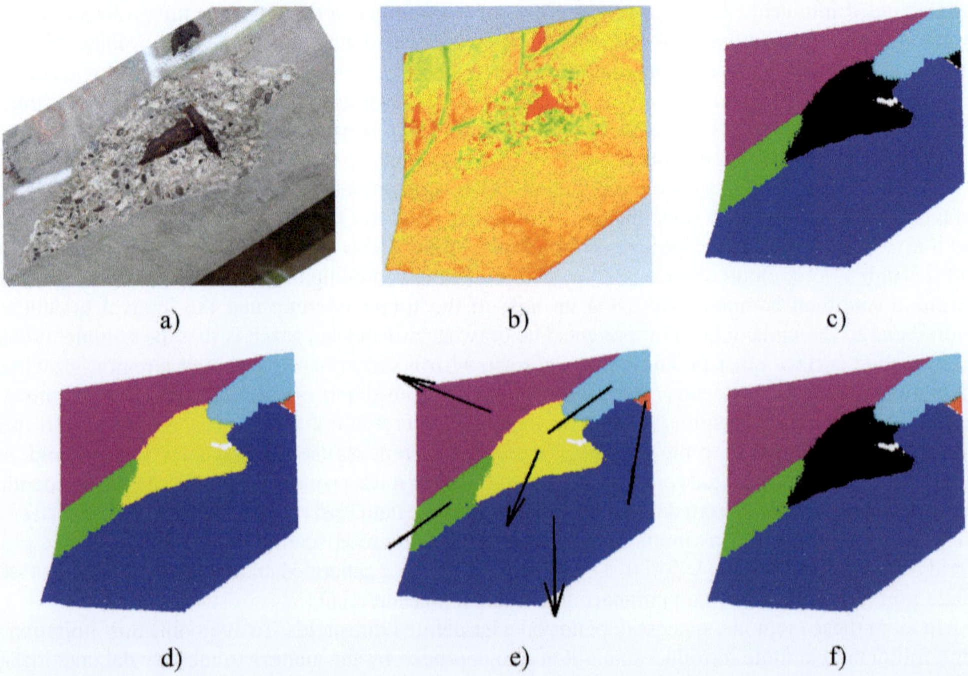

Figure 12.10. A hierarchical method for identifying a point cloud of damaged surfaces: (a) photo of the girder's section containing spalling; (b) point cloud of the surface; (c) resulting segmentation; (d) resulting clustering of unsegmented points; (e) normal vectors of all segments; (f) resulted damage detection.

is based on an observation that the damaged surface is often rougher than the undamaged surface. The roughness of the clusters/surfaces can be described by the residual, which is the root mean square of orthogonal distances from a point cloud to its fitting surface. The fourth condition checks if the damage surface located in a different plane away to the undamaged surface. That implies that the sign is larger than zeros to ensure the damaged surface is further away from the scanner than the undamaged surface (Figure 12.10f). In the case of rust-jacking for steel, the sign condition is likely to be reversed due to the expansive nature of the damage mechanism.

$$\begin{aligned} &Condition1: r_{eqv} \geq r_0 \\ &Condition2: s_c \geq s_s - t\sigma \\ &Condition3: R_c \geq R_s \\ &Condition4: sign\left(\overrightarrow{p_{0s}p_0}, n_s\right) \geq 0 \end{aligned} \tag{8}$$

where, r_{eqv} is an equivalent diameter of the cluster while r_0 is the threshold; s_c is the average point density of the cluster; s_s is the average point density of the surfaces connecting to the cluster, σ is a standard deviation of point density of the surfaces; t is set equal to 1; R_c and R_s are, respectively, the roughness of the cluster and undamaged surfaces adjacent to the cluster; p_0 and p_{0s} are the centroid of the cluster and the projection of p_0 onto the adjacent undamaged surfaces, respectively; and n_s is the normal vector of the undamaged surfaces adjacent to the considered damaged surface.

Finally, the filtering conditions are applied for each cluster derived from the unsegmented points (Figure 12.10d). Results of the filtering are shown in Figure 12.10e, where the black points illustrate the damage cluster. Once the damaged clusters and connecting undamaged surfaces are identified, methods proposed by Olsen *et al.* (2010) and Mizoguchi *et al.* (2013) can be used to calculate spalling depth or volume loss.

In another application, Truong-Hong and Laefer (2015a) proposed a 2D grid combining an angle criterion to determine the areas of holes on the surfaces of metal structures due to corrosion. The method employed an octree-based region-growing method (Vo *et al.*, 2015) to extract a point cloud of a damaged surface containing the holes. Next, the point cloud was projected on its fitting surface and then aligned to a plane parallel to the xy-plane in the global coordinate system. A 2D grid with a predefined cell size was used to divide a bounded, 2D region of the aligned point cloud into uniform cells. The cell size was set equal to 3 times of a sampling step, as the hole was assumed to appear at locations where at least two consecutive data points in x and y directions could not be achieved. Cells in the 2D grid were classified as "*full*" if the cells contained at least one data point; otherwise, the cell was labelled as "*empty*". Empty cell clusters inside the surface were assumed to represent to the holes. Next, the angle criterion (Truong-Hong *et al.*, 2012b) was used to determine data points on a boundary of the hole from the data points possessed by full cells connected to the empty cell cluster. Finally, a polygon created from a set of the boundary points was used to calculate the hole's area, which was considered as the surface loss area. Figure 12.11 illustrates the proposed method showing the absolute and relative errors of the detected surface loss area of less than 60 mm^2 and 10%, respectively, when compared to manual measurement.

4.2.2 Chemical attack, leading water bleed, and corrosion

At the locations subjected to water bleed, chemical attack, or corrosion, significant geometric changes may not be presented. A possible solution in extracting such damage is to generate ortho-images from a high dense point cloud of the surface (Kalenjuk and Lienhart, 2017). Subsequently,

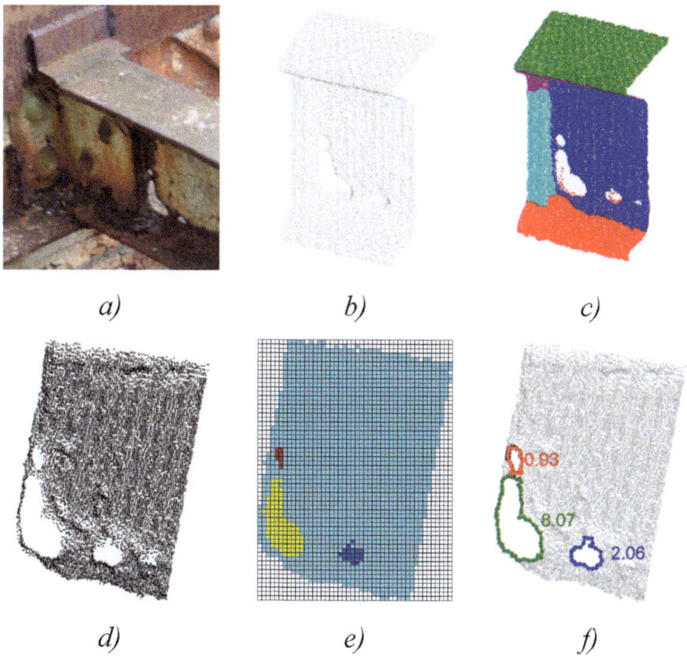

Figure 12.11. A combination of 2D grid and an angle criterion to determine surface loss area: (a) photo of the surface loss due to corrosion; (b) input data of the section of interest; (c) segments of the input data; (d) data points of a surface loss projected on a xy-plane; (e) 2D cell grid and detected holes; (f) boundary points and hole areas (unit in cm^2).

image processing or machine learning methods could be employed to extract damaged regions. For example, Shen *et al.* (2013) proposed a combination of a Fourier transform method and color image processing to identify rusted areas. The method started to determine whether the input image containing contained rusts. The K-means algorithm (Lloyd, 1982) was implemented to distinguish rust defects from other defect types. An accuracy of up to 100% in identifying the rust from 25 steel bridge images was achieved, showing the effectiveness in machine learning methods for identifying such damage.

In the case of working directly with orthoimages generating from attributes of a point cloud, González-Jorge *et al.* (2012) used K-means and Fuzzy C-means for determining biological crusts on concrete surfaces. The study was tested on an orthoimage generated from the intensity values of a point cloud of two pillars of Ribeirino Bridge in Ourence, Spain, acquired by a Riegl LMS Z-390i and Trimble GX200 with a sampling step of 5 mm. K-means and fuzzy C-means algorithms built in Matlab were used to classify the intensity orthoimages of three different regions of the pillars independently, which involves two classes (biological crusts and concrete) and three classes (biological crusts, water, and concrete). Results from unsupervised machine learning methods were compared to the ground truth extracted with support of RGB images. The comparison showed that damaged areas were more easily extractable from the higher-resolution REIGL scan data and resulted in no more than a 20% error. Similarly, González-Jorge *et al.* (2013) used the K-means algorithm for efflorescence detection using image-based intensity acquired from the MLS system based on regions of lower intensity values.

Application of a supervised machine learning method in the form of support vector machine (SVM) was used by this chapter's authors to extract damaged surfaces caused by water bleeding of sides of a pier and the superstructures of the Blessington bridge connected Kilbride Rd. to Lake Dr Rd., cross over Liffey Lake in County Wicklow, Ireland, was selected as a case study (Figure 12.12a and 12.12b). The bridge part was scanned with a Leica ScanStation P20 from approximately 63.0 m away from the part with the sampling step of 1.6 mm at a measurement range of 10 m. An external camera captured RGB images of the structure, and RGB colors for individual data points were generated within point cloud processing software, likely Trimble RealWorks or Leica Cyclone. The point clouds then had 3D coordinates and attributes including intensity and RGB colors.

To develop a model to predict damaged areas, point clouds of the undamaged and damaged regions were manually extracted to be used as the undamaged and damaged data sets. In a real structure, as the damaged regions are often only a small portion of an entire structure, the number of undamaged data sets is much more plentiful, resulting in an imbalanced training set. A highly skewed data set can cause misclassification and cost-sensitive learning (Weiss *et al.*, 2007). As such, the undamaged data set is randomly split into multiple folds having the number of the points nearly equal to the damaged data set. One fold of the undamaged data sets is then merged into the damaged data set to create the training data.

Since a geometric surface of a damaged region is no different from an undamaged one, coordinates or any geometric properties of the point cloud cannot be good features for the classification. As such, the SVM-based classifier would be done based on features created from attributes of the point cloud, which are intensity and RGB colors. In this example, four different feature combinations are to investigate an importance of features in damage classification, which are (1) Model 1 used raw intensity and RGB colors, (2) Model 2 involves normalized intensity and RGB colors, (3) Model 3 is similar to Model 1 plus two other color spaces called HSV (H, S, and V colors) and Lab (L, La and Lb colors), and (4) Model 4 is similar to Case 3 but normalize intensity used. According to Höfle and Pfeifer (2007), the intensity value depends on a reference range (R_s), a real measurement range (R), an incident angle (α) and atmospheric conditions (η_{atm}), which can be expressed in Equation 9. The reference range R_s was selected equal to 10 m, which is the standard range measurement description the scanning resolution. The incident angle α is the angle between the laser beam and a normal vector of the surface, which can be estimated by using PCA based on the neighbor points, in which the number of the neighbor points by 20 is used in this study.

Moreover, since a bridge section containing surface damage was scanned in a short time (approximately 5 minutes), the atmospheric attenuation can be neglected. All features were normalized into the range of [0, 1] to avoid ill conditions during constructing predicted models.

$$I_{norm} = I\frac{R^2}{R_s^2}\frac{1}{\eta_{atm}cos\alpha} \tag{9}$$

A binary SVM machine is trained to classify the points as the damaged and undamaged classes. To determine the best parameters for constructing a model, the algorithm hyperparameters are chosen using cross-validation and a random search method (Bergstra and Bengio, 2012). This process is splitting the training data set into five sub-data sets and training them with separate SVMs. The process can be repeated to avoid sampling bias. To classify a new point, every SVM is asked to make its prediction, then a majority voting happens to determine the final prediction. Results showed the best parameter as follows: the kernel function is linear, and a regularization parameter, C, is 100. They are used to construct the predictive model.

Results of damage classifications based on SVM were illustrated in Figure 12.12. To evaluate classification accuracy, point clouds in damaged (water bleed) and undamaged regions were manually extracted from the data points of parts of the bridge which are considered as the ground truth. A point-based evaluation strategy was used to evaluate an prediction accuracy by comparing to the ground truth (Truong-Hong and Laefer, 2015b). Although different feature combinations were used, the classification accuracy (ACC) of these cases is mostly the same over 82%, and there was a small difference between cases, which was less than 0.03% (Case 2 vs. 3). Moreover, when the normalized intensity was used, the prediction is given better results, for example, 84.7% (Case 1) vs. 85.4% (Case 2). The demonstration shows a supervised machine learning method, likely SVM, is a prominent solution to extract damaged surfaces directly from attributes of the point cloud.

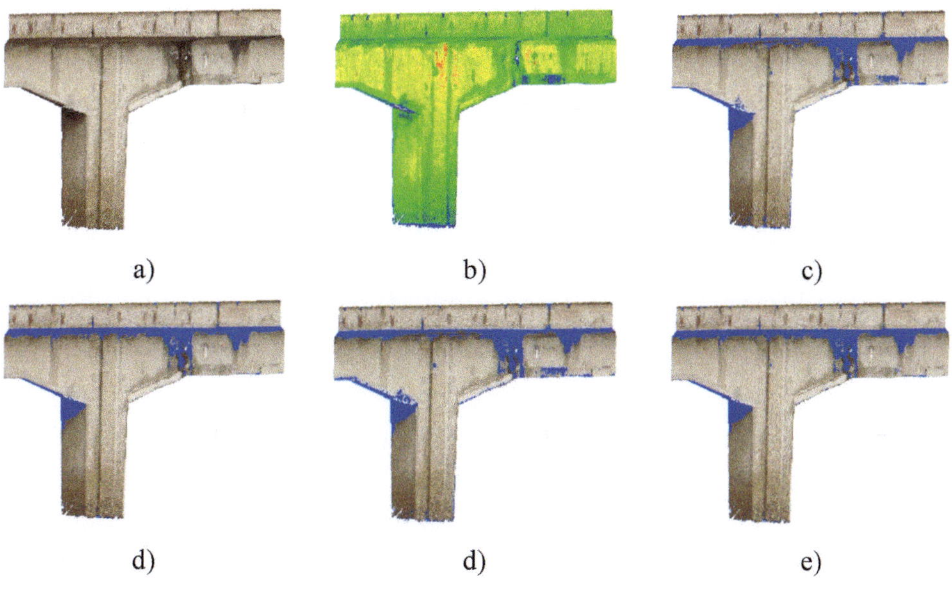

Figure 12.12. Illustration of damage prediction by SVM: (a) a structure rending by RGB; (b) a point cloud of the bridge parts rending by intensity; (c) model 1: ACC = 0.847; (d) model 2: ACC = 0.854; (e) Model 3: ACC = 0.822; (f) model 4: ACC = 0.841.

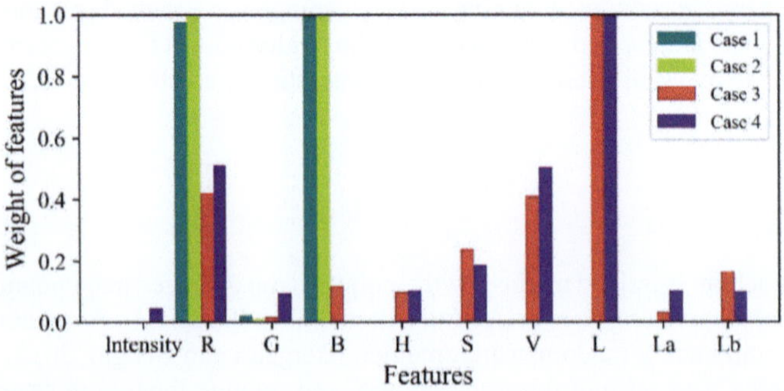

Figure 12.13. Feature importance.

Importantly, the weight of features in constructing a model for damage detection shows that the intensity value is high sensitivity (Figure 12.13). Additionally, the colors (R, B, S, V, L, and Lb) are top important features for this task. There is a small difference of accuracy classification derived from the classifier model using a raw intensity against that based on normalized intensity, for example Model 1 vs. Model 3 and Model 3 vs. Model 4.

4.2.3 Crack detection

Laefer *et al.* (2014) established fundamental mathematics to determine the minimum crack width detectable by laser scanning, and addressed that a minimum visible crack can be detectable if there are at least two reflected laser beams return the scanner from a back face of the crack. That implies the smallest crack width to be detected should be larger than 2 times of a sampling step. The minimum detectable crack width was expressed in Equation 10, when an elevation of the crack cross-section differs level to the scanner's mount.

$$w \geq (D' + d')\tan(\theta + 2\Delta\theta) - H \tag{10}$$

where notations D', d', H are shown in Figure 12.14, while θ and Δθ are respectively a scanning angle and an interval scanning angle.

However, as data points on a surface of the structure component lie in different planes from the back face of the crack, the crack width should be a distance between two data points located on the left and right edges of the crack after eliminating the data points of the back face (Figure 12.14). The minimum detectable crack width would be expressed in Equation 11.

$$w = D\left[\tan(\theta + 3\Delta\theta) - tan\theta\right] \tag{11}$$

A series of laboratory tests have been conducted to determine a minimum detectable crack width (CW), various vertical crack widths (1.10 mm, 3.14 mm, 5.33 mm, and 7.22 mm), range measurements (5.0 m, 7.5 m, 10.0 m, and 12.5 m), and scanning angles (0°, 15°, and 30°). The crack was scanned by a Trimble GS200 (Trimble Navigation Limited, 2003) having horizontal and vertical angular by 0.0018° and 0.0009°. The crack width was manually determined as a distance between the data points on the crack's edges. The experimental test showed the minimum absolute errors of the crack width-based TLS is 0.43 mm (CW = 1.10 mm), 0.81 mm (CW = 3.14 mm), 0.02 mm (CW = 5.33 mm), and 0.31 mm (CW = 7.22 mm) when the crack was measured from an orthogonal distance by 5.0 m with the scan angle by 0° except for CW = 7.22 mm measured under

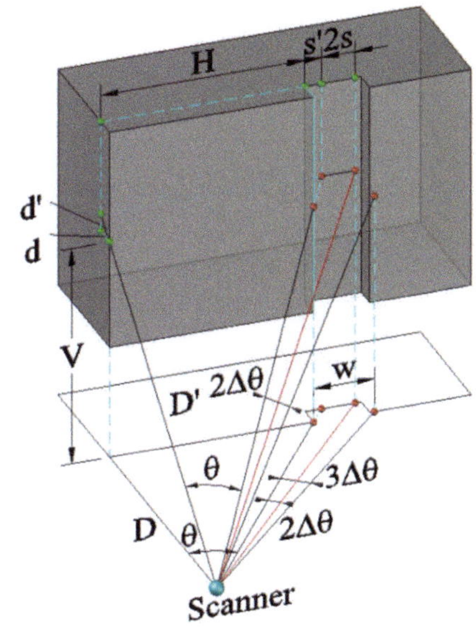

Figure 12.14. Diagram of laser beams and crack different level from the scanner's mount.

the angle by 30°. Results showed that the best performance of crack width can be achieved from laser scanning points when the crack width is larger than 5 mm to be scanned with the scanning range and angle respectively less than 7.5 m and 15°. Interestingly, the study also finds that the orthogonal scan distance impacts an error of the detected crack width significantly larger than the scanning angle.

The study concludes the laser beam's footprint and the sampling step on a target surface are two parameters dominated the smallest crack width detectable. Both parameters are proportion of the orthogonal distance (D) from the scanner to the surface and the scanning angle (θ) or incidence angle. When the footprint is enlarged greater than the sampling step, the laser's footprints of two consecutive pulses can be overlapped, which leads to benefit of the correlated sampling does not fully deploy (Lichti and Jamtsho, 2006). Moreover, ideally, by using the assumption that the minimum detectable crack width must be equal or greater than 2 times of the sampling step, the data points on the crack edges can be extracted. However, in practice, noise and mixed data points cannot be inevitable because of enlarging footprint of the laser pulse and incidence angle (Soudarissanane *et al.*, 2011). Thus, the small crack width can be difficult to automatically determine from laser scanning points (Truong-Hong *et al.*, 2016).

Truong-Hong *et al.* (2016) suggested using RGB colors of the point cloud support to identify crack's characteristics (e.g. length and width). That is because the crack is hard to identify directly from the point cloud. The RGB colors can be obtained by mapping a high-resolution image taken from an external camera onto the point cloud. The authors developed a tool to allow users to measure length and width of the crack by measuring distance between data points on the crack edges. The process of a crack's width and length determination was illustrated in Figure 12.15.

In an attempt to develop an automatic method for crack identification, Valença *et al.* (2017) combined laser scanning data and image processing. The laser scanning data were used to (i) identify discontinuity of the surface as a crack location where the gap between adjacent data points was larger than others and (ii) to convert the image in an image model to a real model, which

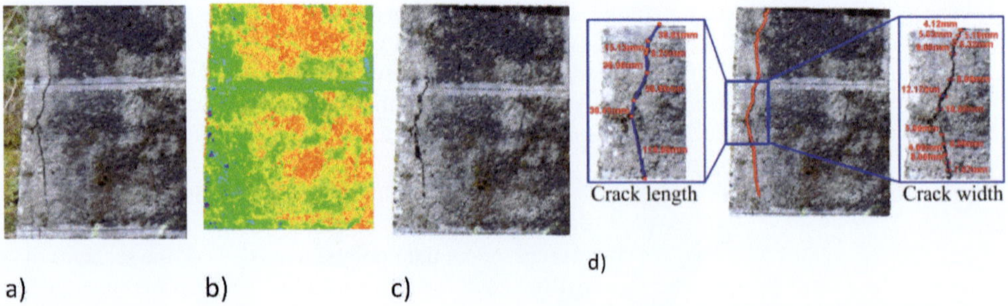

Figure 12.15. Crack detection by incorporating high-resolution image and point cloud: (a) photo of a crack; (b) point cloud rendering by intensity; (c) point cloud rendering by RGB from the image; (d) identification of crack's width and length

allows one to obtain crack characteristics in real coordinates instead of in image coordinates. The proposed method first removed data points inside a crack in a region of interest (ROI) by using intensity filtering. Second, the data points of the crack's face were continuously extracted as the data points having the orthogonal distance larger than 2s of the orthogonal distances of the point cloud in ROI to its fitting surface determined by using PCA. Unfortunately, when TLS captures the crack surface in the north abutment of IC2 viaduct at Rio Maior and EN114, Portugal, from a long distance, it produced a large sampling step of the point cloud on the surface, which leads to us being unable to identify the surface discontinuity. Thus, the crack characteristics (i.e. length, width, and location) were mainly identified from the image process technique (Valença *et al.*, 2012). Next, the image was orthorectified by the point cloud of the ROI, which allows one to retrieve the crack characteristics in a real coordinate system. Results showed that the error of crack width detection is less than 10%.

5 LESSONS AND CHALLENGES

Current TLS units have the flexibility to acquire 3D topographic and texture information of structures' surfaces quickly and accurately. Incomplete data often occur in the data captures because of (i) complexity of the bridge structure causing self-shadowing, as well as obstructions from the terrain surrounding the bridge and (ii) limited access to set up scan stations to capture all parts of the bridge. In such scenarios, the number of scan stations must be increased to minimize missing data. Moreover, additional equipment may also be required to set up the scanner. For example, for a long span bridge over a river, the scanner may have to be set up on a boat to scan parts of the bridge that are not visible or cannot capture from the river's banks. Such solutions-based TLS to maximum data coverage can increase the cost of data collection. An alternative method is to use a drone with an integrated camera, providing flexible and low-cost data collection. However, critical issues can occur: (i) shadow, image noise, and environmental conditions can affect the quality of images for a point cloud generation, (ii) errors in a point cloud generated from images captured from the drone may exceed the error budget, and (iii) a drone can crash into the structures when a special flight path requires the drone flights close to the structure to capture the surface with high resolution. Recent drones with laser sensors can overcome the first two issues. Finally, irrespective of the complementary methods used to enrich the data, incomplete data is inevitable, which can cause failures in reconstructing 3D geometric models and/or reporting quantitative deficiencies in the existing bridge. Thus, acquiring sufficient data of the structures' surfaces is an important key for success of bridge inspection-based laser scanning.

Surface damage (i.e. spalling/scaling, corrosion, ruts, or cracks) appeared at different scales, shapes, and orientations. High accuracy in documentation can occur if the structure is captured with sufficient quantity and quality of data. In practice, the scanner is often located away from the bridge, which may prevent the inspector/surveyor from identifying damage location. Furthermore, the damage may not be readily identifiable directly from a point cloud collected for geometric modelling. In that case, subscans of the damaged region may not be captured. One approach to this is to supplement high-resolution images to support the inspectors/surveyors to identify the damage location.

As TLS can capture a structure's surfaces as spatial data points, real edges of the surfaces may not be acquired when they are located between two consecutive scan lines. Moreover, mixed pixels are inevitable when the laser beam hits the edge. Such errors especially affect the accuracy of edges of representing surface edges, which can lead to the deviation of incorrect dimensions. Accurate edges can be obtained through an intersection of two fitting surfaces.

Adverse weather conditions, for example foggy and windy, can also cause noise in data and strongly impact the backscatter energy of laser beams and colors of the point cloud. Wind or traffic vibration can introduce error in TLS-based data capture. These problems reduce the quality of the point cloud and require significant effort to remove low-quality data points and can consequently cause difficulty in generating accurate 3D models and/or extracting damage.

One of the big challenges in 3D geometric modelling is to efficiently extract a point cloud of each individual structural component from the massive point cloud of the bridge because of a complexity of the structures in size, shape, and orientation. Recent efforts have proposed an automatic method for this task, but it is limited to simple slab bridges (Lu and Brilakis, 2017). In popular geometric modelling approaches, the structures are modelled as planar surfaces or by extruding polygons of cross-sections, which are fitted through a point cloud of undamaged parts of the structures. A 3D geometric model may not include an appearance of volumetric loss of the structure and a local deformation of the structure. Thus, the 3D geometric model can be the embodiment of the structure when surface meshes are directly generated from a point cloud (Salman *et al.*, 2010). However, this approach requires a large memory to store the mesh and visualization, which may not be applicable for bridge information modelling and bridge management.

While the hierarchical method was demonstrated to be successful in automatically detecting surface defects (e.g. spalling, scaling, or surface loss), segmentation of undamaged surfaces is an important key to obtain accurately quantitative damage. The segmentation method often requires users to tune parameters to obtain expected results (Vo *et al.*, 2015). However, in practice, a decision of resulted segments was often based on visualization evaluation, which imposes subjective and subsequently gives low accuracy of damage measurement. Development of a complex algorithm improving the segmentation process and decision-making could lead to more accuracy of location and quantity of damage. An alternative solution, a machine learning method (e.g. a deep learning method), can be used for detecting surface defects. However, training such models may require a huge amount of data and dealing with imbalance in training sets, because damaged regions are often a small portion compared to undamaged regions in the structure. Finally, noise data can be problematic for computing correct values of the features for training.

Demonstration of extracting surface defects due to chemical attack, corrosion, and water bleed through texture information of a point cloud proves machine learning–based methods can detect such defects with high accuracy. However, attributes of the point cloud including backscatter energy (or intensity value) and RGB colors depend on various factors, for example, a laser scanning unit (e.g. a wavelength or divergence of a laser beam) and a camera, properties of an object's surface (e.g. material, color, and roughness of the surface), scanning range and incident angle, and atmospheric conditions (e.g. humidity, temperature, and wind). As such, values of intensity and RGB colors of damaged surfaces vary in a wide range, since the structure's surface is to be scanned in different periods of time. Clearly, these values also change in different structures.

As such, the machine learning methods can give reliable results once a training data set is large enough and sufficient features are to be prepared. That implies a numerous surface damage to be captured, and feature engineering may require huge labor time to label the point cloud as either "damaged" or "nondamaged". A robust method should be developed to obtain invariant values of the features from various sources of laser scanning data. As only two main features are intensity and RGB colors of the point cloud derived directly from the TLS, these features may not be enough for constructing a good generalization model. Additional features used in the training model can improve the performance in term of classification accuracy of the predicted model. Automated feature learning can be achieved by using a deep learning method, but this learning method requires huge training data.

Finally, the biggest challenge of using TLS for bridge inspection is to identify cracks in structures. The crack length can be detected directly from the point cloud where the point cloud of the crack appears in slightly different geometry and texture compared to the point cloud surrounding the crack (Valença *et al.*, 2017). Recent work can determine the crack length with a relative error of no more 3% when the minimum crack width by 6 mm scanned in a range 7.5 m and an incident angle less than 10° (Cabaleiro *et al.*, 2017). However, the crack width is still difficulty as quantity and quality of the laser scanning data insufficient representation of crack edges. Several laboratory tests have concluded the sapling step, noise level, and positional accuracy of the point cloud play out critical roles (Cabaleiro *et al.*, 2017; Laefer *et al.*, 2014; Anil *et al.*, 2013). Anil *et al.* (2013) stated the smallest diagonal crack width by 1.25 mm cannot be detected from a point cloud acquired from Leica ScanStation C10 with the sampling step by 0.5 mm, which exceeds the crack width corresponding to the moderate crack in reinforced concrete structure (American Association of State Highway and Transportation Officials, 2011). Success of crack detection from image processing [e.g. (Barazzetti and Scaioni, 2009)] can offer a hybrid solution to identify crack characteristics (width, length, and orientation) by mapping the high-resolution image to a point cloud (Truong-Hong *et al.*, 2016; Valença *et al.*, 2017). That solution allows one to determine the crack characteristics in a real scale instead of those in an image scale and provide geodetic information of the crack for bridge management. This work required an automatic, complex algorithm allowing the user to align and scale the images with the point cloud.

6 CONCLUSIONS

In summary, laser scanning technology and state of the art of a point cloud processing can overcome many of the drawbacks of manual bridge inspection. Key advantages involve the collection of an objective data set in a truncated schedule that is able to support the creation of both bridge inventories and structural defects for both assessment and management. To acquire a point cloud of sufficient quantity and quality for geometric modelling and surface deficiency detection remains a challenge with TLS. Coming years are likely to see a more integrated use of TLS with hand-held, MLS, and those that are UAV mounted to gain a more comprehensive documentation of bridge structures. As part of this, a real-time system should be introduced to check the quantity and quality of the data on site to avoid having to rescan a bridge due to incomplete data sets. Finally, robust, efficient algorithms need to be further developed for greater abilities to automatically identify various types of damages. Machine learning–based approaches are emerging as alternative solutions to object-based methods, but obtaining the required large training data sets still poses its own set of challenges.

REFERENCES

American Association of State Highway and Transportation Officials. (2011) *AASHTO Guide Manual for Bridge Element Inspection.* 1st ed. American Association of State Highway and Transportation Officials.

Anil, E.B., Akinci, B., Garrett, J.H. & Kurc, O. (2013) Characterization of laser scanners for detecting cracks for post-earthquake damage inspection. *International Symposium on Automation and Robotics in Construction and Mining (ISARC), 11–15 August*. Montreal, QC, Canada.

Anil, E.B., Sunnam, R. & Akinci, B. (2012) Challenges of identifying steel sections for the generation of as-is BIMs from laser scan data. *Gerontechnology*, 11(2), 317. https://doi.org/10.4017/gt.2012.11.02.266.00.

Ayala, D., Brunet, P., Juan, R. & Navazo, I. (1985) Object representation by means of nonminimal division quadtrees and octrees. *ACM Transaction on Graphics*, 4(1), 41–59. https://doi.org/10.1145/3973.3975.

Bae, K.-H. & Lichti, D.D. (2008) A method for automated registration of unorganised point clouds. *ISPRS Journal of Photogrammetry and Remote Sensing*, 63(1), 36–54.

Barazzetti, L. & Scaioni, M. (2009) Crack measurement: Development, testing and applications of an automatic image-based algorithm. *ISPRS Journal of Photogrammetry and Remote Sensing*, 64(3), 285–296.

Bergstra, J. & Bengio, Y. (February 2012) Random search for hyper-parameter optimization. *Journal of Machine Learning Research*, 13, 281–305.

Cabaleiro, M., Lindenbergh, R., Gard, W.F., Arias, P. & van de Kuilen, J.W.G. (2017) Algorithm for automatic detection and analysis of cracks in timber beams from LiDAR data. *Construction and Building Materials*, 130, 41–53. https://doi.org/10.1016/j.conbuildmat.2016.11.032.

Ester, M., Kriegel, H.-P., Sander, J. & Xu, X. (1996) A density-based algorithm for discovering clusters in large spatial databases with noise. *Second International Conference on Knowledge Discovery and Data Mining, 2–4 August*. Portland, Oregon, USA.

Fujino, Y. & Abe, M. (2003) Structural health monitoring in civil infrastructures–research activities at the bridge and structure laboratory of the university of Tokyo. *Struct Health Monitor Intell Infrastruct*, 1, 39–50.

Gibson, S.F.F. (1995) Beyond volume rendering: Visualization, haptic exploration, and physical modeling of voxel-based objects. *Visualization in Scientific Computing'95, 3–5 May*, Chia, Italy.

Goldman, R. (2009) *An Integrated Introduction to Computer Graphics and Geometric Modeling*. CRC Press, Boca Raton, FL, USA.

Gong, J., Zhou, H., Gordon, C. & Jalayer, M. (2012) Mobile terrestrial laser scanning for highway inventory data collection. *Computing in Civil Engineering (2012)*, 545–552.

González-Jorge, H., Gonzalez-Aguilera, D., Rodriguez-Gonzalvez, P. & Arias, P. (2012) Monitoring biological crusts in civil engineering structures using intensity data from terrestrial laser scanners. *Construction and Building Materials*, 31, 119–128. https://doi.org/10.1016/j.conbuildmat.2011.12.053.

González-Jorge, H., Puente, I., Riveiro, B., Martínez-Sánchez, J. & Arias, P. (2013) Automatic segmentation of road overpasses and detection of mortar efflorescence using mobile LiDAR data. *Optics & Laser Technology*, 54, 353–361. https://doi.org/10.1016/j.optlastec.2013.06.023.

Hinks, T., Carr, H. & Laefer, D.F. (2009) Flight optimization algorithms for aerial LiDAR capture for urban infrastructure model generation. *Journal of Computing in Civil Engineering*, 23(6), 330–339.

Hinks, T., Carr, H., Truong-Hong, L. & Laefer, D.F. (2012) Point cloud data conversion into solid models via point-based voxelization. *Journal of Surveying Engineering*, 139(2), 72–83.

Höfle, B. & Pfeifer, N. (2007) Correction of laser scanning intensity data: Data and model-driven approaches. *ISPRS Journal of Photogrammetry and Remote Sensing*, 62(6), 415–433.

Kalenjuk, S. & Lienhart, W. (2017) Automated surface documentation of large water dams using image and scan data of modern total stations. *FIG Working Week 2017: Surveying the World of Tomorrow-From Digitalisation to Augmented Reality, 29 May – 2 June*. Helsinki, Finland.

Kang, Z., Li, J., Zhang, L., Zhao, Q. & Zlatanova, S. (2009) Automatic registration of terrestrial laser scanning point clouds using panoramic reflectance images. *Sensors*, 9(4), 2621–2646.

Laefer, D.F. (2013) Planning infrastructure documentation with aerial laser scanning. *IABSE Symposium Report, 6–8 May*, Rotterdam, The Netherlands.

Laefer, D.F., Fitzgerald, M., Maloney, E.M., Coyne, D., Lennon, D. & Morrish, S.W. (2009) Lateral image degradation in terrestrial laser scanning. *Structural Engineering International*, 19(2), 184–189.

Laefer, D.F. & Truong-Hong, L. (2017) Toward automatic generation of 3D steel structures for building information modelling. *Automation in Construction*, 74, 66–77.

Laefer, D.F., Truong-Hong, L., Carr, H. & Singh, M. (2014) Crack detection limits in unit based masonry with terrestrial laser scanning. *NDT & E International*, 62, 66–76.

Leica Geosystems. (2017) *Leica Cyclone V9.1.3*. Available from: http://hds.leica-geosystems.com/en/Leica-Cyclone_6515.htm [accessed 14th February 2018].

Lichti, D.D., Gordon, S.J., Stewart, M.P., Franke, J., Tsakiri M. (2002) Comparison of digital photogrammetry and laser scanning. *Proceedings CIPA WG 6 International Workshop on Scanning Cultural Heritage Recording*, Corfu, Greece, 1–2 September 2002, pp. 39–44.

Lichti, D.D. & Jamtsho, S. (2006) Angular resolution of terrestrial laser scanners. *The Photogrammetric Record*, 21(114), 141–160.

Liu, W. & Chen, S.-E. (2013) Reliability analysis of bridge evaluations based on 3D light detection and ranging data. *Structural Control and Health Monitoring*, 20(12), 1397–1409.

Liu, W., Chen, S.-E. & Hasuer, E. (2012) Bridge clearance evaluation based on terrestrial LIDAR scan. *Journal of Performance of Constructed Facilities*, 26(4), 469–477. https://doi.org/10.1061/(ASCE)CF.1943-5509.0000208.

Lloyd, S. (1982) Least squares quantization in PCM. *IEEE Transactions on Information Theory*, 28(2), 129–137.

Lovas, T., Barsi, A., Detrekoi, A., Dunai, L., Csak, Z., Polgar, A., . . . Szocs, K. (2008) Terrestrial laser scanning in deformation measurements of structures. *ISPRS Congress, 3–11 July*. Beijing, China.

Lu, R. & Brilakis, I. (2017) Recursive segmentation for as-is bridge information modelling. *Joint Conference on Computing in Construction (JC3), July 4–7*. Heraklion, Greece.

Mailhot, G. & Busuioc, S. (2006) Application of long range 3D laser scanning for remote sensing and monitoring of complex bridge structures. *7th International Conference on Short and Medium Span Bridges, 23–16 August*. Montreal, Canada.

Metni, N. & Hamel, T. (2007) A UAV for bridge inspection: Visual servoing control law with orientation limits. *Automation in Construction*, 17(1), 3–10.

Minehane, M.J., O'Donovan, R., Ruane, K.D. & O'Keeffe, B. (2014) The use of 3D laser scanning technology for bridge inspection and assessment. *Structural Health Monitoring (SHM)*, 13, 14.

Mizoguchi, T., Koda, Y., Iwaki, I., Wakabayashi, H., Kobayashi, Y., Shirai, K., . . . Lee, H.-S. (2013) Quantitative scaling evaluation of concrete structures based on terrestrial laser scanning. *Automation in Construction*, 35, 263–274. https://doi.org/10.1016/j.autcon.2013.05.022.

New York State Department of Transportation. (2014) "Bridge Inspection Manual." Available from: https://www.dot.ny.gov/divisions/engineering/structures/repository/manuals/inspection/nysdot_bridge_inspection_manual_2017.pdf

Olsen, M.J., Kuester, F., Chang, B.J. & Hutchinson, T.C. (2010) Terrestrial laser scanning-based structural damage assessment. *Journal of Computing in Civil Engineering*, 24(3), 264–272. https://doi.org/10.1061/(ASCE)CP.1943-5487.0000028.

Pakrashi, V., O'Brien, E. & O'Connor, A. (2011) A review of road structure data in six European countries. *Proceedings of the ICE-Urban Design and Planning*, 164(4), 225–232.

Park, S.W., Oh, B.K. & Park, H.S. (2017) Terrestrial laser scanning-based stress estimation model using multi-dimensional double layer lattices. *Integrated Computer-Aided Engineering*, 24(4), 367–383.

Pauly, M., Keiser, R. & Gross, M. (2003) Multi-scale feature extraction on point-sampled surfaces. *Computer Graphics Forum*, 22(3), 281–289.

Phares, B.M., Washer, G.A., Rolander, D.D., Graybeal, B.A. & Moore, M. (2004) Routine highway bridge inspection condition documentation accuracy and reliability. *Journal of Bridge Engineering*, 9(4), 403–413.

Pulli, K., Duchamp, T., Hoppe, H., McDonald, J., Shapiro, L. & Stuetzle, W. (1997) Robust meshes from multiple range maps. *International Conference on Recent Advances in 3-D Digital Imaging and Modeling, 12–15 May*. Ottawa, Canada.

Rabbani, T., van den Heuvel, F. & Vosselmann, G. (2006) Segmentation of point clouds using smoothness constraint. *International Archives of Photogrammetry, Remote Sensing and Spatial Information Sciences*, 36(5), 248–253.

Riveiro, B., González-Jorge, H., Varela, M. & Jauregui, D.V. (2013) Validation of terrestrial laser scanning and photogrammetry techniques for the measurement of vertical underclearance and beam geometry in structural inspection of bridges. *Measurement*, 46, 784–794. https://doi.org/10.1016/j.measurement.2012.09.018.

Salman, N., Yvinec, M. & Mérigot, Q. (2010) Feature preserving mesh generation from 3D point clouds. *Computer Graphics Forum*, 29(5), 1623–1632.

Schnabel, R., Wahl, R. & Klein, R. (2007) Efficient RANSAC for point-cloud shape detection. *Computer Graphics Forum*, 26(2), 214–226.

Schneiders, R. (1996) A grid-based algorithm for the generation of hexahedral element meshes. *Engineering With Computers*, 12(3–4), 168–177.

Schneiders, R., Schindler, R. & Weiler, F. (1996) Octree-based generation of hexahedral element meshes. *5th International Meshing Roundtable, 10–11 October*, Pittsburgh, PA, USA.

Shen, H.-K., Chen, P.-H. & Chang, L.-M. (2013) Automated steel bridge coating rust defect recognition method based on color and texture feature. *Automation in Construction*, 31, 338–356. https://doi.org/10.1016/j.autcon.2012.11.003.

Soudarissanane, S., Lindenbergh, R., Menenti, M. & Teunissen, P. (2011) Scanning geometry: Influencing factor on the quality of terrestrial laser scanning points. *ISPRS Journal of Photogrammetry and Remote Sensing*, 66(4), 389–399.

Tang, P., Akinci, B. & Huber, D. (2009) Quantification of edge loss of laser scanned data at spatial discontinuities. *Automation in Construction*, 18(8), 1070–1083. https://doi.org/10.1016/j.autcon.2009.07.001.

Tang, P., Huber, D., Akinci, B., Lipman, R. & Lytle, A. (2010) Automatic reconstruction of as-built building information models from laser-scanned point clouds: A review of related techniques. *Automation in Construction*, 19(7), 829–843.

Tarsha-Kurdi, F., Landes, T. & Grussenmeyer, P. (2007) Hough-transform and extended RANSAC algorithms for automatic detection of 3d building roof planes from lidar data. *ISPRS Workshop on Laser Scanning 2007 and SilviLaser 2007, 12–14 September*, Espoo, Finland.

Teza, G., Galgaro, A. & Moro, F. (2009) Contactless recognition of concrete surface damage from laser scanning and curvature computation. *NDT & E International*, 42, 240–249. https://doi.org/10.1016/j.ndteint.2008.10.009.

Trimble Navigation Limited. (2003) *Trimble GS200*. Available from: http://49.128.56.163/hydronav/templates/rhuk_milkyway/gg/pdf/TrimbleGS200Spec_0.pdf [accessed 11th April 2014].

Truong-Hong, L., Falter, H., Lennon, D. & Laefer, D.F. (2016) Framework for bridge inspection with laser scanning. *EASEC-14 Structural Engineering and Construction, 6–8 January*, Ho Chi Minh City, Vietnam.

Truong-Hong, L. & Laefer, D.F. (2014) Using terrestrial laser scanning for dynamic bridge deflection measurement. *IABSE Istanbul Bridge August 11–13, 2014*, Istanbul, Turkey.

Truong-Hong, L. & Laefer, D.F. (2015a) Documentation of bridges by terrestrial laser scanner. *IABSE Symposium Report, 13–15 May*, Nara, Japan.

Truong-Hong, L. & Laefer, D.F. (2015b) Quantitative evaluation strategies for urban 3D model generation from remote sensing data. *Computers & Graphics*, 49, 82–91.

Truong-Hong, L., and Laefer, D.F. (2017) Tunneling appropriate computational models from laser scanning data. *IABSE Symposium Vancouver 2017: Engineering the Future, September 19–23*, Vancouver, Canada.

Truong-Hong, L., Laefer, D., Hinks, T. & Carr, H. (2012a) Flying voxel method with Delaunay triangulation criterion for façade/feature detection for computation. *ASCE, Journal of Computing in Civil Engineering*, 26(6), 691–707.

Truong-Hong, L., Laefer, D.F., Hinks, T. & Carr, H. (2012b) Combining an angle criterion with voxelization and the flying voxel method in reconstructing building models from LiDAR data. *Computer-Aided Civil and Infrastructure Engineering*, 28(2), 112–129.

Truong-Hong, L., Laefer, D.F., Hinks, T. & Carr, H. (2013) Combining an angle criterion with voxelization and the flying voxel method in reconstructing building models from LiDAR data. *Computer-Aided Civil and Infrastructure Engineering*, 28(2), 112–129. https://doi.org/10.1111/j.1467-8667.2012.00761.x.

Valença, J., Dias-da-Costa, D. & Júlio, E.N.B.S. (2012) Characterisation of concrete cracking during laboratorial tests using image processing. *Construction and Building Materials*, 28(1), 607–615.

Valença, J., Puente, I., Júlio, E., González-Jorge, H. & Arias-Sánchez, P. (2017) Assessment of cracks on concrete bridges using image processing supported by laser scanning survey. *Construction and Building Materials*, 146, 668–678.

Vo, A.-V., Truong-Hong, L., Laefer, D.F. & Bertolotto, M. (2015) Octree-based region growing for point cloud segmentation. *ISPRS Journal of Photogrammetry and Remote Sensing*, 104, 88–100. https://doi.org/10.1016/j.isprsjprs.2015.01.011.

Walsh, S.B., Borello, D.J., Guldur, B. & Hajjar, J.F. (2013) Data processing of point clouds for object detection for structural engineering applications. *Computer-Aided Civil and Infrastructure Engineering*, 28(7), 495–508. https://doi.org/10.1111/mice.12016.

Wang, J., Oliveira, M.M., Xie, H. & Kaufman, A.E. (2005) Surface reconstruction using oriented charges. *Computer Graphics International, 22–24 June*, New York, NY, USA.

Weinmann, M., Jutzi, B. & Mallet, C. (2014) Semantic 3D scene interpretation: A framework combining optimal neighborhood size selection with relevant features. *ISPRS Annals of the Photogrammetry, Remote Sensing and Spatial Information Sciences*, 2(3), 181.

Weiss, G.M., McCarthy, K. & Zabar, B. (2007) Cost-sensitive learning vs. sampling: Which is best for handling unbalanced classes with unequal error costs? *DMIN*, 7, 35–41.

Williams, K., Olsen, M., Roe, G. & Glennie, C. (2013) Synthesis of transportation applications of mobile LIDAR. *Remote Sensing*, 5(9), 4652.

Yan, L., Tan, J., Liu, H., Xie, H. & Chen, C. (2017) Automatic registration of TLS-TLS and TLS-MLS point clouds using a genetic algorithm. *Sensors*, 17(9), 1979.

Yan, Y., Guldur, B. & Hajjar, J.F. (2015) Automated structural modelling of bridges from laser scanning. *Structures Congress 2017, 6–8 April*, Denver, CO, USA.

Zogg, H.-M. & Ingensand, H. (2008) Terrestrial laser scanning for deformation monitoring – Load tests on the Felsanau Viaduct. *ISPRS Congress, 3–11 July*, Beijing, China.

Laser Scanning – Riveiro and Lindenbergh
© 2020 ISPRS, ISBN 978-1-138-49604-0

Chapter 13

Laser scanning data for inverse problems in structural engineering

Belén Riveiro, Manuel Cabaleiro, Borja Conde, Mario Soilán and Ana Sánchez-Rodríguez

ABSTRACT: This chapter is presenting a review of applications of laser scanning technology into structural analysis. In the first part of the chapter, the reader will be presented with the state of the art of successful applications where laser scanning provided the basic data to build geometric models that are required as input to the structural models that are subsequently used for structural behavior simulation. This first part will review not only the creation of 3D models using existing CAD software but also the latest trends in creating customized geometric models using image processing tools adapted to 3D point clouds. The second part of the chapter focuses on demonstrating how laser scanning is a technology that not only provides accurate and detailed geometric information for further 3D modelling but also can provide the necessary data to solve inverse problems such as estimating the value of the most influential material properties of structures subjected to loading, or even contributing in the forensic analysis of existing structures. Two different applications, at laboratory scale and in an existing masonry bridge, are presented in order to illustrate these last applications of laser scanning in solving inverse problems in structural analysis.

1 INTRODUCTION

This chapter presents a revision of structural engineering applications in which laser scanning technology has proven to be a very useful tool in structural analysis problems of existing structures. The first sections are devoted to making a review of successful applications of laser scanning in the process of structural evaluation of existing civil engineering structures and buildings, including creation of geometrical models suitable for advanced structural analysis (including finite element method [FEM], discrete element method [DEM], or more simple analysis using limit analysis theory) in a manual manner; conversion of point clouds into models suitable for structural analysis in a semi- or fully automated manner; other procedures to automatically assist structural evaluation of existing constructions. The second part of this chapter is devoted to presenting some applications where laser scanning not only contribute to create the basis model for further analysis, but it also provides the real scenario, or final state of a structural member, so this condition allows the calibration of unknown parameters during the structural simulation; this is the so-called inverse analysis problem in structural engineering by using laser scanning data. For this second part, two methodologies are shown: the first one is based on the estimation of material properties of a beam at laboratory scale; and the second one is focused to understand the causes that provoked a pathological condition in a real structure, in this case an arch bridge.

2 SUCCESSFUL APPLICATIONS OF LASER SCANNING IN STRUCTURAL ENGINEERING PROBLEMS

During the last decade, the use of terrestrial laser scanning in the structural engineering domain has significantly evolved, and thus, many different applications have been reported in the literature. Moreover, terrestrial laser scanners are nowadays common tools in structural engineering laboratories to document geometrical condition in both pre- and postmechanical tests, monitoring construction processes, accurately determine displacements during experiments, etc. These operations can be done both using existing commercial tools, or developing customized tools that automate the analysis of the structural condition. In this section a review of existing approaches that uses laser scanning data as a basic source to create models for structural evaluation is presented, attending to the following classification:

- 3D modelling of existing constructions for further structural analysis.
- Automated conversion of point clouds into structural models.
- Procedures to enrich structural models based on automated processing of point clouds.

2.1 *3D modelling of existing constructions for further structural analysis*

Many authors have published works in which the point clouds collected with laser scanners result in a very convenient data source to derive geometric models that can be used for structural simulations. Thus Riveiro *et al.* (2016b) describe the most common procedures to convert point clouds into such models (alignment, filtering, cleaning, etc.). A point cloud is composed of discrete and isolated points that cannot be directly used by the software applications used for structural analysis, so it needs to be converted into a continuous model. However, there can be different approaches when converting the optimized point clouds into the solid models required by the software applications performing the numerical simulation: using digital surface models, using primitives and volumetric shapes, using NURBS.

2.1.1 Digital surface model

A digital surface Model (DSM) is computed by triangulation or another interpolation method. Delaunay condition is commonly used to produce the triangulation. This triangulation requires defining the maximum edge length of the triangle, the maximum angle between the two beams originating in the optical center of the scanner and terminating in the vertices of a triangle edge, thereby limiting the length of the edge. This type of continuous model is also very convenient when a photorealistic model is to be created by projecting RGB images, or other textures, over the surface with the aim of subsequently produced orthophotos.

This modelling approach has important limitations such as the presence of gaps, excessive sharp angles, etc., which may cause problems during the mesh generations in the finite element analysis (FEA) simulation software and therefore prevent convergence in the solution. So, besides the DSM creation, a subsequent processing involves the re-edition of the mesh model into an intermediate CAD (computer-aided design) software. From the continuous model (mesh) already created, cross-sections can be produced so the extracted feature lines describing the object or structural member are then used as the basis to use extrusion operations. Compared to other approaches, this process is not computationally demanding (Tang *et al.*, 2010) but can lead to smoothed geometries, losing relevant information of deformed geometries.

2.1.2 3D modelling using volumetric shapes

When a detailed three-dimensional model is required for the structural analysis of a given structure, including all the relevant features about the complex geometry, it can be created by using the 3D modelling tools (primitive fitting tools, cross-section methods, extrusion techniques, etc.) of CAD platforms. This approach consists of using directly the original or improved point cloud into the CAD software, avoiding the intermediate step of mesh generation. However, the CAD software must incorporate specific modules for point cloud visualization when dense point clouds need to be handled (Riveiro *et al.*, 2015). Alternatively, it would be needed to draw the contour of the object to be modelled from different orthogonal views in the point cloud editing software to be subsequently imported into the CAD software and use the aforementioned modelling tools to create the model. By repeating the operation for all parts and properly defining the assembly of all parts, the entire 3D model is created.

Even though such a modelling approach can be the most suitable in terms of compatibility with FEA simulation software, sometimes the detailed geometry measured by the laser scanner becomes partially idealized. This may result in the loss of some relevant information such as out-of-plane deformations, lack of material, etc.

2.1.3 3D modelling using NURBS

This last method consists of creating nonuniform rational B-splines (NURBS) on a given mesh. One of the main advantages of using NURBS is that you can rely on its accuracy when modelling complex geometries and the simple exportation to FEA software. Additionally, this method has the advantage of requiring less manual intervention by the operator. The main drawback, compared to the previous approaches, is that the computational resources required are higher.

Independently of the method used, which is normally a decision of the operator performing the further structural analysis, laser scanning data has been traditionally used as a data source to derive the geometrical models. The selection of one approach or another may be constrained by the type of structural analysis to be performed afterward.

Riveiro *et al.* (2011) proposed a workflow to use point clouds collected by laser scanning in order to develop the geometrical models needed to perform the structural analysis of masonry arches based on limit analysis theory using the rigid blocks method. For their investigation, they converted the point clouds into continuous surface models where RGB texture was incorporated. Thus, the orthophotos created allowed them to prepare the geometrical models of the arches to be used during the structural analysis.

Sánchez-Aparicio *et al.* (2014) conducted an in-depth investigation of the geometric state of a church in Portugal. The aim of their work was to improve the numerical model used previously in Lourenço *et al.* (1999) by having a more detailed geometry (not idealized), that incorporates also the pathological condition (main cracks). In their approach, the authors propose converting the point cloud into a solid model in two steps: mesh generation (DSM) describing main geometric details and further processing and parameterization for the conversion into a solid CAD model. The model obtained could be discretized into elements compatible with the requirements of the FEA software.

McInerney *et al.* (2012), proposed a methodology to convert laser scanning point clouds into geometrical models that could be used for structural analysis based on discrete element modelling (DEM). They developed and tested their method in two masonry vaulted structures in Cambridge (United Kingdom).

Figure 13.1 shows geometric models created from laser scanning data, for different types of structural analysis:

2.2 *Automated conversion of point clouds into structural models*

The previous section showed different suitable approaches to convert point clouds into structural models. In most cases, the data processing required the intervention of the human operator, who must also be trained in the mentioned processes. This fact has motivated that, in the last years, an intense activity on developing tools for the automated data processing has been reported in the literature. As shown in most of the chapters of the present book, most authors work on point cloud semantic segmentation, object recognition, 3D modelling, conversion of point clouds to building information models (BIMs), etc. In the particular case of structural engineering applications, many authors have presented approaches for structural evaluation of existing constructions based on structural health monitoring tools, or direct conversion of point clouds into FEM or DEM.

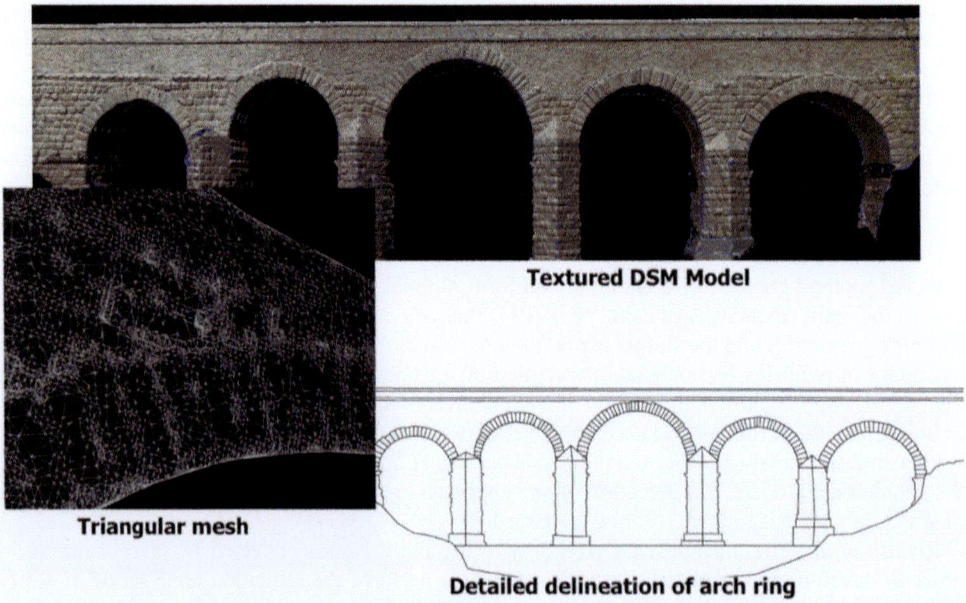

Figure 13.1. View of the DSM of a masonry bridge to be subjected to a structural analysis based on limit analysis theory.

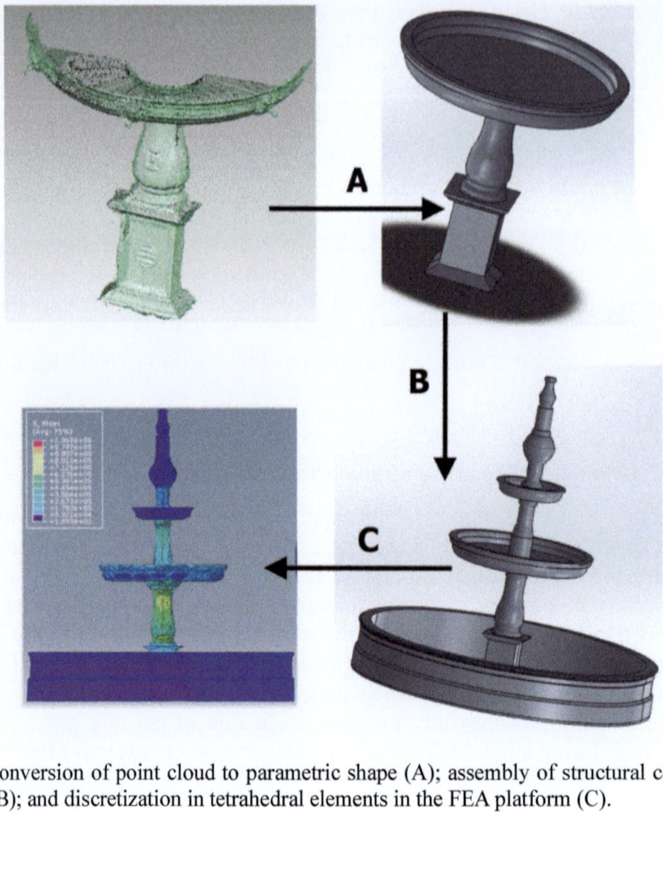

Figure 13.2. Conversion of point cloud to parametric shape (A); assembly of structural components in the CAD software (B); and discretization in tetrahedral elements in the FEA platform (C).

To avoid redundancy in the contents related to automated processing of point clouds that are already presented in other chapters of this book, this section is just summarizing some relevant works dealing with automated creation of structural models.

One of the pioneer works regarding the automated creation of CAD models from point clouds was presented by Truong-Hong *et al.* (2013). Their approach proposes the segmentation of façade point clouds based on the voxelization method, which also permits one to identify the mayor features from the point cloud.

Cabaleiro *et al.* (2014) presented a new method for the detection and parameterization of connections in metallic structures that ended up with the automatic 3D modeling of the detected connections. One of the main advantages of their method is that once the joint is modeled, the entire portal frame can be created. The method projects the 3D point clouds into a virtual 2D plane where density images are created (using rasterization), so using the Hough Transform, the main features of the metal joint can be derived. As a result of this methodology, the coordinates of the center of the joint, the composition (profiles, size and shape of the haunch) and the direction of the profiles are estimated so the semantic characterization of every connection is completed. With this information, most common structural simulation FEA software can define the geometric model corresponding to the connections and entire frame.

Castellazzi *et al.* (2015) recently presented a new method that can convert point clouds of complex structures into FEM models by using voxelization. Their method has been validated into a historic masonry castle where the model obtained was compared to a 3D CAD model of the construction. They concluded that the two models provide comparable results, and the proposed model is created faster and with a high level of automation.

Riveiro *et al.* (2016c) have also presented a work for the automatic data processing of lidar point clouds to automatically derive structural models of masonry walls. In their approach, based on heuristic rules, they use rasterization and the subsequent application of image processing tools to create the delineation of masonry structures. This way they are able to automatically perform morphologic analysis of the structures after identifying every single stone in the point cloud of the wall and its main geometric parameters.

Other examples where point clouds are automatically processed with the aim of performing the structural evaluation of existing bridge structures is presented by Riveiro *et al.* (2016a). In their work, the combination of voxelization, together with image processing tools adapted to the 3D space, following heuristic rules, permitted the semantic classification of bridge point clouds. At the end of the process, the method provides a partitioned point cloud corresponding to every structural element (spandrel walls, arches, cutwaters, pathway, etc.), which are spatially related. This was the input for a subsequent work in which Sánchez-Rodríguez *et al.* (2018) developed a diagnosis procedure that can detect and quantify a pathological condition in piers. This is achievable by means of further segmentation and parameterization of pier components, which are all spatially related, so the faults in piers can be also related to other anomalies in other elements of the bridge.

A closely related field to the automated creation and monitoring of structural elements is the creation of BIM models from laser scanning data. As will be shown in the last chapter of this book, important contributions have been published focusing on converting laser scans into BIM and, further, BIM into structural models. Barazzetti *et al.* (2015) proposed a two-step method, "cloud-to-BIM-to-FEM," where scans of a historical construction are modelled within the historical (H)-BIM logic; and later, a tetrahedron solid mesh is automatically generated from the volume enclosed by 2D surfaces of the HBIM. For further details on BIM, refer to Chapter 15.

2.3 *Procedures to enrich structural models based on automated processing of point clouds*

In addition to the applications summarized in sections 2.1 and 2.2, laser scanning has been largely used for the detection and parameterization of elements or damages that may affect the structural performance of constructions. To avoid including redundant contents in this section, the reader is encouraged to see other chapters in this book: Chapter 11 about integral diagnosis and structural analysis of historical constructions using TLS and Chapter 12 about the application of TLS for bridge inspection.

3 LASER SCANNING DATA FOR INVERSE PROBLEMS

The methodologies and applications shown in this chapter so far involve the use of laser scanning data in a direct manner, this is, point clouds are the source data to build the geometric models that would be later used for the structural analysis. However, laser scanning has proven to be also an excellent source of data for documenting the deformation state of structures. Thus, laser scanning becomes a suitable technology to provide the required data for calibration of FEA models or similar problems that involve uncertainties that need to be solved through inverse analysis procedures. In this sense, this section is presenting some experiences where the dense point clouds collected using laser scanning, combined with automated data processing algorithms, allowed us to solve inverse problems in structural engineering at different scales. The first subsection presents a methodology for the estimation of material properties of a beam in a laboratory test. The second subsection summarizes a methodology that allowed us to predict the causes that led a masonry arch bridge to a certain pathological condition.

3.1 *Estimation of material properties of FEM models using laser scanning data*

As previously said, in this experience, dense point clouds acquired by laser scanning were explored as a source of data for the calibration of FEM models. The final aim is to determine the values of the most influential elastic parameters in the structural behavior of the tested structural members. The experiment where the methodology has been tested involves a steel beam subjected to the combined effect of bending and torsion. The structural member was subjected to different loads, and a point cloud of the deformed state was recorded. An algorithm for the automated analysis of the point cloud was developed so that it is possible to automatically monitor the deformation process in the real point cloud, as well as to compare these measured surfaces with those ones obtained using FEA simulation software. By minimizing differences between the deformations obtained both by simulation and by real measurement, it was possible to calibrate the elastic parameters.

3.1.1 Point clouds

The point clouds used during this investigation were collected with a terrestrial laser scanner FARO Focus 3D. The scanning instrument was placed in a fixed position at a distance of 2.6 m from the tested beam. The detailed scans ensured that the lower resolution was less than 5 mm in the further areas and up to 1 to 2 mm in the denser areas (corresponding to the beam center). These point clouds were manually segmented in order to isolate the structural member from the other elements in the scene.

On the other hand, simulated point clouds were produced from a numerical model of the beam (whose geometrical model was built using the unloaded beam). To obtain the point clouds corresponding to every load of the experiment, a FEM analysis was performed with Abaqus. Since this simulation is part of the optimization process, the process has to be iteratively repeated for every value of the input parameters (elastic properties being calibrated). Thanks to its Python scripting interface, it was possible to access the model characteristics from Matlab to change the values of the input parameters and thus to obtain a simulated point cloud where the differences with the real point clouds were minimized.

The methodology started with the definition of the reference system for both the one measured by a terrestrial laser scanner and the one obtained as a result of the FEM simulation, as well as the method for the establishment of correspondences. Next, point normals were calculated in order to determine the effects of bending or torsion in every deformed surface evaluated. By using the theories of beam deformation and adopting the elastic parameters calibrated at the end of the methodology, normal and shear stresses at the surface of analysis are going to be estimated.

3.1.2 Automated segmentation and filtering of the point cloud to be evaluated

After isolating the beam points using CloudCompare software, the automated processing of the point cloud started. The algorithms presented here were implemented using Matlab software.

Further filtering operations were implemented in Matlab to remove undesired points such as noise, mixed pixels, etc. Additionally, since several parts of the beam would contain occlusions, the points defining the upper flange were selected as a reference surface in the comparison between real and simulated point clouds.

The next operation consisted in the rasterization of this reduced point cloud, since this regular structure was chosen as the link between the different point clouds being compared. Finally, over this raster structure, the stress values will be calculated based on the geometric parameters of every pixel: the already mentioned point normals. As a summary, the segmentation of the beam flange was performed in the following steps:

- Normals calculation
- Removing points that do not belong to planar entities
- Coarse filtering using connected components method
- Definition of image limits (flange limits)
- Point cloud smoothing

3.1.2.1 Normals calculation

The points of interest (upper flange) are those that define a flat or slightly curved surface (after deformation). Point normals can provide robust geometric descriptors to segment and classify points. In our case, using principal components analysis was used to determine the orientation of every single point having into account its neighborhood, defined as points with a radius depending on points density. The first step of a principal components analysis is to calculate covariance matrix:

$$\Sigma = \begin{pmatrix} \sigma_x^2 & \sigma_{xy} & \sigma_{xz} \\ \sigma_{xy} & \sigma_y^2 & \sigma_{yz} \\ \sigma_{xz} & \sigma_{yz} & \sigma_z^2 \end{pmatrix} \tag{1}$$

Where variance in each axis x_i or covariance in each pair of axis $x_i x_j$ are calculated as:

$$\sigma_{xi}^2 = \frac{1}{n} \Sigma_{k=1}^{n} (xi_k - \overline{xi})^2 \tag{2}$$

$$\sigma_{xixj} = \frac{1}{n} \Sigma_{k=1}^{n} (xi_k - \overline{xi})(xj_k - \overline{xj}) \tag{3}$$

Where n is the number of points, xi_k is the xi coordinate of k point, and $\overline{xi}$ is the mean of all xi coordinates.

The matrix eigenvectors represent the three principal directions, the three orthogonal axes for which covariances are null. The diagonal elements get maximum or minimum values, that can be seen as a Cartesian coordinate system whose axes correspond to maximum and minimum data deviation. The third eigenvector represents the normal direction to the plane that best fits the points.

3.1.2.2 Removing points that do not belong to planar entities

As said, the points corresponding to the flange surface can be assumed to belong to a planar surface at their local neighborhood, so dimensionality descriptors become very useful for our point classification. For that reason, the Gressin method (Gressin *et al.*, 2013) for point classification based on their dimensionality was used. This method assumes that points and their corresponding neighborhoods represent an ellipsoid. The ellipsoid diameters are oriented according to eigenvectors, and sizes are directly related with eigenvalues. Based on this information, points can be classified as being contained in ellipsoids with linear predominance (a_{1D}), planar predominance (a_{2D}), and volumetric predominance (a_{3D}).

Let σ_1, σ_2, and σ_3 be typical deviations of points with respect to eigenvectors 1, 2, and 3, respectively, so that $\sigma_1 \geq \sigma_2 \geq \sigma_3$. Dimensional coefficients are calculated as:

$$a_{1D} = \frac{\sigma_1 - \sigma_2}{\sigma_1}; \; a_{2D} = \frac{\sigma_2 - \sigma_3}{\sigma_1}; \; a_{3D} = \frac{\sigma_3}{\sigma_1} \tag{4}$$

From this classification, we kept the planar points (a_{2D}) that correspond to flange surface and linear points (a_{1D}) that define the flange ends.

3.1.2.3 Coarse filtering using connected components method

Once points were classified according to their dimensionality, the subset of classified points were voxelized, in order to be subsequently subjected to an analysis that cluster voxels according to their connectivity to adjacent voxels. In graph theory, connected components are any subgraph which is connected to other one by a path and is not connected to graph vertex. After applying a connected components algorithm, those sets of connected elements composed of a small number of voxels were removed, so only big groups were kept as those representing the point cloud of interest.

3.1.2.4 Definition of image limits

After coarsely removing part of the point clouds, the ends of the flange will not be homogeneous along the beam length. Consequently, a density-based filtering when projecting points in the x and y directions allowed us to set the minimum and maximum y and x coordinates, respectively, for the xy plane containing the flange points. Once these limits were defined, the point cloud could be rasterized.

3.1.2.5 Point cloud smoothing

When measuring the deformed surface, the measurement error of the scanner provokes that the point cloud defining such a surface is composed of fuzzy points whose thickness normally approximates to the nominal error of the scanner. On the other hand, with previous knowledge about beam mechanics, the deformed state of a beam can be parameterized. Even though this last model is not perfect (is just an idealization), it can help define some parameters to fit the fuzzy point cloud to polynomial surfaces, and thus, noisy points can be removed. The criteria to remove such points was to set a threshold for residuals so that points with higher residuals were removed from the adjustment.

Finally, with the purpose of having a smoother point cloud, a moving average filter was applied in order to remove those remaining points that prevent the point cloud be a continuous and smooth surface. The filter consisted in replacing the z coordinate of every point by the averaged z value obtained from its neighborhood, according to formula 5.

$$\overline{z_i} = \frac{1}{n_i} . \sum_{j=1}^{n_i} z_{i,j} \tag{5}$$

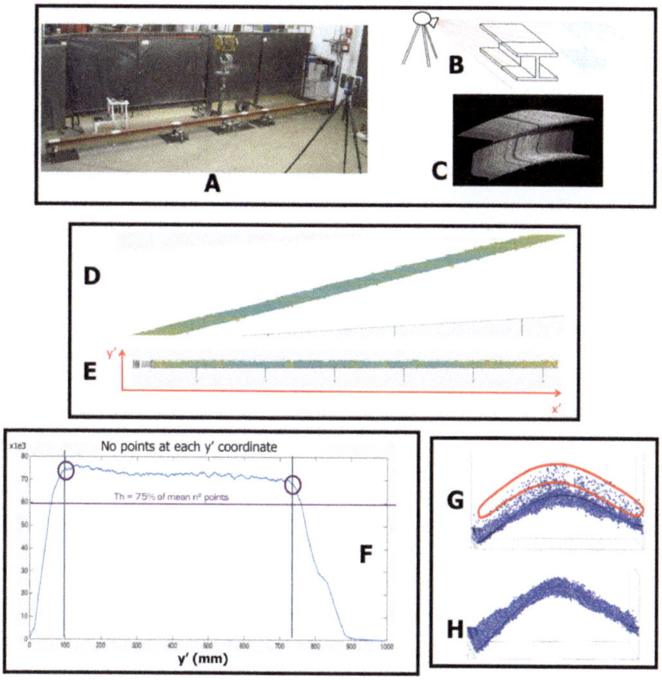

Figure 13.3. Point cloud acquisition and preprocessing: (a) experiment in the laboratory; (b) orientation of scanner to measure beam points; (c) point cloud of the beam measured with the FARO Focus 3D; (d) perspective view of flange points after coarse cleaning and filtering; (e) top view of flange points after coarse cleaning and filtering oriented to the coordinate system adopted for the analysis; (f) definition of image limits in the y' direction; (g) flange points before fine filtering using beam mechanics theory (elevation view of the beam); (h) flange points after fine filtering (elevation view of the beam).

Where, $\overline{z_i}$ the averaged z coordinate of point i, n_i is neighborhood population of point i and $z_{i,j}$ is the z coordinate of point j.

Figure 13.3 summarizes the processing of real point clouds, from acquisition to fine filtering.

Once the point cloud was segmented and the area of analysis was defined and finely filtered, every point cloud was rasterized so an image of deflections in z coordinate was obtained. These images represent the real deformation images that were compared with the simulated deformed images created using a numerical model that reproduces the conditions of the test performed in the laboratory, as explained in the next section.

3.1.3 Simulated point clouds and parameter identification

As presented in section 3.1.1, point clouds created with FEA simulation software were compared with the point clouds acquired in every deformation stage of the beam being analyzed in this investigation. The purpose is to define a numerical model that minimizes differences between the real and the simulated point clouds. Thus, we are facing an optimization problem where the most influential elastic parameters of the material and the boundary conditions are to be estimated by minimization of differences between the simulated point cloud produced by the FEA simulation software and the real point cloud produced in section 3.1.2. Both point clouds are linked through the raster structure defined in section 3.1.2.4. Figure 13.4 summarizes the process for the calibration of the numerical model.

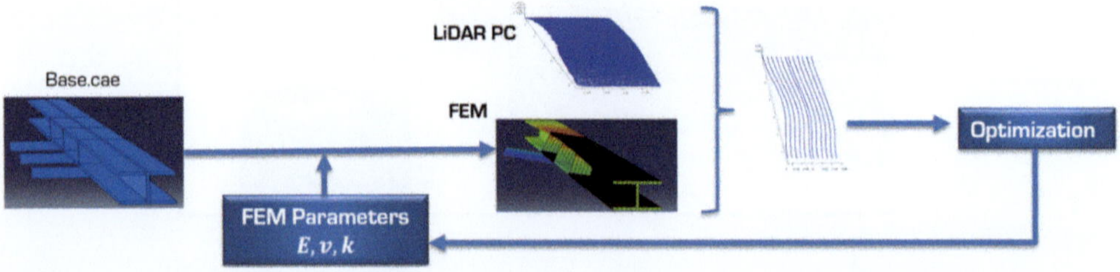

Figure 13.4. Workflow for the parameter identification using both real point clouds and FEM simulated point clouds.

3.1.3.1 Definition of the numerical model

Abaqus software was used for the FEA simulation, and using its Python scripting interface, it was possible to update the value of the input parameters. For every simulation performed, the displaced mesh nodes were exported as simulated point clouds. These point clouds were also subjected to the segmentation of the flange and rasterization operations explained in section 3.1.2 (note that filtering operations were not needed here because there is not the noise of the scanner).

It should be noted that the uncertainties of the numerical model can be due to different parameters, so the following variables were submitted to a sensibility analysis in order to identify which of them are relevant influential parameters: supports rigidity, Young's modulus, Poisson's ratio, and steel density.

The sensitivity analysis was based on Pearson's correlation coefficients, which assess influence of each input parameter on the result of a weighting function. The employed function calculates the root mean square error (RMSE) between the real and the simulated point cloud:

$$RMSE = \sqrt{\frac{1}{n}\Sigma_{i=1}^{n}(z_{\mathrm{Re}\,fi} - z_{Abi})^2} \tag{6}$$

Where Z_{Refi} is i z value of the pixel in the raster image created from the real point cloud, z_{Abi} is i z value of the pixel in the raster image created from the simulated point cloud in Abaqus image, and n is number of pixels. Figure 13.5 summarizes the workflow of sensitivity analysis implemented in Matlab connected to the Abaqus kernel.

As a result of this sensitivity analysis, only Young's modulus was found to be an influential input parameter. For that reason, this was the only parameter submitted to the optimization process.

3.1.3.2 Optimization process

The optimization process was executed using a genetic algorithm. The genetic algorithm makes a population evolve by random actions, simulating the biological action through crossover and natural selection in order to adapt to the environment. The adaptation is measured using a weighting function. In our problem, the aim is to find that value of the Young's modulus corresponding to a global minimum of the weighting function. Thus the optimization started with a population of 25 individuals that includes those whose RSME was minimum in the previous sensibility analysis. For every individual of every generation, a simulated point cloud is created from the FEM analysis and compared with the real point cloud. As output, the best individual of the latest generation is stored as solution.

This method allowed us to estimate as Young's modulus that minimized differences between simulated and real point clouds a value of 237.82 GPa, whilst the value of Young's modulus typically given in the standards is 200 to 210 GPa).

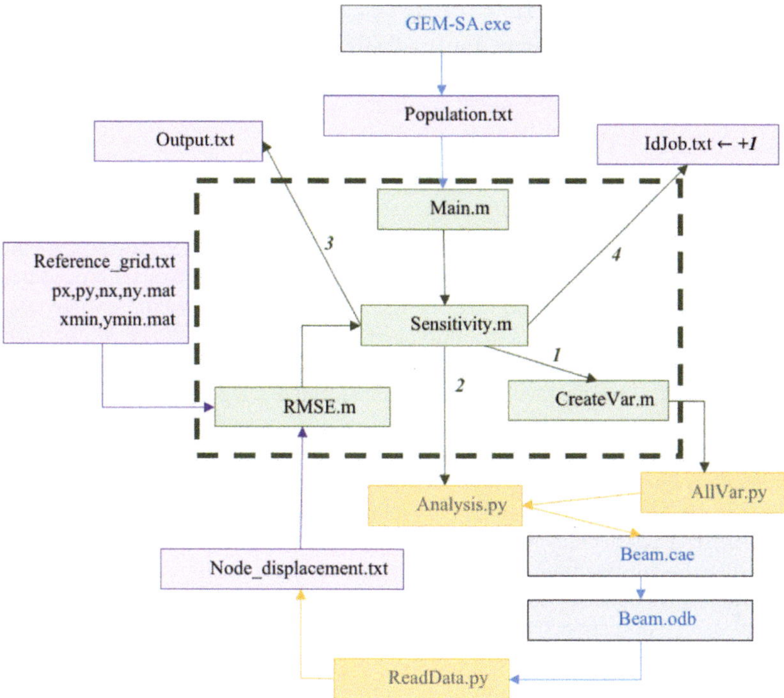

Figure 13.5. Workflow for the sensibility analysis implementation.

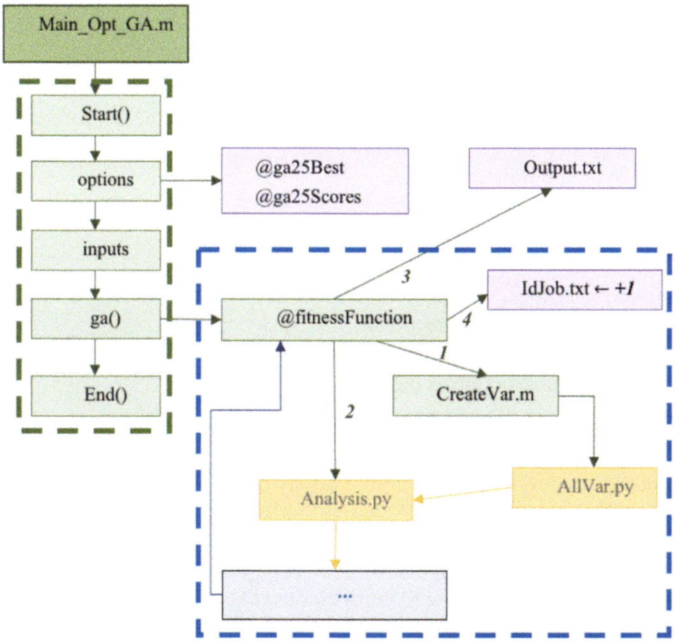

Figure 13.6. Workflow for the optimization process using a genetic algorithm implemented in Matlab connected to the Abaqus kernel.

Finally, it can be concluded that the detail and accuracy of the point cloud provided by terrestrial laser scanners allowed us to calibrate some unknow parameters in simulated structural models. Thus, laser scanning becomes a tool that not only provides geometric details but also allows one to solve inverse problems when it is properly combined with automated processing methods.

3.2 *Automated processing of TLS point clouds to understand the pathological condition of masonry arch bridges*

Conde *et al.* (2016) presented an approach for the investigation of the causes that might lead a masonry arch bridge to a pathological condition by using inverse analysis procedures. In their research, they used the damaged condition of a real bridge located in Kakodiki (Greece) as a starting point and searched for the causes that provoked the bridge become to be in a critical structural condition. Unfortunately, the bridge collapsed in January 2018.

In their approach, the damage identification is transformed in a parameter identification problem which is tackled by comparing the predictions obtained by numerical modelling with those actually measured in the deformed (damaged) bridge. Similarly to the previous methodology, a global optimization approach was adopted within the inverse problem, using a genetic algorithm, aiming at calibrating the parameters of the numerical model of the structure. Again, an error function that measures the differences between the numerically predicted damage pattern and the deformation state of the bridge had to be minimized. Since the numerical modelling and the optimization process of this bridge have been already published (Conde *et al.*, 2016) and a problem of optimization was already presented in section 3.1, this part of the investigation will be omitted here. This section will focus instead in presenting a methodology for the automated processing of laser scanning data so that the geometrical models, needed for the comparison of the real deformed state of the bridge with the numerically predicted damage state, can be obtained.

In essence, the geometric model here refers to the delineation of the arch, since a 2D finite element discretization was adopted for the computational models. The variables to be estimated during the optimization problem have to be extracted from the measured point cloud of the arch, namely rise at mid span, rise at quarters of span, horizontal, and vertical displacements at both abutments. Also, the fit of both predicted and real geometries is done using the singular points placed in the intrados of the arch at the joints. Thus the challenge for the automated processing of laser scanning data is to obtain the coordinates of the real joints. These coordinates are compared with those resulting from using a theoretical initial geometry plus the structural displacements experimented with in the nonlinear analysis.

The methodology for the automatic creation of the geometric models can be summarized in the following steps:

- Segmentation of the bridge point cloud into main structural components: isolation of the point cloud corresponding to the spandrel walls.
- Rasterization of the point cloud: creation of an intensity image of the spandrel wall.
- Extraction of main geometric parameters.
- Detection of joints and computation of coordinates at the intrados.

3.2.1 Segmentation of the bridge point cloud into main structural components

This type of segmentation adapted to the geometric characteristics of masonry arch bridges was developed by Riveiro *et al.* (2016a). In their methodology, the authors propose a method for the automated segmentation of large point clouds of masonry arch bridges that decompose the point clouds into spatially related and organized point clouds that contain the relevant data of a bridge's components (pier, arch, spandrel wall, cutwaters, etc.). The strategy to get these segments was based in the combination of a heuristic approach (following logical and topological constraints) and image processing tools adapted to voxel structures.

The relevant segmented point cloud for the delineation of arch geometry is the spandrel wall that would be used to create the geometric model that is being compared with the predicted geometrical output of the numerical model. Both spandrel walls can be used in case the survey is to be done at both barrel ends.

3.2.2 Rasterization of the point cloud

The point cloud of the spandrel wall was rasterized, and the intensity attribute was used to color this image. The value of intensity of each pixel was computed by using the observed data (point cloud sample) contained in the region of the pixel. The pixel values correspond to the infrared spectrum (1540 ηm in the Riegl LMS Z-390i scanning system).

3.2.3 Extraction of main geometric parameters

The intensity image (Figure 13.7a) created above is subjected to two different operations: first, the image is binarized, labelling empty pixels as 0 values. Second, the intensity image will be postprocessed as explained in section 3.2.4.

In the binary image, many pixels are empty due to the typical occlusions that may happen in the point cloud, or areas of low resolution. Thus, the image was subjected to morphological operations to fill empty pixels within the surface of the bridge and also removing noisy points outside of bridge surface. Next, the boundary of this image is computed, so the lower boundary contains the line of intrados whose ends need to be calculated. To do this operation, the line of lower boundary is subjected to a principal component analysis (PCA), where the second eigenvalue allowed to clearly identify those points where the intrados of the arch ends (Figure 13.7b).

As previously said, the geometric variables to be estimated during the optimization problem refer to basic geometric measures at some particular locations of the arch: rise at different points, span (related to displacements at the abutments). The estimation of arch radius is also relevant, because it is the first geometric value that will allow distinguishing the arch typology (semicircular, pointed, segmental, etc.). For that purpose, the previous intrados contour is partitioned in two halves (at the keystone). For each half, the center of a circumference is estimated. This operation is done by randomly selecting three points and interpolating a center; this operation is repeated several times. When all the centers closely lie into a single point, the arch is determined to be segmental or semicircular; however, if there is more than one clear center (e.g. one for the points of each half intrados), then the arch is determined to be pointed. After the type of arch has been identified, the main geometric measures are obtained using logical reasoning.

3.2.4 Detection of joints and computation of coordinates at the intrados

Using the previously created intensity image, it needs to be postprocessed in order to emphasize the pixels corresponding to joints between voussoirs. For that purpose, the standard deviation of the intensity value of the eight nearest neighbors of each pixel was computed, creating an image were pixels contain the value of standard deviation (standard deviation image). Thus, as shown in Figure 13.7c, one can clearly distinguish pixels corresponding to joints from those corresponding to masonry where the texture would be more homogeneous.

In a previous investigation, Riveiro *et al.* (2016c) proposed the segmentation of individual stones in quasi-periodic masonry walls. In their approach, they iteratively operated with the value of pixels according to orthogonal directions (which is the case of periodic masonry), so that joints can be detected according to those rectangular directions. However, this approach is not suitable for the identification of joints between the voussoirs of an arch, because the joints are oriented with the radius of the arch. Thus, a customized approach has been developed here using the geometric information extracted in section 3.2.3.

By using the position of the arch center (one or two depending on the type of arch) as reference point, and the ends of intrados, the coordinates of pixels were converted to polar coordinates (radius and angle). This conversion was done with the purpose of having pixels sorted by angle and radius in the standard deviation image: first, the pixels are sorted by angle, and later, the mean of standard deviation of all pixels with the same radius value is calculated. At the extrados line, the average standard deviation should be higher because it denotes the joint between the arch voussoirs and the masonry of the spandrel wall. Thus, the radius with higher value of averaged standard deviation is assumed to be the extrados, and it provides information about the averaged arch thickness (radius of extrados minus radius of intrados). To avoid including nonrelevant parts of the spandrel wall, the set of pixels delimited by the intrados and extrados is used to calculate the location of every joint between adjacent voussoirs in the arch. Here the operation is simple and consists of calculating the accumulated sum of standard deviations of pixels with the same angle. Those angles with higher accumulated sum denote the presence of joints. The coordinates of every joint at the intrados (Figure 13.7d) provide the necessary information to proceed with the comparison with those coordinates predicted by numerical modelling, within the optimization process explained at the beginning of section 3.2.

4 SUMMARY

Laser scanning data has proven to be very useful not only as source data for the creation of geometric models needed for structural analysis operations of real structures in a direct manner, but also it has been demonstrated how the data acquired by these instruments can provide the necessary data (in terms of detail and accuracy) when inverse problems need to be solved. For this second application, customized methodologies based on image processing tools and different heuristic rules were the key to solving the problem in an automatic manner avoiding the human operator intervention; this ensures a more productive procedure, without subjective decisions, and facilitating the adoption of the technology by technicians and engineers from outside of the geomatics domain.

One of the methodologies presented in this chapter to illustrate the application of laser scanning to solve inverse problems focused on the calibration of FEM models by determining the value of those most influential elastic parameters in the structural behavior at a laboratory test. For this

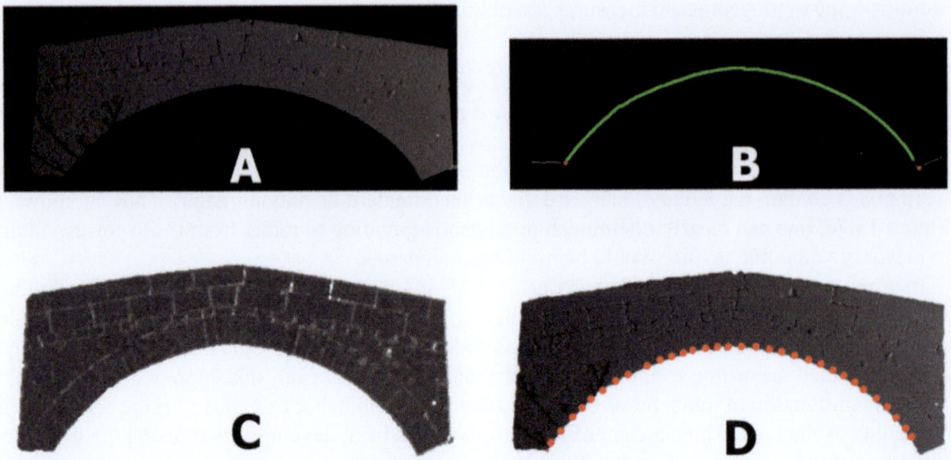

Figure 13.7. Automated detection of arch joints at the intrados line: (a) intensity image created from the point cloud of the spandrel wall; (b) line of intrados extracted from the lower contour in the binarized image; (c) standard deviation image; (d) location of arch joints at the intrados line.

methodology, using an algorithm that is able to automatically monitor deformations in the real beam and to minimize the differences between these deformations and those predicted by the FEA simulation software, it was possible to estimate the value of those searched elastic parameters.

In the second methodology, it was shown how laser scanning data can also contribute to determine the causes that led a masonry arch bridge to a pathological condition. In this case, the damage identification problem was tackled similarly to the previous methodology, where a global optimization approach was adopted, using a genetic algorithm, aiming at minimizing the differences between the numerically predicted damage pattern and the deformation state of the bridge. In this case, a customized automated processing of laser scanning data has been presented so that the geometrical models needed for the comparison of the real bridge arch deformation could be compared with the numerically predicted damage state.

Additionally to these specific applications where laser scanning data has been successfully applied to solve inverse problems, the first part of the chapter provides readers with a review of applications of laser scanning to structural analysis problems in a direct manner; this is contributing to create 3D models for further structural analysis or applications.

REFERENCES

Barazzetti, L., Banfi, F., Brumana, R., Gusmeroli, G., Previtali, M. & Schiantarelli, G. (2015) Cloud-to-BIM-to-FEM: Structural simulation with accurate historic BIM from laser scans. *Simulation Modelling Practice and Theory*, 57, 71–87.

Cabaleiro, M., Riveiro, B., Arias, P., Caamaño, J.C. & Vilán, J.A. (2014) Automatic 3D modelling of metal frame connections from LiDAR data for structural engineering purposes. *ISPRS Journal of Photogrammetry and Remote Sensing*, 96, 47–56.

Castellazzi, G., D'Altri, A.M., Bitelli, G., Selvaggi, I. & Lambertini, A. (2015) From laser scanning to finite element analysis of complex buildings by using a semi-automatic procedure. *Sensors*, 15(8), 18360–18380.

Conde, B., Drosopoulos, G.A., Stavroulakis, G.E., Riveiro, B. & Stavroulaki, M.E. (2016) Inverse analysis of masonry arch bridges for damaged condition investigation: Application on Kakodiki bridge. *Engineering Structures*, 127, 388–401.

Gressin, A., Mallet, C., Demantké, J. & David, N. (2013) Towards 3D lidar point cloud registration improvement using optimal neighborhood knowledge. *ISPRS Journal of Photogrammetry and Remote Sensing*, 79, 240–251.

Lourenço, P.B. & Ramos, L.F. (1999) *Investiçasão sobre as Patologias do Santuário de São Torcato*. Final Report. University of Minho; 99-DEC/E-5.

McInerney, J., Trzcinski, I. & DeJong, M.J. (2012), Discrete element modelling of masonry using laser scanning data. *Proceedings of the 8th International Conference on Structural Analysis of Historical Constructions*, Wroclaw, Poland.

Riveiro, B., Conde-Carnero, B. & Arias-Sánchez, P. (2015) Laser scanning for the evaluation of historic structures. In: *Handbook of Research on Seismic Assessment and Rehabilitation of Historic Structures*. IGI Global, Hershey, PA, USA, pp.765–793.

Riveiro, B., DeJong, M.J. & Conde, B. (2016a) Automated processing of large point clouds for structural health monitoring of masonry arch bridges. *Automation in Construction*, 72, 258–268.

Riveiro, B., González-Jorge, H., Conde, B. & Puente, I. (2016b) Laser scanning technology: Fundamentals, principles and applications in infrastructure. *Non-Destructive Techniques for the Evaluation of Structures and Infrastructure*, 11, 7.

Riveiro, B., Lourenço, P.B., Oliveira, D.V., González-Jorge, H. & Arias, P. (2016c) Automatic morphologic analysis of quasi-periodic masonry walls from LiDAR. *Computer-Aided Civil and Infrastructure Engineering*, 31(4), 305–319. https://doi.org/10.1111/mice.12145.

Riveiro, B., Morer, P., Arias, P. & De Arteaga, I. (2011) Terrestrial laser scanning and limit analysis of masonry arch bridges. *Construction and Building Materials*, 25(4), 1726–1735.

Sánchez-Aparicio, L.J., Riveiro, B., Gonzalez-Aguilera, D. & Ramos, L.F. (2014) The combination of geomatic approaches and operational modal analysis to improve calibration of finite element models: A case of study in Saint Torcato Church (Guimarães, Portugal). *Construction and Building Materials*, 70, 118–129.

Sánchez-Rodríguez, A., Riveiro, B., Conde, B. & Soilán, M. (2018) Detection of structural faults in piers of masonry arch bridges through automated processing of laser scanning data. *Structural Control and Health Monitoring*, 25(3), e2126.

Tang, P., Huber, D., Akinci, B., Lipman, R. & Lytle, A. (2010) Automatic reconstruction of as-built building information models from laser-scanned point clouds: A review of related techniques. *Automation and Construction*, 19(7), 829–843.

Truong-Hong, L., Laefer, D.F., Hinks, T. & Carr, H. (2013) Combining an angle criterion with voxelization and the flying voxel method in reconstructing building models from LiDAR data. *Computer-Aided Civil and Infrastructure Engineering*, 28(2), 112–129.

Laser Scanning – Riveiro and Lindenbergh
© 2020 ISPRS, ISBN 978-1-138-49604-0

Chapter 14

Construction site monitoring based on laser scanning data

Daniel Wujanz

ABSTRACT: Erecting structures has always been challenging in physical, financial and organisational terms. While the building sector is one of the least developed ones in terms of digitization, it can benefit from current developments in 3D data acquisition by means of laser scanning and spatio-temporal data analysis in order to improve its level of productivity. At first, this chapter has a look at various aspects of construction sites in general, followed by considerations of data acquisition and processing. The third section finally discusses several strategies for construction site monitoring.

1 THE INHERENT NATURE OF CONSTRUCTION SITES

The construction of structures has become far less challenging over the last decades in terms of physical effort due to the support that machines, computers and contemporary tools provide. Nevertheless, it still remains a logistically demanding, costly and time-consuming process. Sydney's opera house, for instance, is undeniably one of the most iconic structures ever built. However, looking at the building solely from the perspective of construction site management leaves a bitter taste. This impression can be justified by an estimated cost of 7 million AUD in the early stages of the project, which finally added up to 49.4 million AUD as well as a delay in construction of six years (Murray, 2003, p. xii). These issues are of course widespread across the globe (Cantarelli *et al.*, 2012) and not a peculiarity of the Australian building sector. An extensive study by Kostka and Fiedler (2016) analysed 119 large infrastructure projects in Germany, which revealed a mean cost overrun of 73% for already finished projects and 41% for ongoing projects while it can expected that these numbers will increase. Several reasons for cost overrun and delays of construction were identified such as pioneer risks in the light of novel materials or untested construction methods – an aspect that clearly applies to the aforementioned Sydney opera house. Other explanations include aspects correlated to project management (Shenhar and Dvir, 2007) as well as lack of external controlling and monitoring. The latter explanation contains several aspects such as monitoring craftsman-like, temporal and geometrical concerns. Typical tools in the architecture, engineering and construction (AEC) industries for monitoring are visual inspection, photographic documentation, geometrical spot-checks by usage of plumbs, tape measures or handheld rangefinders and geodetic techniques such as levelling or tacheometry. All in all, these techniques lead to a random and heterogeneous documentation that do not satisfy demands of digital archiving in the light of a structures' lifecycle. According to Manyika *et al.* (2015), the degree of digitisation in construction is one of the least developed sectors in the U.S., which hinders its improvement in productivity and performance (Barbosa *et al.*, 2017). Laser scanning is an emerging and well-suited sensor technology that could undeniably contribute to construction site monitoring on several levels, such as structural progress and productivity tracking, quality assurance, digital inventories and archiving to dimensional compliance.

The necessity of monitoring from a monetary point of view is depicted in Figure 14.1, which illustrates the inherent relation between the chance of making necessary corrections during

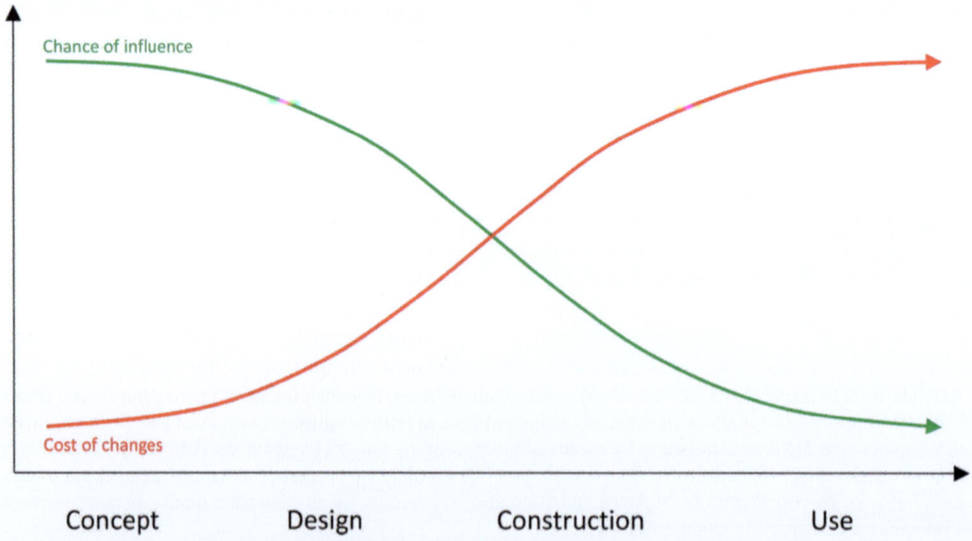

Figure 14.1. Relation between the chance of making necessary changes and the required costs to correct recognised issues (based on Project Management Institute, 2008, p. 17).

construction and the corresponding costs (Schoonwinkel *et al.*, 2016). It is obvious that the chance of influencing the final outcome is greatest at the very beginning of the project, while the required monetary effort is at the lowest point. Ultimately, it can be concluded that it is advisable to identify inadequate components or lots, preferably soon in order to avoid cost overruns and a delay of the entire project.

This chapter will discuss various aspects related to monitoring of construction sites. At first, the focus is set on data acquisition and referencing of point clouds in section 2 which forms a vital prerequisite for monitoring. Different forms of construction site monitoring (CSM) based on terrestrial laser scanning are of interest in section 3, while the last section summarises the content of this chapter.

2 DATA ACQUISITION AND REFERENCING OF POINT CLOUDS

Working on construction sites brings some peculiarities to light that usually cannot be found in other problem domains where a surveyor can navigate freely through the area of interest, which will be the subject of section 2.1. The detectability of changes with TLS is discussed in section 2.2. A prerequisite before conclusions about the construction progress can be drawn is the necessity to reference datasets into a common coordinate system and hence that stands in the focus of section 2.3.

2.1 *Challenges regarding data acquisition on construction sites*

Construction sites are challenging environments for geodetic data acquisition for numerous reasons. One of the most demanding circumstances is the fact that measurements usually have to be carried out during construction work on the bustling site in order to avoid delays of the construction process. Consequently, people and machines are moving through the area of interest, yielding occlusions, which may have to be compensated by additional scans extending the time on site. The issue of laser safety regulations can be particular cumbersome when scanners above laser class 1 (IEC, 2001) are being used, since it requires barricading certain areas of the site which hence means additional work as well as a potential source for conflicts. Heavy machines such as plate vibrators or pneumatic hammers are frequently used at construction sites, leading to notable vibrations of the ground even at greater distances to the TLS that may interfere with the geometric quality of the current scan. Large amounts of dust are released at every construction site as a consequence of, e.g., drilling, smoothing or mixing floor pavement or concrete. These particles rapidly settle, resulting in a thin layer on the instrument's surface. Dust on the deflection mirror yields in signal deterioration and hence has a negative impact on the measurements. Thus, the tidiness of the mirror should be frequently checked and cleaned if necessary. The most challenging aspect of scanning at construction sites is linked to the ever-changing geometry on site during data acquisition that may be mistaken with construction process. These changes can be for instance provoked by machines or scaffolds that are frequently moved into different places, building material which has been temporarily placed or simply modifications on the structure itself. Hence, it is desirable to identify and track secondary and temporary objects at construction sites. Turkan *et al.* (2014) presented a solution for this problem in the context of structural concrete works by incorporating 3D-model knowledge.

A typical scene is depicted in Figure 14.2 that shows an intensity image of a construction site. A scaffold can be seen in the centre that results in inevitable occlusions on the structure's surface. Pink boxes highlight building material that is temporarily stored at the site, yielding additional occlusions. Green boxes depict areas where rubble or trash was placed, while blue boxes mark building material and equipment that is currently in use.

2.2 *Detectability of changes or compliance on construction sites*

The suitability of a sensor for monitoring always depends on the geometric magnitude of changes which have to be documented. As for the case of construction sites, a large spectrum of possible magnitudes can be expected, ranging from earthworks for volumetric estimates (Slattery *et al.*, 2011) with low demands in terms of precision to monitoring e.g. the thickness of plaster, which may be fairly close to the precision of most scanners on the market. In brief, the detectability of geometric changes that are observed over time mainly depends on two ingredients, namely the magnitude or signal and the noise, which is a probabilistic or stochastic measure. The

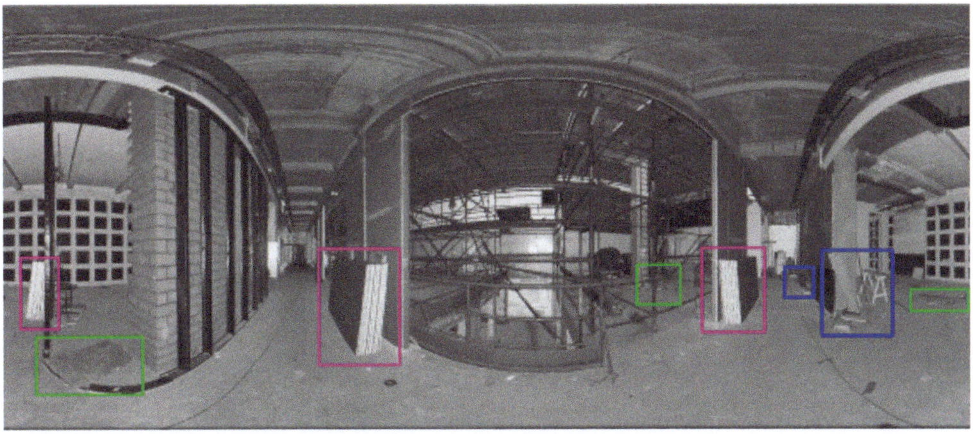

Figure 14.2. Intensity image of a construction site.

relationship among these two quantities can be expressed in the so-called signal-to-noise ratio, abbreviated by SNR

$$SNR = \frac{M_{signal}}{M_{noise}} \tag{1}$$

where M_{signal} denotes the magnitude of the signal, for instance a geometric modification of a building component as a consequence of construction, and M_{noise}, which represents the magnitude of noise. In general, the SNR should be a potentially large number, while geometric modifications of a certain magnitude are undetectable with a low SNR. This equation shows that it may be sufficient to use a less precise scanner in order to detect large-scale changes, while an instrument with a quite low level of noise is required to identify fairly small geometric phenomena. The description of stochastic effects is quite complex and is not just limited to the applied TLS (Soudarissanane, 2016; Wujanz *et al.*, 2017) but also has to consider the influence of registration as well as the polymorphic nature of terrestrial laser scanning (Wujanz *et al.*, 2016). Polymorphic means in this context that every point cloud of an identical object leads to another yet different geometric description. Apart from considering random errors, it is also vital to ensure that the applied scanner is calibrated regarding its rangefinder as well as its mechanical components, as extensively discussed in Chapter 6. If this requirement is not satisfied, wrong conclusions regarding construction progress or compliance with specified standards are unavoidable. A comparative study in terms of accuracy and level of completeness among photometric techniques and terrestrial laser scanning for the sake of documenting infrastructure was published by Dai *et al.* (2012) on the example of a bridge.

2.3 *On the importance of referencing*

The most precarious preprocessing step before progress monitoring can be conducted is referencing several datasets into a common coordinate system. The criticality of this task can be justified by the fact that biases of the computed transformation parameters yield in erroneous magnitudes of construction progress or false conclusions regarding compliance to construction specifications. In general, two strategies can be distinguished, namely georeferenced and registration approaches. The peculiarity of the first method is that the common coordinate frame is described by a reference coordinate system, which is established by geodetic means, whereas several different techniques can be applied. The second methodology works solely on the acquired data. In general, the following scenarios likely arise in construction monitoring:

- In all cases: merging single scans captured at a certain point in time to a connected epoch.
- Data-driven CSM: referencing epochs captured at different points in time.
- Model-driven CSM: referencing of a certain epoch into a model's coordinate system.

In the following, several strategies are briefly discussed and finally assessed from the perception of construction monitoring.

2.3.1 Target-based registration

A quite common way of transforming several scans into a reference coordinate system is by usage of artificial targets (Chow *et al.*, 2010). Therefore, spherical or printed planar targets are used, which have to be placed within the scene. The target locations must be chosen in a way that they are visible from at least two viewpoints that should be registered. After data acquisition the target centres have to be determined either by computing the centre point of a sphere by means of adjustment calculation or by using appropriate algorithms (Abmayr *et al.*, 2008). Regarding the target configuration, it is important to mention that the targets should be located in a way that they are distributed in all cardinal directions and do not form a line. After extraction of the target centres point to point correspondences can be established between individual points from different viewpoints. Based on this information, transformation parameters among the two viewpoints can be computed and finally applied to the corresponding point clouds, yielding a registered dataset.

Several disadvantages can be associated to this strategy such as (i) the effort of distributing the targets in the area of interest, (ii) the limited extent of the placed targets within the region of interest and (iii – optionally) the required survey of the targets in order to determine their 3D-coordinates in a reference system. A general assumption of this approach is that the majority of targets remain geometrically stable over the course of the survey campaign. However, this might not be the case in the ever-changing environment of a construction site. A geodetic principle in the context of computing transformation parameters is that corresponding points should surround the ROI. If this is not the case, extrapolative effects occur quite likely, which falsify the outcome. The reason why targets are still commonly used in practice despite their numerous drawbacks can be found in the fact that all well-established operational procedures for total stations are directly transferable to TLS.

2.3.2 Surface-based registration

The most versatile registration method uses redundantly captured regions of two point clouds and forms the family of surface-based registration algorithms. A substantial advantage of this strategy over the remaining ones in this chapter is the actual use of the information present in the overlapping region of two or more point clouds. A drawback of these surface-based algorithms is their dependence to a sufficient pre-alignment of two datasets. Three general options can be used to satisfy this requirement, namely (i) manual determination of a few correspondences, (ii) measurement of the individual location and orientation of two scans – see section 2.3.4 for details – and (iii) use of pre-alignment algorithms such as Aiger *et al.* (2008).

After coarse alignment of two datasets relative to each other has been carried out, fine matching can be conducted. The necessity of conducting coarse matching before fine matching can be explained by the non-convexity of the optimisation function that leads to a convergence region in which the two point clouds need to fall into in order to avoid local minima of the target function during parameter estimation. One of the first surface-based registration algorithms, named iteratively closest point algorithm (ICP), has been proposed by Besl and McKay (1992), where point-to-point correspondences are established. This step has to be rated critical due to the fact that non-repeatable points are captured by TLS while this issue can be compensated by establishing point-triangle correspondences (Chen and Medioni, 1991). As measurements in general are always subject to erroneous impacts, a certain positional accuracy results for the case of laser scanners. In order to address this issue, Bae and Lichti (2008) proposed a weighted variant of Chen and Medioni's (1991) algorithm. An overview of several variations of the ICP can be found in Rusinkiewicz and Levoy (2001). It is well known that ICP-based approaches are sensitive against deformed regions, which yields erroneous registrations (Wujanz *et al.*, 2016). Since notable geometric differences usually arise on construction sites between epochs, it is of vital importance to reveal these regions within the captured datasets and to exclude them from the computation of transformation parameters.

2.3.3 Feature-based registration

Point clouds captured by terrestrial scanners also contain radiometric information (Höfle and Pfeifer, 2007), which is also referred to as intensity, in addition to the geometric content. Intensity values are assigned to individual points and are based on the strength of a reflected signal. If the topology of points within a dataset is known, this information can be used to convert a point cloud into an image where intensity values represent the brightness of individual pixels. By doing this, well-established techniques from the field of image matching can be applied to 3D datasets. As a first step, so-called keypoints have to be extracted, which are distinct radiometric features within a local neighbourhood in terms of their grey scale values. After keypoint extraction, descriptors are used to establish correspondences between keypoints from different datasets (Lowe, 1999). Based on this information, transformation parameters can be computed (Böhm and Becker (2007).

Apart from using radiometric features, geometric information can be used in the form of geometric primitives, such as planes (Bosché, 2012; Previtali *et al.*, 2014; Förstner and Khoshelham, 2017; Technet, 2019). As a first step, planar segments have to be extracted from the original point clouds (Vosselman *et al.*, 2004), while for each segment, plane parameters are adjusted. Then correspondences among identical planes are computed instead of matching single points between scans such as in radiometric approaches. By using adjusted planes instead of points, the precision of the resulting transformation parameters notably increases. Geometric features in the form of planes can be numerously found in nearly all structures and are hence particularly suitable for registration of data captured on construction sites, especially for the documentation of the interior.

2.3.4 Direct geo-referencing

Instead of computing transformation parameters based on established correspondences, the relation between different local TLS coordinate systems can also be achieved by observing their location and orientation within a superior coordinate frame. Methods for direct georeferencing of TLS were initially solely of scientific interest, e.g. Paffenholz *et al.* (2010) while lately several commercial systems emerged (Riegl, 2019; Zoller and Fröhlich, 2019). A significant drawback of direct georeferencing is the notable extension of a scanner's error budget (Soudarissanane, 2016), due to the use of additional positioning- and orientation sensors such as GNSS-equipment or electronic compasses. With increasing scanning range the impact onto the relative rotation between two point clouds also increases, which is a result of the limited accuracy of compasses. This could for instance be an issue for monitoring the progress of multi-storey buildings or dams where the range between object and scanner can sum up to several hundred metres. An issue related to data acquisition of interiors is that GNSS solutions for determination of the scanner's location fail, since the signals are shielded by the structure.

2.3.5 Assessing referencing strategies for construction monitoring

The previous subsections discussed several referencing approaches of point clouds, while Table 14.1 summarises important characteristics in the light of construction monitoring. For this, an assessment scheme is used that includes the classes "–", which stands for "not advisable", "o" that represents acceptable solutions or "+" highlighting recommendable methods. As discussed in section 2.2, the error budget of monitoring is not limited to the accuracy of the scanner but also involves influences for instance provoked by post-processing such as referencing. The most versatile referencing solutions are surface-based approaches such as the ICP, which can be found in nearly every commercial software solution. The only drawback is its vulnerability against

Table 14.1 Overview about different referencing approaches

	Target-based registration	*Surface-based registration*	*Feature-based registration*	*Direct-geo-referencing*
Error budget	–	o	Radiometric: o Geometric: +	–
Model-to-scan registration	No	Yes	Radiometric: No Geometric: Yes	No
Registration among epochs	–	+	Radiometric: – Geometric: +	o
Applicability indoors	Yes	Yes	Yes	No

deformation in combination with non-transparent quality measures (Wujanz, 2012). Since the location of corresponding points and their magnitude of residuals are not known, it is not clear if the computed set of transformation parameters has been biased. Another recommendable method for referencing is based on geometric features in the form of planes due to its favourable error budget as well as its suitability for usage in built environments. Despite its popularity in practice, it is not recommended to solely use artificial targets for referencing on construction sites. The first reason for this argument is the inability to compute transformation parameters of as-built scans into a model-coordinate system. Furthermore, the likelihood that targets are being damaged, manipulated or even removed is quite high on construction sites, which increases the risk of erroneous registrations or none in the worst possible case. Similar issues apply to radiometric features, since the surface characteristics on construction sites are subject to various sources of change, for instance due to pargetting, painting, concreting or plastering. Direct-geo-referencing is not suitable to be solely used however may be helpful to compute approximate solutions for other methods.

3 METHODS OF CONSTRUCTION SITE MONITORING

The term "construction site monitoring" is ambiguous and can be interpreted in several ways. A first interpretation of CSM is an exclusively data-driven monitoring approach of a construction site without any previous knowledge about the sequential order of lots at the building site. Therefore, the site is frequently surveyed or at specified points in time, referred to as epochs, which can be optionally defined by the construction manager. This strategy will be discussed in greater detail in section 3.1. A second way of understanding CSM is correlated to the question whether a lot or element has been produced or not at a given point in time according to a design intent. This spatio-temporal problem requires specific information about the construction schedule as well as details regarding the building plan and is the subject of section 3.2. Apart from the circumstance that a lot has been built, it is vital to validate its compliance with given construction standards. This obviously requires construction standard specifications that are compared to individual parts of the captured point cloud, as explained in section 3.3. In summary, CSM can be used to draw conclusions about a lot or component in terms of:

- timeliness against a construction schedule,
- construction progress between epochs,
- correctness in comparison to a design or plan and,
- its professional implementation.

Table 14.2 summarises characteristics among the CSM categories. Data-driven CSM, as its name says, solely works on captured data and hence does not require any previous knowledge. For that, point clouds captured in at least two epochs are required, which have to be registered into a common coordinate system. A common way to illustrate the outcome of construction monitoring is to colourise individual points of the point cloud, for instance in dependence to the magnitude of geometric difference. Model-driven CSM can be carried after acquisition of at least one epoch, which is then compared to a building plan in geometric terms and/or to a work schedule that allows drawing conclusions about the timeliness of construction. Again, it is vital to register both geometric datasets into a common coordinate system that allows drawing geometric conclusions. Assessing the compliance of an individual construction component with a given standard specification can be carried out after one visit of the corresponding site. Before data acquisition, it is important to level the applied scanner and to activate the inclinometer or compensator in order to ensure that conclusions about the horizontality or verticality can be drawn. The outcome of this CSM-process is a Boolean statement about the compliance of the component. In this context, it is important to point out that tolerances, which are the most common quality measure in construction,

Table 14.2 Overview about different CSM strategies

	Data-driven CSM	*Model-driven CSM*	*Compliance*
Least required data	> = 2 epochs	> = 1 epoch(s)	1 epoch
Required previous knowledge	None	Work schedule & building plan	Construction standard
Registration required	Yes	Yes	No (levelled)
Outcome	Colour-coded point cloud or Boolean conclusion (geometry)	Boolean conclusion (geometry, time)	Boolean conclusion (geometry)

are deterministic measures, while the specified precision and accuracy values of a scanner are of stochastic nature. A straightforward solution to this problem is to compute statistical boundaries based on the stochastic information (e.g. Pukelsheim, 1994) and to compare the resulting values with the given construction standards.

Common actions yielding geometric modification of the construction site are:

- Construction or demolition of a lot or component.
- Modification of a given lot – for instance, plastering or insulation.

Apart from the aforementioned typical actions temporarily stored building material, installation or removal of moulds as well as piles of rubble or trash also lead to a geometric modification of a construction site and require precautions in order to avoid falsification of the referencing process. Due to the aforementioned occlusions, it is usually not possible to capture the entire surface of a construction site. As a consequence, conclusions about the status of a building component can only be partially drawn.

3.1 *Data-driven construction site progress monitoring*

A straightforward form of construction site monitoring is purely data driven and generally coincides with methods for deformation monitoring based on laser scans, as already discussed in Chapter 8. While the majority of academic research focuses on model-based CSM, this procedure is quite common in practice. This can be explained by the fact that site managers or architects usually do not pass digital construction plans on to the surveyor. The other way around, experts from the AEC industry do not possess required software and/or the necessary expertise to perform progress monitoring. In order to be able to quantify geometric changes that occurred in between epochs, correspondences between the scans have to be established. This problem is identical to surface-based registration where point-to-point correspondences are determined by means of the ICP (Besl and McKay, 1992) or point-to-triangle correspondences. Equivalents in deformation monitoring are for instance Girardeau-Montaut *et al.* (2005) or Cignoni *et al.* (1998). Building progress is interpreted as geometrical differences between point clouds or derived measures captured in at least two different epochs.

Figure 14.3 illustrates different strategies on the example of a single room that was captured in two epochs. The first epoch is shown Figure 14.3a and features a matrix of windows on the lower left, a wall that surrounds supply circuits in the upper right, steel columns for drywalling as well as some building material. Two weeks after, bricklayers have closed the wall, drywalls were installed and the window frames were plastered as depicted in Figure 14.3b. This example highlights two critical issues related to this strategy of construction site progress monitoring. The first problem is that it is not advisable to register the two given datasets based on the depicted parts of the point clouds, since all components were either built or modified apart from the floor.

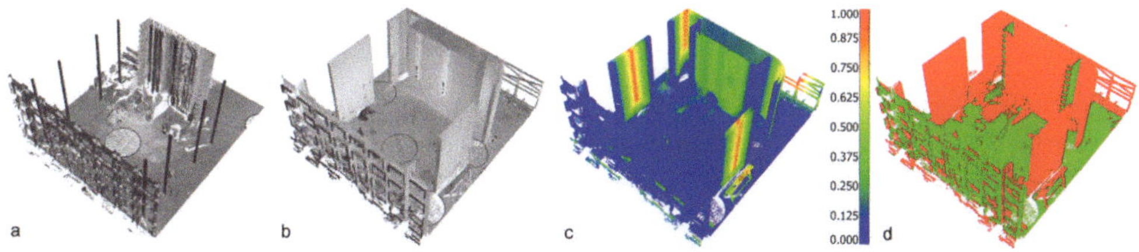

Figure 14.3. Point cloud of a room before (a) and after drywalling and plastering (b). Building progress according to C2C-inspection given in metres (c) and thresholding (d).

If one would nevertheless do so, the bias of transformation parameters would not allow us to correctly identify changes that occurred at the site. In the given case, the registration was carried out by using unaltered regions on site that are not depicted in the figure. The second issue becomes apparent in Figure 14.3c that shows the outcome of a C2C-inspection [m] based on the two point clouds. The colour-coding on various parts of the point cloud leads to the conclusion that progress was made at the construction. However, the magnitude of change does not correspond to the construction progress. This is clearly visible in regions of the point cloud where components were erected, for instance on the upper left side of the room. The colour pattern leads to the conclusion that non-planar modification occurred in between epochs. This effect stems from the determination of correspondences based on the closest point principle. Since there was no wall in the first epoch, point correspondences were established between the gypsum panels and the steel column. Therefore, the colour-coding represents the distance from a certain point on the panel to the column. The generated result indicates that this strategy is suitable to monitor construction progress that is subject to changes with smaller magnitudes such as plastering. To detect recently constructed components, the result can simply be converted into a binary colour scheme. Therefore, a threshold has to be specified, so that all points with larger deviations between the two point clouds are considered to be construction progress. Figure 14.3d illustrates such a binary representation where the threshold was set to 1 cm. This visualisation allows one to distinguish changes that occurred on site, highlighted by red points, and unaltered parts, which are tinted in green. Note that both results were generated with CloudCompare (Girardeau-Montaut, 2015). An alternative procedure to compute a binary coloured point cloud would be to decompose the datasets into voxels of equal size (Samet, 2006, p. 211). Every voxel that contains points from just one dataset can be considered to be construction progress. Data-driven approaches were used, for instance, by Shih *et al.* (2004) and are implemented in the commercial solution BuildIT Construction (BuildIT, 2019).

Even though the solutions described above are widespread in the scientific community, they nevertheless are subject to several disadvantages. One of them is the use of deterministic thresholds as a quality measure that inherently assumes equal precision for all points within a captured dataset, which is clearly not the case, as extensively discussed in Chapter 6. Wujanz *et al.* (2018) address these issues by carrying out plane-based monitoring. Therefore, the original point clouds are segmented into planar fragments. Monitoring is performed by comparing individual plane parameters from one epoch for identity to ones captured at a different point in time. If a plane has been redetected in between epochs, it is considered to have remained unaltered while changes to the original plane parameters or until-then-undetected planar patches are interpreted as construction progress. The decision on geometrical identity is carried out by a statistical test deploying a comprehensive error budget that contains effects provoked by the applied sensor as well as registration. By using adjusted planes, a higher level of precision is achieved compared to individual points at the cost of spatial resolution. Figure 14.4 illustrates the same scene depicted in Figure 14.3 in the form of intensity images. The left part shows the captured room before (left) and after drywalling and

Figure 14.4. Intensity image with detected planes (red patches) of a room before (left) and after drywalling and plastering (right).

plastering. The red patches highlight areas that have been detected as being planar that are hence used for progress monitoring.

3.2 *Construction site monitoring deploying model information*

A current trend in managing construction sites is to use building information models (BIM) instead of computer aided design (CAD) models, since it embodies a far more extensive description of the design intent (Turkan *et al.*, 2012; Bieńkowska, 2017). BIM includes, apart from dimensional specifications, component-wise schedules and prevailing tolerances as well as details regarding the implementation in terms of material etc. for individual building components. Despite its indisputable advantages, it is recommended to consider critical aspects of BIM; see for instance Hosseini *et al.* (2016) or Dainty *et al.* (2017).

In contrast to data-driven approaches, building monitoring is performed by comparison of a dataset captured at a certain point in time against a designated state by considering knowledge from a building model. In general, two cases can be distinguished, namely, model-wise and component-wise strategies (Boukamp and Akinci, 2004). As for the first case, the assumption is made that a structure's design is identical to its as-built model in geometric terms. Discrepancies between the two input instances are identified by comparison of a deviation against a global threshold (Akinci *et al.*, 2006). However, this strategy is subject to several difficulties provoked by e.g. missing components in either the as-built data or the design-model, secondary and temporary objects in the scene, and misplaced components at the construction site. This will lead to issues during registration (Wujanz, 2012) as well as consequently lead to false correspondences between parts of the point cloud and individual building components. In addition, use of a global threshold does not make sense from the perception of construction standardisation since different components require different tolerances. Hence, it is desirable to perform the actual monitoring process at component level for individual objects. A crux in this context is to decompose the entirety of the captured point cloud into subsets that represent individual components, since no method for 3D-data acquisition does naturally generate data at object level (Turkan *et al.*, 2012). Generally, this task corresponds to the segmentation procedures described in Chapter 7. Bosché (2010) tackles this hurdle by performing a two-tiered registration procedure. At first, a point cloud describing the as-built state of a construction site is registered into the coordinate system of a given model by using surface-based registration comparable to the ICP. Subsequently, this procedure is repeated on object level for individual components between model and point cloud, yielding refined registrations. As a result, discrepancies among the as-designed and as-built state can be revealed. The aforementioned issue of object occlusion is considered by defining minimum thresholds of coverage (Bosché *et al.*, 2009). Thus, objects are considered to be built when this threshold is surpassed. Bosché (2010) assumed that the shape of every as-built entity complies with specified tolerances, which is reasonable for prefabricated components. However, this assumption usually does not hold for the majority of cases leading to biased transformation parameters and ultimately false conclusions.

Recent developments on the market also revealed some noteworthy commercial solutions. Clearedge (2019) introduced software that is capable of verifying individual elements and building components in terms of geometric compliance and punctuality in comparison to a given construction plan, which is comparable to the implementation of Bosché (2010). Information about the planned status of the construction site is exchanged through an integrated interface to Autodesk's Navisworks (Autodesk, 2019). The workflow of the software is initiated by a transformation of the point clouds captured in the field into the coordinate system of the given 3D model. By aligning the point cloud with the 3D design model, each design element is compared against its true position as indicated by the point cloud. Subsequently, the point cloud is decomposed into segments that describe parts of or entire building components. This subset of points is then associated with the corresponding component. A duplicate of the component under investigation of the model is then registered to the point cloud subset via ICP, yielding local transformation parameters. The geometric compliance to the building plan is then derived based on the discrepancy among the planned building component and its registered duplicate. This information can then be extracted and shared amongst the other stakeholders of the project. Figure 14.5 illustrates the inspection of a set of stairs. The left side of the figure shows the point cloud as well as the planned staircase, that is tinted in blue, in Navisworks. On the right of the figure, the outcome of the inspection is depicted in form of a heat map in Verity. The colour-coding is directly dependent to the specified tolerance, which was 15 mm in relation to the construction plan in this case. Red regions, especially on the lower stairs, indicate erroneous parts of the component. Consequently, the staircase receives the status "out of tolerance".

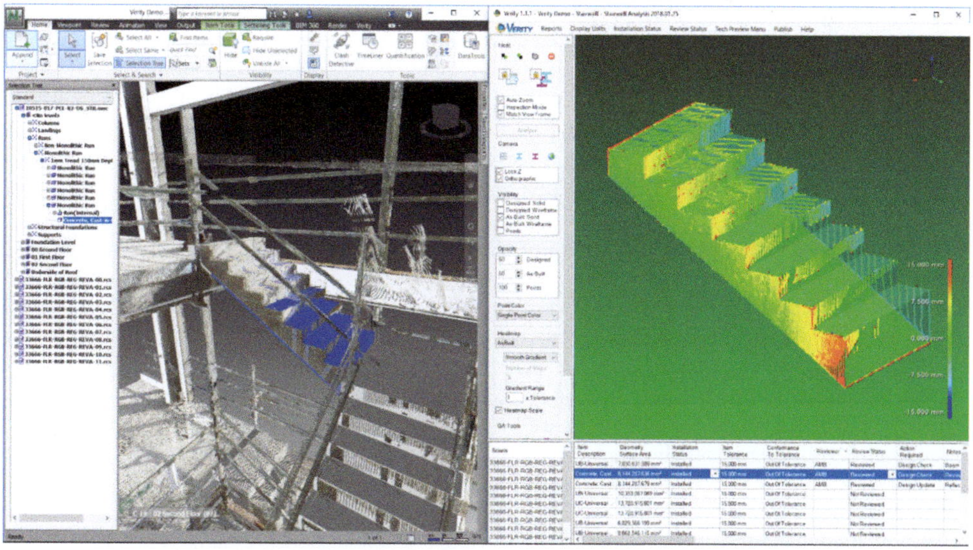

Figure 14.5. A 3D design model of a staircase and a captured point cloud in Navisworks is depicted on the left. The right part of the figure illustrates the result of an inspection generated by Verity.

3.3 *Compliance of individual components with construction standards*

Tolerances in the context of construction define numerical boundaries or thresholds that parts of building components have to satisfy. A typical example would be a specified planarity of a wall's face, while related standards are e.g. ASTM E1155, ISO 7976-1 or the German DIN 18202. Conventional tests for planarity compliance according to DIN 18202 are usually performed as spot-checks or in the form of grid measurements, where a larger surface is systematically tested. Typically, planarity is analysed by use of a straightedge and a rule. At first, the straightedge is placed on the wall under investigation, where it has two mechanical contact points and thus defines a reference line. Then, the perpendicular distance between the straightedge and its largest deviation to the wall is measured. If this value falls below the specified tolerance, the wall complies with the assumed construction standard. It is obvious that the detected deviations are directly dependent on the chosen location of the straightedge and hence are subjective. Due to their high spatial resolution, compliance checks based on TLS point clouds can notably decrease the level of subjectivity. Other desired characteristics in architecture are for instance the alignment of several columns, the horizontality of a floor or the vertical orientation of a wall. For the last two tests, it is vital to thoroughly level the applied TLS or to activate the instrument's inclination sensor. Assuming the case that compliance is tested based on data that was previously registered and hence stems from more than one viewpoint, it is important to restrict the rotational degrees of freedom in order to avoid vertical misalignments in the registration process.

In the following, a practical compliance check is exemplified on a brick wall that is depicted in Figure 14.6a. Since the wall was surveyed from one viewpoint, the error budget does not contain influences provoked by registration. Based on the captured data and usage of an intensity-based stochastic model (Wujanz *et al.*, 2017), a colour-coded visualisation of the 3D precision (3σ) for individual points was computed that is illustrated in Figure 14.6b. It can be seen that the precision of individual points is quite heterogeneous, reaching from approximately 1.5 to 3.5 mm. Since the effective tolerance for a wall of this kind is 15 mm according to DIN 18202, it can be established that the precision of the given data is sufficient to test the wall for compliance. A practical difference between conventional compliance checks and inspection by TLS is that the magnitudes of the computed deviations differ. This effect stems from the fact that deviations to the straightedge are based on a mechanical realisation, while this is not the case for laser scanning data. A commonly chosen reference for compliance checks based on TLS point clouds is plane parameters that are estimated by means of least squares adjustment. It is crucial to exclude points that do not physically belong to the plane from the adjustment, as it would otherwise bias the estimated plane parameters and consequently the compliance check. Finally, perpendicular distances for all points in a laser scan to the reference plane are computed. Since the reference plane has been computed based on a least squares adjustment, positive and negative discrepancies arise. Hence, the tolerance threshold has to be divided by 2 and finally compared against absolute deviations to the computed plane. The outcome for the brick wall according to DIN 18202 is depicted in Figure 14.6c. Several red regions are visible that indicate that the corresponding deviations to the reference plane are larger than the tolerances. For the given case, the chosen wall does not comply with the legal construction standard. Another practical solution is to combine different compliance tests. This can be achieved by introducing constraints to the least squares adjustment. Assuming the case that planarity and verticality should be checked for compliance, the constraint has to be introduced that the face normal of the adjusted plane is perpendicular to the vertical axis of the scanner's coordinate system. The outcome of this test is depicted in Figure 14.6d. A comparative look at Figures 14.6c and 14.6d shows notably different deviation pattern. Yet again, the red regions indicate that the wall does not comply with the effective construction standards.

An algorithm comparable to the discussed course of action is implemented in the commercial software BuildIT Construction (BuildIT, 2019), which is able to perform flatness and floor levelness checks according to the ASTM E1155 standard. Interested readers on the subject are referred to Bosché and Guenet (2014), who present different implementations of straightedge methods on

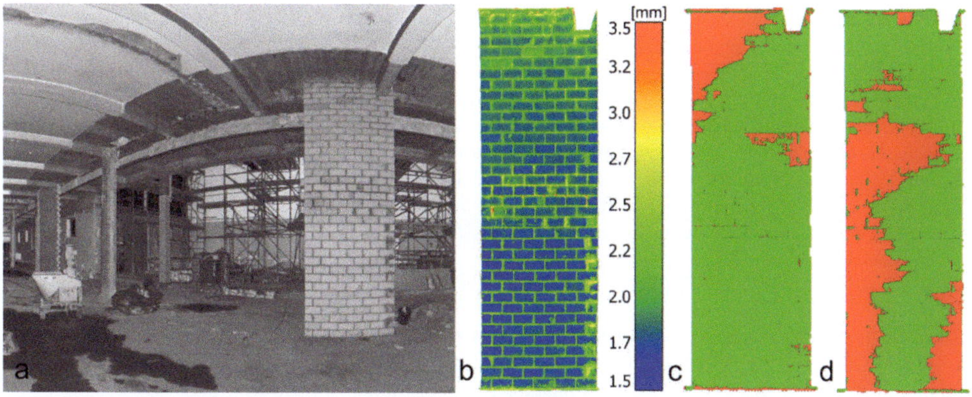

Figure 14.6. Compliance check of a brick wall (a) under the assumption of planarity according to DIN 18202. 3D precision (3σ) of individual points (b). The remaining colour coded inspection maps test planarity (c) as well as planarity and verticality (d).

terrestrial laser scans in compliance to ASTM E 1155–96 as well as grid-based solutions for flatness control.

4 SUMMARY

This chapter argued for the necessity of monitoring of construction sites while terrestrial laser scanners are a suitable sensor technology for it. TLS should be seen as an additional tool that adds value to existing and established methods in terms of digital archiving, information density and geometric accuracy. In section 2, it was shown that construction sites are a challenging environment for data-acquisition by usage of laser scanners and that demands regarding the precision of laser scanners depend on the magnitude of geometric changes of interest. A critical component of construction site monitoring is linked to registration among point clouds or to an existing model. Biases of the computed transformation parameters consequently lead to an immediate falsification regarding the subsequent construction monitoring process. Section 3 discussed several strategies for construction monitoring with and without given model information as well as compliance control to given construction standards. A topic that requires more scientific attention in the future is the identification of individual building components from acquired point clouds. Until now this process is exclusively based on geometric information, while prospective research should also incorporate radiometric information that may be provided by a novel generation of terrestrial laser scanners (Eitel *et al.*, 2016). Another issue worth considering is the exploitation of a comprehensive error budget for the entire process chain of construction site monitoring ranging from data acquisition to post-processing. This should supersede the naive assumption of equal precision within point clouds that is inherently postulated in the form of global thresholds.

REFERENCES

Abmayr, T., Härtl, F., Hirzinger, G., Burschka, D. & Fröhlich, C. (2008) A correlation based target finder for terrestrial laser scanning. *Journal of Applied Geodesy*, 2(3), 131–137.

Aiger, D., Mitra, N.J. & Cohen-Or, D. (2008) 4-points congruent sets for robust pairwise surface registration. *Transactions on Graphics*, 27(3), Article 85.

Akinci, B., Boukamp, F., Gordon, C., Huber, D., Lyons, C. & Park, K. (2006) A formalism for utilization of sensor systems and integrated project models for active construction quality control. *Automation in Construction*, 15(2), 124–138.

Autodesk. (2019) *Navisworks – Project Planning and Review Software for AEC professionals*. Available from: www.autodesk.co.uk/products/navisworks/overview [accessed 22nd July 2019].

Bae, K.-H. & Lichti, D.D. (2008) A method for automated registration of unorganised point clouds. *ISPRS Journal of Photogrammetry and Remote Sensing*, 63(1), 36–54.

Barbosa, F., Woetzel, J., Mischke, J., Ribeirinho, M.J., Sridhar, M., Parsons, M., . . . Brown, S. (2017) Reinventing construction: A route to higher productivity. *McKinsey Insights Report*, McKinsey & Company and McKinsey Global Institute.

Besl, P.J. & McKay, N.D. (1992) Method for registration of 3-D shapes. *Robotics-DL Tentative*, International Society for Optics and Photonics. pp.586–606.

Bieńkowska, E. (2017) *Handbook for the Introduction of Building Information Modelling by the European Public Sector*, s.l., EUBIM Taskgroup.

Böhm, J. & Becker, S. (2007) Automatic marker-free registration of terrestrial laser scans using reflectance. *Proceedings of 8th Conference on Optical 3D Measurement Techniques*, Zurich, Switzerland. pp.338–344.

Bosché, F. (2010) Automated recognition of 3D CAD model objects in laser scans and calculation of as-built dimensions for dimensional compliance control in construction. *Advanced Engineering Informatics*, 24(1), 107–118.

Bosché, F. (2012) Plane-based registration of construction laser scans with 3D/4D building models. *Advanced Engineering Informatics*, 26(1), 90–102.

Bosché, F. & Guenet, E. (2014) Automating surface flatness control using terrestrial laser scanning and building information models. *Automation in Construction*, 44, 212–226.

Bosché, F., Haas, C.T. & Akinci, B. (2009) Automated recognition of 3D CAD objects in site laser scans for project 3D status visualization and performance control. *Journal of Computing in Civil Engineering*, 23(6), 311–318.

Boukamp, F. & Akinci, B. (2004) Towards automated defect detection: Object-oriented modeling of construction specifications. *International Conference on Computing in Civil and Building Engineering*, Bauhaus-Universität Weimar, Germany.

BuildIT. (2019) *BuildIT Construction – Validate, Evaluate, Position and Monitor*. Available from: www.builditsoftware.com/products/buildit-construction [accessed 22nd July 2019].

Cantarelli, C.C., Flyvbjerg, B. & Buhl, S.L. (2012) Geographical variation in project cost performance: The Netherlands versus worldwide. *Journal of Transport Geography*, 24, 324–331.

Chen, Y. & Medioni, G. (1991) Object modeling by registration of multiple range images. *Proceedings of the International Conference on Robotics and Automation*, Sacramento, CA, USA. pp.2724–2729.

Chow, J., Ebeling, A. & Teskey, B. (2010) Low cost artificial planar target measurement techniques for terrestrial laser scanning. *Proceedings of the FIG Congress 2010: Facing the Challenges–Building the Capacity*. Sydney, Australia. 1–3 May. 6 p. (on CD-ROM).

Cignoni, P., Rocchini, C. & Scopigno, R. (1998) Metro: Measuring error on simplified surfaces. In: *Computer Graphics Forum*. Vol. 17, No. 2. Blackwell Publishers, Hoboken, NJ, USA. pp.167–174.

Clearedge (2019) *Verity – Construction Verification Software*. Available from: www.clearedge3d.com/products/verity [accessed 22nd July 2019].

Dai, F., Rashidi, A., Brilakis, I. & Vela, P. (2012) Comparison of image-based and time-of-flight-based technologies for three-dimensional reconstruction of infrastructure. *Journal of Construction Engineering and Management*, 139(1), 69–79.

Dainty, A., Leiringer, R., Fernie, S. & Harty, C. (2017) BIM and the small construction firm: A critical perspective. *Building Research & Information*, 45(6), 696–709.

Eitel, J.U., Höfle, B., Vierling, L.A., Abellán, A., Asner, G.P., Deems, J.S., . . . Mandlburger, G. (2016) Beyond 3-D: The new spectrum of Lidar applications for earth and ecological sciences. *Remote Sensing of Environment*, 186, 372–392.

Förstner, W. & Khoshelham, K. (2017) Efficient and accurate registration of point clouds with plane to plane correspondences. *Proceedings of the IEEE Conference on Computer Vision and Pattern Recognition*. pp.2165–2173. https://ieeexplore.ieee.org/document/8265463

Girardeau-Montaut, D. (2018) *Cloud Compare – 3D Point Cloud and Mesh Processing Software*. Open Source Project.

Girardeau-Montaut, D., Roux, M., Marc, R. & Thibault, G. (2005) Change detection on points cloud data acquired with a ground laser scanner. *International Archives of Photogrammetry, Remote Sensing and Spatial Information Sciences*, 36(part 3), W19.

Höfle, B. & Pfeifer, N. (2007) Correction of laser scanning intensity data: Data and model-driven approaches. *ISPRS Journal of Photogrammetry and Remote Sensing*, 62(6), 415–433.

Hosseini, M.R., Banihashemi, S., Chileshe, N., Namzadi, M.O., Udaeja, C., Rameezdeen, R. & McCuen, T. (2016) BIM adoption within Australian Small and Medium-sized Enterprises (SMEs): An innovation diffusion model. *Construction Economics and Building*, 16(3), 71–86.

IEC (2001) *International Standard IEC 60825-1: Safety of Laser Products – Part 1: Equipment Classification, Requirements and User's Guide*. IEC Geneva, Switzerland.

Kostka, G. & Fiedler, J. (2016) *Large Infrastructure Projects in Germany: Between Ambition and Realities – Working Paper 1*. Available from: www.hertie-school.org/infrastructure [accessed 15th March 2018].

Lowe, D. (1999) Object recognition from local scale-invariant features. Computer vision, 1999. *The Proceedings of the Seventh IEEE International Conference On Computer Vision*, vol. 2. pp.1150–1157. https://ieeexplore.ieee.org/document/790410

Manyika, J., Ramaswamy, S., Khanna, S., Sarrazin, H., Pinkus, G., Sethupathy, G. & Yaffe, A. (2015) *Digital America: A Tale of the Haves and Have-Mores*. McKinsey Global Institute. Available from: http://www.mckinsey.com/industries/high-tech/our-insights/ digital-america-a-tale-of-the-haves-and-have-mores [accessed 22nd July 2019].

Murray, P. (2003) *The Saga of Sydney Opera House: The Dramatic Story of the Design and Construction of the Icon of Modern Australia*. Routledge, Abingdon-on-Thames, UK.

Paffenholz, J.A., Alkhatib, H. & Kutterer, H. (2010) Direct geo-referencing of a static terrestrial laser scanner. *Journal of Applied Geodesy*, 4(3), 115–126.

Previtali, M., Barazzetti, L., Brumana, R. & Scaioni, M. (2014) Scan registration using planar features. *The International Archives of Photogrammetry, Remote Sensing and Spatial Information Sciences*, 40(5), 501.

Project Management Institute. (2008) *A Guide to the Project Management Body of Knowledge. ANSI/PMI 99-001-2008*. Project Management Institute, Philadelphia, PA, USA.

Pukelsheim, F. (1994) The three sigma rule. *The American Statistician*, 48(2), 88–91.

Riegl (2019) *Riegl VZ 2000i Data Sheet*. Available from: http://www.riegl.com/uploads/tx_pxpriegldownloads/RIEGL_VZ-2000i_Datasheet_2019-05-28.pdf [accessed 22nd July 2019].

Rusinkiewicz, S. & Levoy, M. (2001) Efficient variants of the ICP algorithm. *Third International Conference on 3-D Digital Imaging and Modeling*. https://doi.org/10.1109/IM.2001.924423.

Samet, H. (2006) *Foundations of Multidimensional and Metric Data Structures*. Kaufmann, San Francisco, CA, USA. 1024 p.

Schoonwinkel, S., Fourie, C.J. & Conradie, P.D.F. (2016) A risk and cost management analysis for changes during the construction phase of a project. *Journal of the South African Institution of Civil Engineering*, 58(4), 21–28.

Shenhar, A.J. & Dvir, D. (2007) Project management research-the challenge and opportunity. *Project Management Journal*, 38(2), 93.

Shih, N.J., Wu, M.C. & Kunz, J. (2004) The inspections of as-built construction records by 3D point clouds. *Center for Integrated Facility Engineering (CIFE), Working Paper # 90*, Stanford University.

Slattery, K.T., Slattery, D.K. & Peterson, J.P. (2011) Road construction earthwork volume calculation using three-dimensional laser scanning. *Journal of Surveying Engineering*, 138(2), 96–99.

Soudarissanane, S. (2016) *The Geometry of Terrestrial Laser Scanning; Identification of Errors, Modeling and Mitigation of Scanning Geometry*. Doctoral dissertation. Delft University of Technology.

Technet (2019) *Scantra – Beyond Clouds*. Available from: www.technet-gmbh.com/en/products/scantra/#c1263 [accessed 22nd July 2019].

Turkan, Y., Bosché, F.T., Haas, C.T. & Haas, R. (2012) Automated progress tracking using 4D schedule and 3D sensing technologies. *Automation in Construction*, 22, 414–421.

Turkan, Y., Bosché, F.T., Haas, C.T. & Haas, R. (2014) Tracking of secondary and temporary objects in structural concrete work. *Construction Innovation*, 14(2), 145–167.

Vosselman, G., Gorte, B.G., Sithole, G. & Rabbani, T. (2004) Recognising structure in laser scanner point clouds. *International Archives of Photogrammetry, Remote Sensing and Spatial Information Sciences*, 46(8), 33–38.

Wujanz, D. (2012) Towards transparent quality measures in surface-based registration processes: Effects of deformation onto commercial and scientific implementations. *Proceedings of the XXII Congress of the International Society of Photogrammetry and Remote Sensing*, Melbourne, Australia.

Wujanz, D., Burger, M., Mettenleiter, M. & Neitzel, F. (2017) An intensity-based stochastic model for terrestrial laser scanners. *ISPRS Journal of Photogrammetry and Remote Sensing*, 125, 146–155.

Wujanz, D., Gielsdorf, F., Romanschek, E. & Clemen, C. (2018) Ebenenbasiertes Baufortschrittsmonitoring unter Verwendung von terrestrischen Laserscans. *Proceedings of the 18th Oldenburger 3D-Tage: Optische 3D-Messtechnik - Photogrammetrie - Laserscanning*, Oldenburg, Germany.

Wujanz, D., Krueger, D. & Neitzel, F. (2016) Identification of stable areas in unreferenced laser scans for deformation measurement. *The Photogrammetric Record*, 31(155), 261–280.

Zoller + Fröhlich (2019) *Data Sheet*. Available from: Z+F IMAGER 5010X. www.zf-laser.com/fileadmin/editor/Datenblaetter/Z_F_IMAGER_5010X_System_Requirements_E_FINAL.pdf [accessed 22nd July 2019].

Laser Scanning – Riveiro and Lindenbergh
© 2020 ISPRS, ISBN 978-1-138-49604-0

Chapter 15

Integrated modelling and management information system (MMIS) for SCAN-to-BIM projects

Fabrizio Banfi and Luigi Barazzetti

ABSTRACT: Building information modelling (BIM) is an innovative tool that produces real benefits for the architecture, engineering and construction (AEC) industry, automating and innovating different phases of the construction process. In recent years, various studies have demonstrated that BIM is an important resource also for existing buildings, becoming a useful method for renovation and building management over time. This research reflects the experience gained in some SCAN-to-BIM projects and proposes a modelling management information system (MMIS) able to convert point clouds into as-built BIM (AB-BIM) by novel grades of generation (GOG) and information (GOI). The integration of specific advanced modelling techniques (AMT) in BIM application has solved generative modelling issues and has permitted a reduction in terms of both time and cost of the 3D reconstruction. The need to adapt the AB-BIM to different types of analyses, such as finite element analysis (FEA) and monitoring, has led to the implementation of a novel automatic verification system (AVS) able to check the metric accuracy between point clouds and 3D objects, giving a numerical indicator of the quality of the whole AB-BIM.

1 INTRODUCTION

In recent years, the construction market has favoured the exponential growth of new building causing uninhabited areas and disadvantaging the preservation and rehabilitation of existing buildings. Besides, the lack of resources has prevented the formation of experts able to manage the built heritage through new technologies, penalizing the re-engineering of the construction sector. Fortunately, countries like the United States, United Kingdom, Scandinavia, Germany, Singapore, France, China and South Korea are trying to provide innovative solutions to improve productivity and reduce cost through the use of building information modeling (BIM) in the AEC sector (Azhar, 2011). Some countries have introduced (in national regulations) the wide-scale adoption of BIM, obtaining impressive results for the renovation and conservation of historical and existing buildings (Volk *et al.*, 2014).

In addition, scientific research and information and communications technologies (ICT) are proposing new market solutions, showing how the integration of 3D survey data and BIM (SCAN to BIM) can also be used for various scopes. Examples are finite element analysis (Fedorik *et al.*, 2016), infrastructure design (Jackson *et al.*, 2016), construction management (Sacks *et al.*, 2010), energy analysis (Kim *et al.*, 2016), MEP (Bosché *et al.*, 2015), mixed reality and virtual reconstruction (Barazzetti and Banfi, 2017).

In these new fields, BIM is not only a virtual representation or innovative tool to support the geometric aspects of buildings. It is a holistic process by which an accurate digital three-dimensional model promotes the dissemination of knowledge during the life cycle assessment (LCA) of the building.

1.1 *Problem statement*

The latest results of research studies carried out in different countries (the reader is referred to Murphy *et al.* (2013), Brumana *et al.* (2017) and Fai and Sydor (2013), among others), show that SCAN-to-BIM adoption for existing buildings is able to improve specific analyses during the rehabilitation project. The use of 3D surveying instruments, the use of laser scanning, photogrammetric and advanced modeling techniques have helped in the survey and management process of existing buildings, reducing costs of rehabilitation projects.

On the other hand, a significant amount of data collected during the 3D survey along with structural, geometric and decay analyses has required the development of extensive digital archives able to transmit the complexity of the built heritage to the team of experts involved in the project.

Accordingly, regular 3D objects in BIM libraries are driving the experts to the 'normalization' of complex shapes, simplifying geometrical aspects or the information related to built heritage. In the field of historic BIM (HBIM), instead, the connection between knowledge and complex geometric entities represents one of the first objectives for preservation, conservation and restoration of complex buildings (Banfi, 2016).

Starting from the assumption that built cultural heritage may support the intellectual capital in public and private organizations, the outputs regarding "capability buildings" should be produced by shaping the conservation process accordingly (Della Torre, 2010).

For this reason, BIM should become an essential requirement for project management and information sharing of historic and existing buildings over time.

The primary cause of this 'normalization' is the lack of advanced modeling tools in BIM applications. Standard 3D objects in BIM databases do not allow the accurate three-dimensional generation of complex elements such as those in churches, basilicas, castles, and medieval bridges. The digital production of their irregular geometric conditions is the first barrier to the management of historical constructions in digital environments. The lack of a user-friendly modeling method results in a waste of resources, leading to considerable costs on the economic side.

Therefore, it can be said that the rapid and accurate 3D digital representation of architectural and structural elements plays a vital role to improve the dissemination of information during the life cycle of the building. Advanced modeling techniques should increase the grade of accuracy of every single element, quantifying and precisely identifying their dimension, material and physical–thermal characteristics, especially their shapes and irregularities in the structures.

This information connected to 3D objects is crucial in developing subsequential analyses that focus on the use of digital models. The combination of all these factors has required the development of a novel method to preserve the morphological and typological values of existing buildings. Vaults, irregular walls, arches, and decorative elements are not part of the parametric software database and require high knowledge of advanced modelling techniques (AMT) for their detailed reproduction. Another determinant factor is the level of automation available in the modeling phase.

The National BIM Report, one of the most comprehensive reviews of the state of BIM within the UK construction sector (NBS, 2017), found that Level 3 is not fully utilized yet. Level 3 envisages the management of the entire life cycle of buildings using digital models capable of performing the function of 'unique knowledge repository'. The main advantages should be better costs and general optimization of the project.

On the other hand, traditional generative procedures of AB-BIM also require two-dimensional drawing procedures and generation of profiles (sections, plans) not mandatory for the digital construction of 3D models. It is clear that BIM adoption is hampered by the lack of in-house expertise, lack of modeling training, high cost for the generation of BIM and the lack of standardized tools and protocol for the re-engineering of their traditional sector.

In this chapter, it will be shown how advanced modeling solutions help to increase the grade of accuracy (GOA) and level of detail (LOD) of the single element, defining a new system able to handle SCAN-to-BIM projects with the integration of free-form modeling software and BIM applications. This new method is not applied through a single software. It is optimized through a combination of different geometric entities, new interoperable interchange formats, new levels of transmissible information during the generative process and a high grade of accuracy of the final model.

1.2 *Research objectives*

This study proposes a novel modelling management information system (MMIS) capable of improving the 3D digital construction of complex AB-BIM by grades of generation (GOG) and information (GOI) defined in previous work (Banfi, 2017). This method aims at the quick and accurate creation of three-dimensional parametric models, automating the extraction of geometric primitives from dense point clouds for the production of specific/unique architectural elements.

In particular, the integration of GOGs 9/10 (automatic point clouds slicing and NURBS interpolation) in BIM application allows one to bridge the modelling gap for complex shapes and structural BIM databases with an estimated grade of accuracy (GOA) at millimetre-level (i.e. the deviation value between point clouds and 3D objects). The planning of the project goals at the beginning of the generative process has enabled the definition of an interoperable system and working method which aims to enhance information exchange at different levels, supporting the intrinsic and intangible values of the three examples presented in this chapter. In particular, the research objectives of the proposed method is an improvement of the conversion of point clouds into AB-BIM by:

1 automatic extraction of geometric primitives and generation of non-uniform rational basis-splines (NURBS) surfaces;
2 parameterization of complex NURBS surfaces in parametric objects avoiding remodeling in BIM application;
3 verification of the deviation value of each 3D object generated from point clouds;
4 improvement of the level of interoperability of AB-BIM and exchange formats for FEA, restoration and monitoring projects

Projects' goals have required a system capable of representing and share existing buildings with high LODs, facilitating information exchange between manufacturers, service providers and experts involved in the rehabilitation and conservation processes of historical construction.

Figure 15.1 aims to describe the proposed method addressing the current SCAN-to-BIM challenges. It underlines how the quality of the model increases if the information flow grows during the generative process. Six crucial steps compose the proposed SCAN to BIM method: (1) project goal definition and collection of available sources; (2) 3D survey with laser scanning, photogrammetry; drone and total station; (3) point cloud processing oriented to BIM generation; (4) generative process 'scan to NURBS'; (5) generative process 'NURBS to BIM'; and finally (6) BIM orientation for different types of analysis.

In this novel generative process, transmissibility of data and related development of new technologies play a crucial role in the creation of AB-BIM. Capturing physical dimensions of architectural heritage with a high level of precision and accuracy is the primary goal of 3D data collection. Consequently, multilevel approaches based on BIM, interdisciplinary collaboration among researchers, operators, software manufacturers and users of open-source communities are necessary to improve information management.

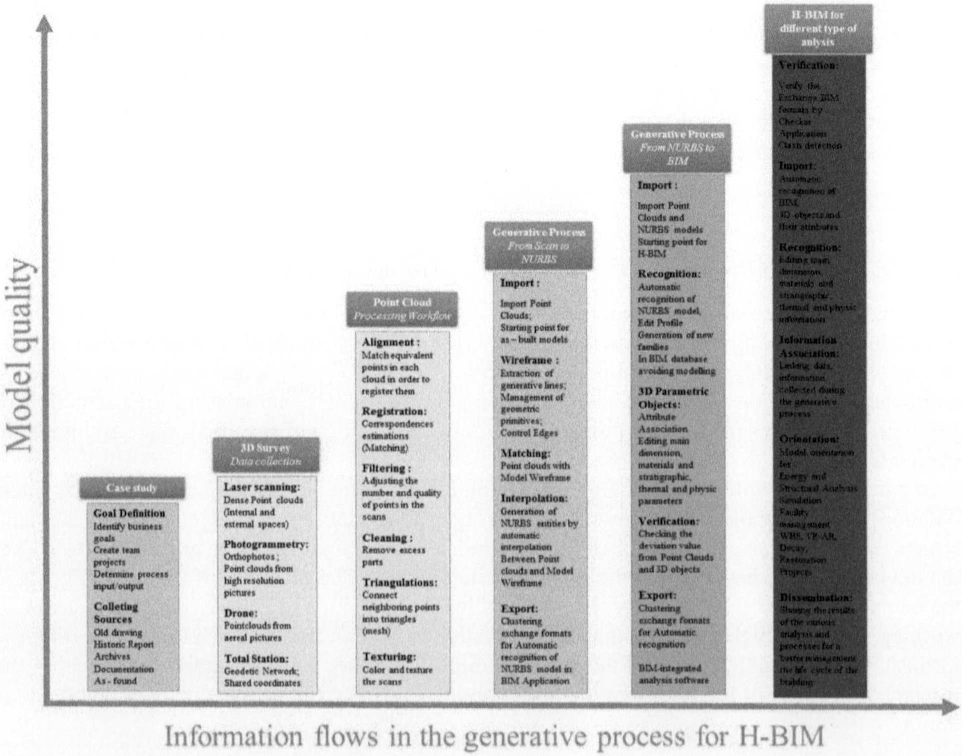

Figure 15.1. The proposed SCAN-to-BIM method

2 DATA ACQUISITION FOR BIM OF LARGE STRUCTURES

As anticipated, 3D data capture is the first step in a generative process for three-dimensional accurate models of existing buildings. The idea is to exploit the survey and management process of built heritage starting from data collected with laser scanning technology (Previtali *et al.*, 2014) and digital photogrammetry (DP; Barazzetti *et al.*, 2010).

Laser scanning technology is an optimal data acquisition technique for AS-BIM project for its metric accuracy and speed. For each proposed case study, laser scanning measured the locations of a huge number of 3D points in a range up to 120–150 meters, which is sufficient to cope with most buildings. The average scan resolution of the three case studies discussed in this chapter was about 44 million points per scan, which is directly related to the point density to ensure reliable 3D reconstruction.

Data processing has been carried out by specific software packages that convert point clouds into formats readable in BIM applications. Experimental investigations were carried out to improve the interoperability level of the exchange format between Autodesk ReCap, McNeel Rhinoceros, and Autodesk Revit. The definition of exchange formats from Bentley ContextCapture/Pointools, Agisoft Photoscan (i.e. photogrammetric software) and Autodesk Recap allowed the direct management of the obtained scans in McNeel Rhinoceros and Autodesk Revit. The proper use of output and input files was the key to reduce point cloud size (in terms of memory consumption and efficient visualization) for a better transmission from their native format in the used modeling applications.

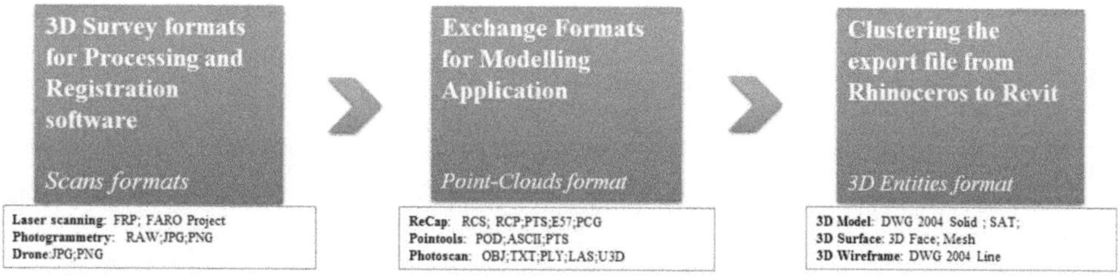

Figure 15.2. The clustering of exchange formats allowed the development of the level of interoperability of data between 3D Survey, NURBS models and BIM application.

PTS and RCS formats have allowed one to import point clouds in McNeel Rhinoceros and Autodesk ReCap. Selecting the best input/output formats and setting a proper NURBS-based modeling solution have permitted users to have point clouds that are not only visual supports during the modeling phase but also active objects to be exploited for automated data processing for the generation of geometric primitives (Figure 15.2).

For instance, the OSnap (Object Snap) function in McNeel Rhinoceros has been used to anchor polylines and slices at specific points with known coordinates. Clustering of the exchange formats (as .pts, .e57, .xyz, .rcp and .ptg) have allowed the transmission of data into the modeling applications without losing geometric information, improving processing, cleaning, editing and measuring phases of measured scans.

The total number of the points handled in this way is significant. This chapter will present the result for some large project, where several laser scanning point clouds were acquired in different days. An overview of the proposed case studies is shown in Figure 15.3. Basically, the laser scanning survey of the Basilica of Collemaggio was made up of 182 scans (8 billion points). Masegra Castel required 176 scans (7,5 billion points). The complexity and size of Visconti bridge required 77 scans (2,5 billion points). The eleven irregular bridge vaults, instead, have required a particular metal bar able to detect the intrados from the road level .

Working with large point clouds required specific strategies and software to preserve the original density of the acquired scans. In fact, the decimation of the point cloud could result in information loss, especially for complex buildings with irregular geometry. One of the aims of the modeling strategy discussed in this chapter is the opportunity to handle huge point clouds, loading only data corresponding to specific constructive elements. Data acquisition has always been carried out considering project requirements, i.e. the creation of a detailed BIM with associated information. Proper planning of the survey (i.e. number of scans, locations, density, time, etc.) cannot be neglected in large SCAN-to-BIM projects.

3 AUTOMATION IN 3D MODELLING THROUGH NOVEL GRADES OF GENERATION (GOG) AND INFORMATION (GOI) FOR SCAN-TO-BIM PROJECTS

Starting from the assumption that the creation of AB-BIM has required the acquisition of dense laser scanning point clouds for modeling in BIM software, this study aims to favour cohesion between advanced modelling techniques (AMT) and parametric applications such as Autodesk Revit and Graphisoft ArchiCAD. As previously stated, the first challenge was to generate irregular and unique architectural and structural elements, improving sharing of information during the life cycle of existing buildings.

3.1 *The application of GOG 9/10: from SCAN to complex NURBS model*

The proposed approach starts from the tools for free-form (pure modeling) software such as McNeel Rhinoceros, Autodesk Maya and 3D Studio Max, in which complex forms can be created.

Figure 15.3. Case studies and the collected TLS data (left) and the orthophoto created by digital photogrammetry technique. Top: the Basilica of Collemaggio; middle: Masegra Castel; and bottom: Azzone Visconti Bridge.

The main characteristic of free-form modeling SCAN to BIM is the possibility to obtain complex shapes through automatic extraction of NURBS geometric primitives (lines and surfaces) from point clouds. Non-uniform rational B-Splines (NURBS) are mathematical functions used in free-form modeling which provide user-friendly and flexible modeling solutions for both regular and irregular surfaces (Piegl and Tiller, 1996). The generative logic of NURBS-based software is different from the parametric modeling logic in BIM application. In the first case, the problem is the generation of geometric entities such as splines, polylines, surfaces, and solids. In the second phase, the modeling logic is based on a limited set of (parametric) modeling tools not able to automatically extract geometric primitives from point clouds, or to create complex surfaces with automatic interpolation algorithms, which are instead available in software such as Autodesk AutoCAD and McNeel Rhinoceros.

The proposed modelling management information system integrates NURBS modelling and BIM generation by GOG 9 and 10. GOG 9 is a procedure that summarizes the technique known as *slicing*. The traditional process of generating 2D CAD models and drawings is essentially based on the manual extraction of cross-sections from point clouds. The sections, usually used for technical drawings, are used as a generative basis for three-dimensional modeling in the digital space. GOG 9, instead, bypass 2D drawings and permits the generation of complex geometric models through the automatic extraction of geometric primitives (Figure 15.5).

Numerous tests have been carried out to identify a novel method capable of automating the generation of complex shapes directly from point clouds, avoiding the slicing technique. Thanks to the identification and targeted application of NURBS interpolation algorithms (Piegl and Tiller, 2012), GOG 10 has been developed. This method provides the automatic generation of complex surfaces directly from point clouds, avoiding the step 2 in picture 5 (internal section curve extraction). GOG

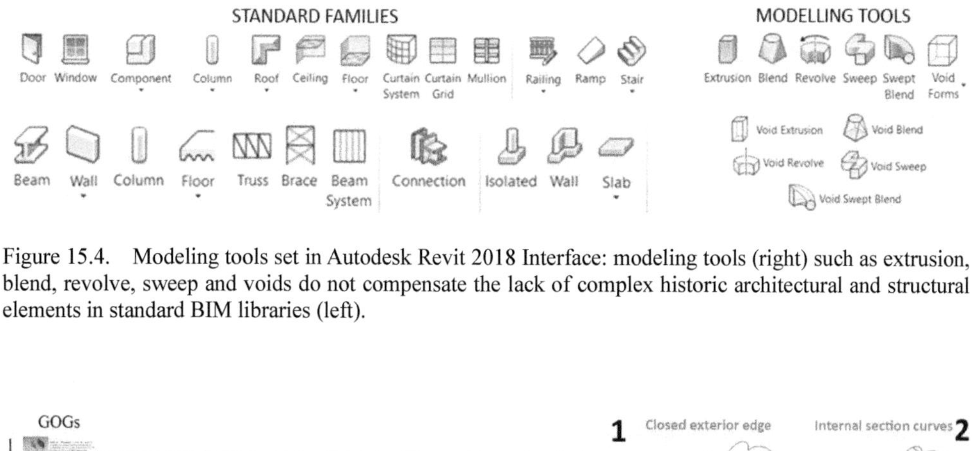

Figure 15.4. Modeling tools set in Autodesk Revit 2018 Interface: modeling tools (right) such as extrusion, blend, revolve, sweep and voids do not compensate the lack of complex historic architectural and structural elements in standard BIM libraries (left).

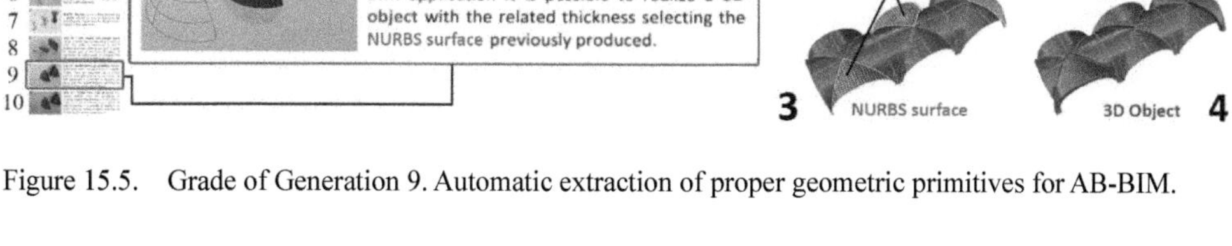

Figure 15.5. Grade of Generation 9. Automatic extraction of proper geometric primitives for AB-BIM.

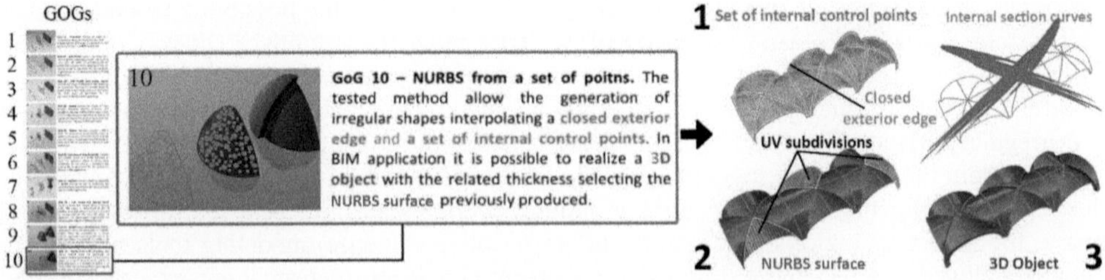

Figure 15.6. Grade of generation 10. Automatic generation of the complex surface by NURBS interpolation for AB-BIM allows one to avoid the generation of internal section curves.

10 is a three-step modeling procedure that has allowed the reduction of time necessary for the generation of AB-BIM (Figure 15.6).

NURBS interpolation algorithms were able to generate complex surfaces that interpolate a large number of points of the scan. Once the point cloud is converted into the NURBS surface by defining specific exchange formats (Banfi *et al.*, 2017), it is possible to parameterize the NURBS surface within Autodesk Revit.

3.2 *AB-BIM Generation: from NURBS model to as-built BIM*

Import tools (import CAD tool) in Autodesk Revit do not allow one to modify the geometric primitives of CAD drawings and geometric models. Once imported via the default command, they are blocked entities without the opportunity to make additional changes. Furthermore, NURBS surface cannot interact with additional information from databases.

For these reasons, it is possible to incorporate NURBS models into the parametric logic and to achieve their parameterization through the tools 'wall by face' and 'roof by face' in Revit. However, the transition from NURBS surfaces to BIM objects depends also on the number of subdivisions U and V. In particular, each NURBS surface required a maximum division limit of 50 × 50, beyond which it is not possible to parameterize them in BIM. If this generative condition is respected, the command 'wall by face' can recognize the components of GOG 9 such as closed exterior edge and internal section curves (Figure 15.7).

This generative procedure, once activated, has permitted the automatic creation of elements capable of accurately following the point cloud, obtaining the NURBS surface in McNeel Rhinoceros and the parametric thickness in Autodesk Revit. The mass command in Autodesk Revit was able to automatically recognize the imported surface. Once correctly introduced through the .sat/.dwg exchange format, it has been possible to create irregular vaults and irregular walls.

The first benefit of this integration between NURBS modeling, 3D scans, and BIM applications is the automation of the generative procedure for complex elements. This procedure has ensured preservation of the geometric anomalies of each architectural and structural element (Figure 15.8). Once the parametric object has been obtained, the second benefit is the possibility to associate new information for different types of analyses and create databases, schedules and sheets in a fully automated way. Figure 15.8 shows the application of GOG 9 and 10 for the creation of a large number of complex elements for the different case studies.

3.3 *Development of grade of information (GOI) in BIM projects for different types of analyses*

BIM objects allow one to connect and edit a large quantity of physical and technical information such as wall stratigraphy, schedules, tables and shared parameters as well as to expand BIM databases with historical information collected during the life cycle of the building, favouring the

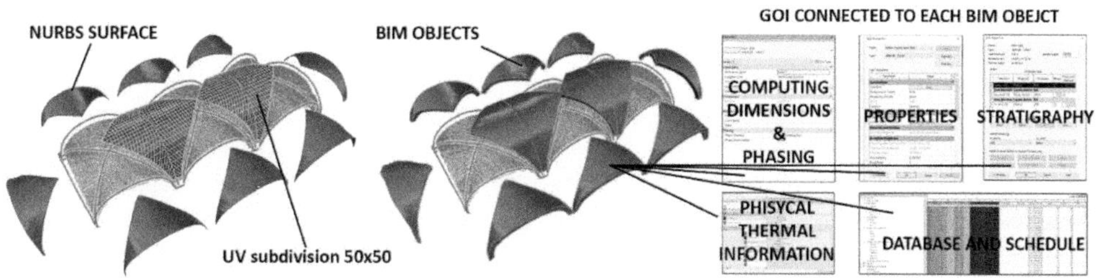

Figure 15.7. The parametrization of NURBS surface in Autodesk Revit. The integration of GOG 9 and 10 and a UV subdivision 50 × 50 has allowed the achieving of high LODs for each BIM object.

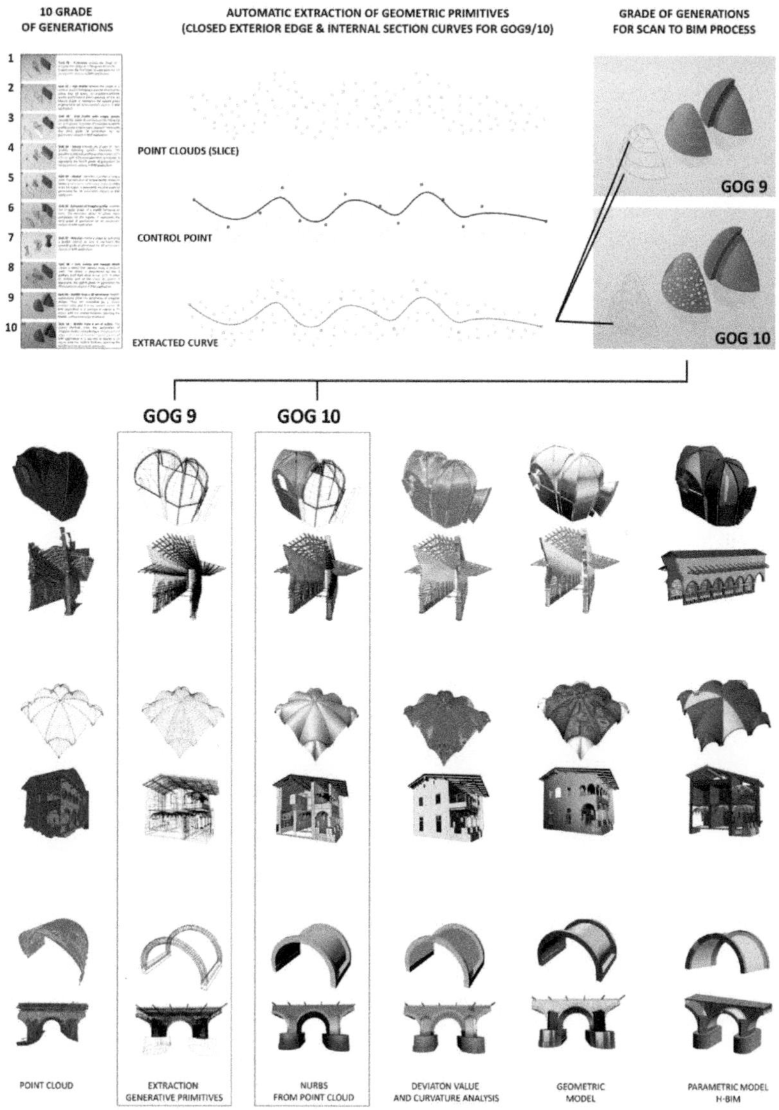

Figure 15.8. The SCAN-to-BIM process applied to complex architectural and structural elements.

dissemination of intangible values through cloud-based digital repositories. Data transfer from free-form modeling to BIM applications has been carried out taking into consideration a high level of detail (LOD) of AB-BIM. NURBS models built according to the construction logic have allowed one to link information and data to complex 3D parametric objects, such as vaults, arches and other complex elements. The identification of the creation logic of different structural 3D objects is fundamental to create an AB-model, which becomes a detailed digital representation of real building conditions. In particular, for each case study, it has been possible to meet the requirements of the rehabilitation projects using GOG 10 and three grades of information (GOI).

A bidirectional relationship in BIM application gives the first grade of information (GOI 1). Thanks to the automatic computing function integrated with Autodesk Revit or Graphisoft ArchiCAD, it is possible to display different types of values and parameters such as volume, area, material, phase, etc. Parametric modeling gives the bidirectional relationship: if the model changes (e.g. geometric dimensions), the numerical values in the database change consequently. This first function facilitates operations such as computing, costs association and the definition of the activities for renovation projects. As shown in Figure 15.9, the GOI 1 of the Basilica of Collemaggio's vaulted system allows one to get the automatic computing (volume and area), the GOG applied for the model generation, the obtained LOD and wall stratigraphy.

The second grade of information (GOI 2) is given by a sub-function in the property window of each BIM object. The definition of the second level of parameters has allowed the dissemination of physical and thermal information such as density, emissivity, permeability, porosity, reflectivity, electrical resistivity, Young's module, Poisson's ratio, linear modulus, yield strength, tensile strength and so on. As shown in Figure 15.10, the bidirectional relationship created between BIM objects and information allows one to edit, import, link and update new descriptive and numeric fields for both the Revit database and external databases.

This type of developed connection has given users the opportunity to investigate a new type of sub BIM objects. GOG 10 applied to the complex vaulted system, and irregular walls of the case studies have allowed 3D BIM mapping of deterioration areas, damages, permanent deformations, decay and accurate quantification in a single visualization.

In particular, the application of GOG 10 has provided corresponding BIM objects for specific deterioration areas. Consequently, this advanced modeling technique has expanded GOG 1, allowing the definition of further parameters (GOG 2) and the creation of related schedules and databases.

Finally, reuse and restoration projects have required BIM orientation for various disciplines. The third grade of information (GOI 3) has been developed with a cloud system. Thanks to Autodesk A360, users with different skills can visualize the AB-BIM and get the GOIs 1 and 2 for each BIM

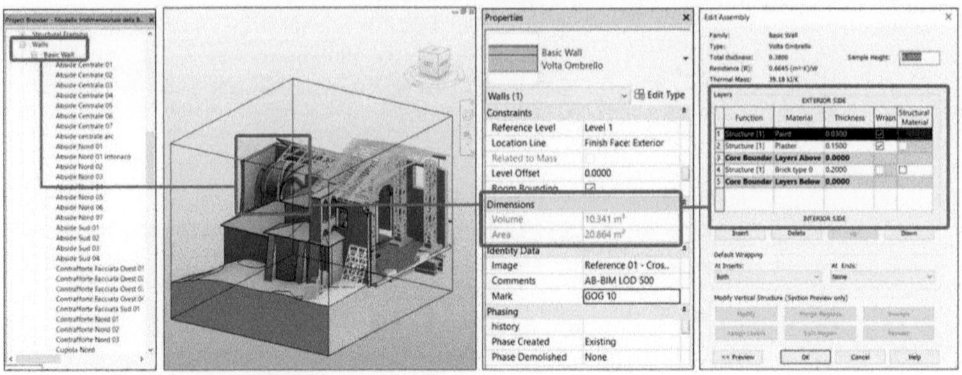

Figure 15.9. The GOI 1 of the complex structural element. The complex vaulted system and the collapsed transept of Basilica of Collemaggio after the earthquake in 2009.

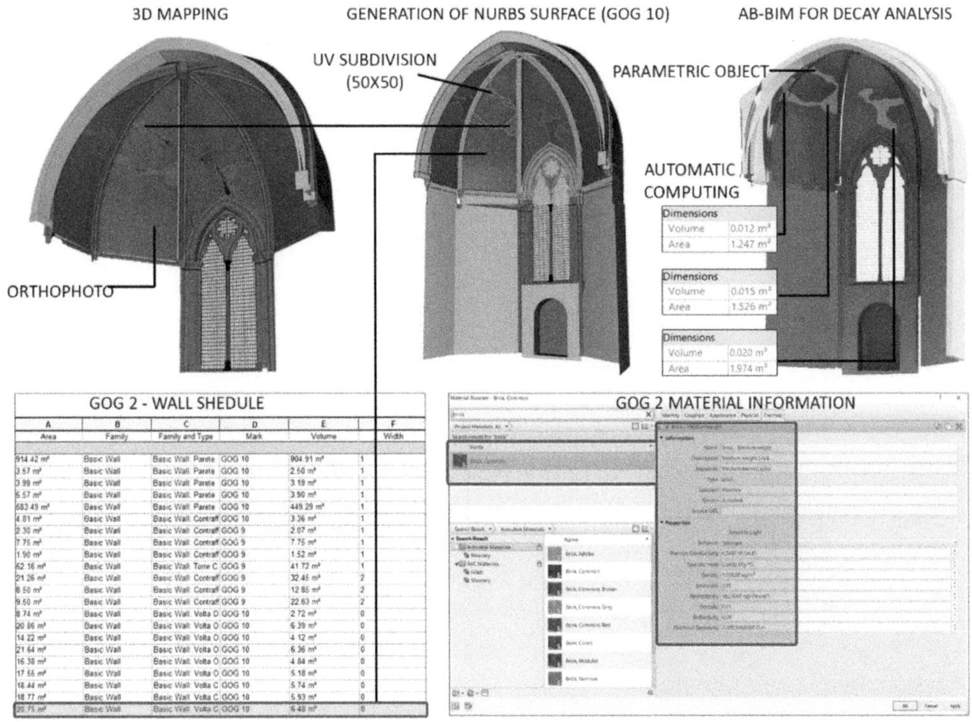

Figure 15.10. The GOI 2: 3D mapping of deterioration areas, NURBS model, AB-BIM, schedules and databases of Basilica of Collemaggio's complex vaulted system.

object. This type of digital information allowed one to improve the connection between different technical disciplines and to update different types of analysis within a common digital repository. The main benefit of GOI 3 is a detailed data usage directly in the browser, supporting more than 50 file formats, including Autodesk, Solidworks, CATIA, Pro-E file formats, Rhino, and NX. Such solutions allow one to share 2D drawings and 3D models via e-mail, or to communicate and get real-time feedback from all the specialists involved in the project, finding various types of linked information from multiple devices (laptops, PCs, mobile platforms).

Figure 15.11 shows the grade of information GOI 1 and two defined in Autodesk Revit and how it is possible to manage the AB-BIM in A360, achieving a cloud-based common data environment (CDE) equivalent to GOI 3. The first benefit of CDE is to share different types of information between various disciplines.

The morphological complexity of the vaulted system and the structural damage of the Basilica were the main reason to refine modeling techniques oriented to the generation of AB-BIM, which has to represent irregular elements starting from a SCAN-to-BIM process.

The objective was to define a high LOD of the historical and irregular parts of the basilica and achieve accurate modeling by the GOG 9 and 10, enabling single BIM elements to become object-based representations. These 3D parametric elements generated from point clouds were created to obtain generative profiles with the integrated use of Autodesk AutoCAD, McNeel Rhinoceros, and Autodesk Revit.

The ability to link information to the model has provided a useful tool for the detailed structural rehabilitation process. The three-dimensional definition of the structural elements has made it possible to identify conservation and restoration techniques for each part of the basilica. The

Figure 15.11. The GOI 3 of AB-BIM: the cloud-based common data environment (CDE) of Masegra Castel allows one to reach new interoperability levels

method focused on adding new type of information such as decay analysis, structural consolidation operations and volumetric quantification of the damaged parts of the apse, transept and pillars of the naves.

Thanks to the automatic volumetric quantification of each BIM object, it has been possible to associate the related cost of the restoration steps. The method has allowed one to create a large number of elements in a short time characterized by cracks analysis, degradation analysis, damage assessment, physical and thermal data of the materials, wall stratigraphy and structural element composition. Finally, the management of schedules and databases based on BIM objects generated by GOG 10 have been crucial for defining a set of restoration goals (Figure 15.12).

GOG 9 and 10 and GOI 1, 2 and 3 for each item have allowed us to find a new way to process the case study of Castel Masegra. Thanks to the conversion system from NURBS objects to meshes, the model of the castle and its irregular elements has been transferred into structural analysis software. Midas FEA, Abaqus and other packages have been able to recognize the detail of the model created without excessive simplification of geometry.

The automatic generation of many sub-elements (triangles and polygons) and the control of geometric primitives has delineated a new level of automation (LOA) in the creation of meshes (useful for FEA) from NURBS models, notwithstanding manual correction has been necessary to have a consistent mesh for FEA. Interoperability of the method has led to the creation of an AB-BIM that can interact with computing applications, running efficient structural simulations for the vaults and irregular walls of the castle.

The deviation value between point cloud and NURBS surface reached the following values: mean distance of 1.0 mm, median distance 0.8 mm, and standard deviation 0.8 mm (Figure 15.13). The development of this method and the optimization of the structural analysis process produced

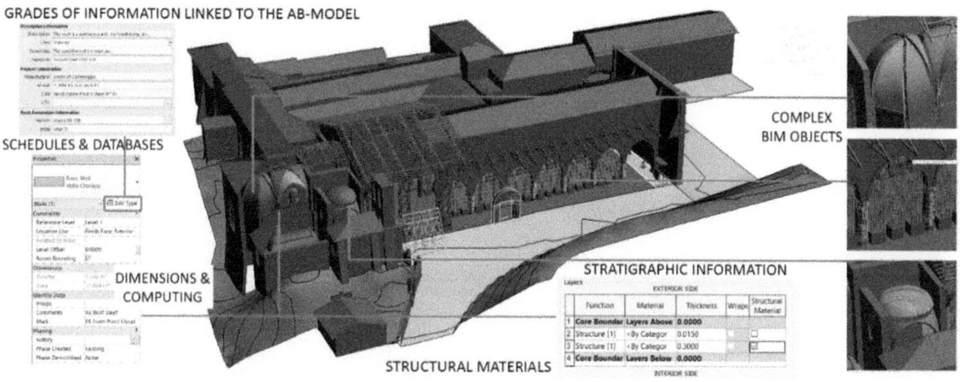

Figure 15.12. AB-BIM Of Basilica of Collemaggio and the GOI 1 and 2 for the restoration project.

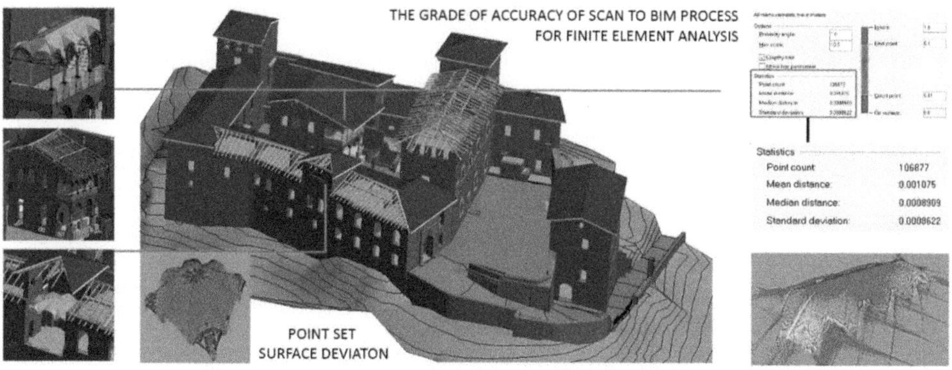

Figure 15.13. AB-BIM of Masegra Castel in Sondrio and the GOA for finite element analysis.

a series of reports and analysis with a high level of modeling precision, providing calculations and quantitative data for better quality controls of the entire medieval complex.

The proposed method has also been used for generating a digital model of a medieval infrastructure like the Azzone Visconti Bridge in Lecco. The generation of 11 irregular arches, the chambers, the stone rings, the vaults, the direct topographical survey of the riverbed, the shape of the pots etc. have achieved a deviation value between model and scans at a millimetre level.

The integration and use of advanced instruments and techniques for the 3D survey and creation of as-built models (such as laser scanner, total station and photogrammetry) combined with NURBS-based and parametric-based modeling have led to the production of an accurate three-dimensional database in Autodesk Revit. Creating novel objects and their associated information inside of the BIM database has led to the adoption of new GOIs (1–2).

The direct benefit has also been an increase in the available information to be exchanged through a cloud-based common data environment (CDE). Surveyors have been able to store the results of their work (3D scans, levelling results, schedules, tables, orthophotos and pictures). This method has allowed one to bypass manual slicing when drawing in CAD applications, as described in the previous paragraph. A high GOA has been achieved for the generation of complex elements by GOG 10.

The proposed generative process of as-built historic models ensures the connection of information during the lifecycle of the infrastructure. The integration between NURBS modeling and BIM application allows the automatic generation of structural elements that are not available in existing BIM databases, also incorporating the information related to management of each component (Figure 15.14).

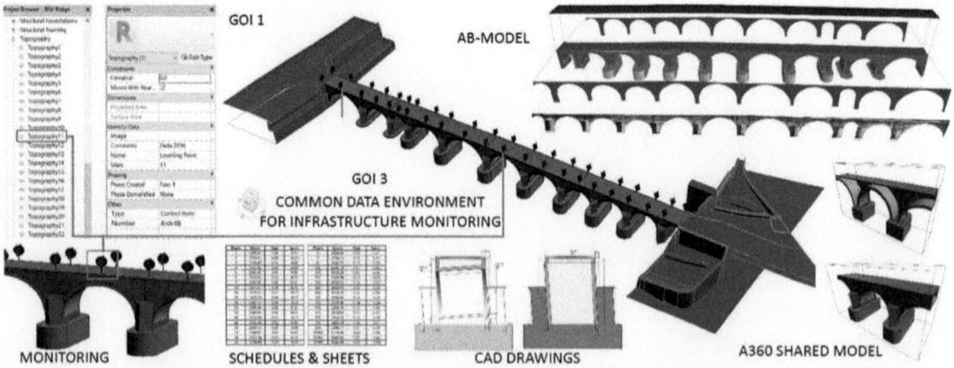

Figure 15.14. AB-BIM of Azzone Visconti bridge in Lecco for cloud-based common data environment (CDE). Surveyors, BIM manager and structural engineering have the same digital repository.

4 AUTOMATIC VERIFICATION SYSTEM (AVS) OF THE GRADE OF ACCURACY (GOA) FOR AS-BUILT BIM

This study showed the possibility of using new tools able to model and manage structures and infrastructures. Advanced modeling techniques from point clouds have a crucial role in reuse, rehabilitation, renovation and conservation, and restoration projects. NURBS-based modeling appears a valid tool to improve the creation of a model for complex and irregular buildings, reducing cost and time related to the BIM generation process. It is possible to improve these aspects and reduce production times by parameterization of NURBS elements in the BIM logic.

The results obtained in the direct development of the proposed SCAN-to-BIM process have been tested by a novel automatic verification system (AVS) able to numerically quantify the grade of accuracy (GOA) of the AB-BIM model.

A quantitative value is provided to quantify the quality of the model produced concerning the acquired point cloud. This provides numeric values that represent the deviation between point clouds and the whole model. Such solution is not only an overall indication, but it can be reused to improve model quality. For instance, Figure 15.9 shows the process for improving the quality of the model. In a first step, AVS has allowed us to reveal large discrepancy values (red points in the second picture in the upper part of the page). Thanks to the application of GOG 9 and GOG 10, it has been possible to improve the NURBS generation of the element and its related deviation value with the identification of geometric primitives (points and external border) with a lower level of metric accuracy (Figure 15.15). The used values are:

Point count: the total number of points n used in the comparison;

Mean distance: average distance between the model and the point cloud ($\mu = \frac{1}{n}\sum_{i=1}^{n} r_i$). This value should be similar to zero, so that systematic errors (biases) do not affect the 3D model;

Median distance: a robust index used to evaluate the presence of gross errors, when compared to the other indexes. The median separates the higher half of the r_i values from the lower half;

Standard deviation: index obtained from the squared distances between model and point cloud $\sigma = \sqrt{\sum_{i=1}^{n}(r_i - \mu)^2 / (n-1)}$. The value should be similar to the precision of the laser scanner used.

The term r_i in the previous relationships represents the distance between a single point and the surface.

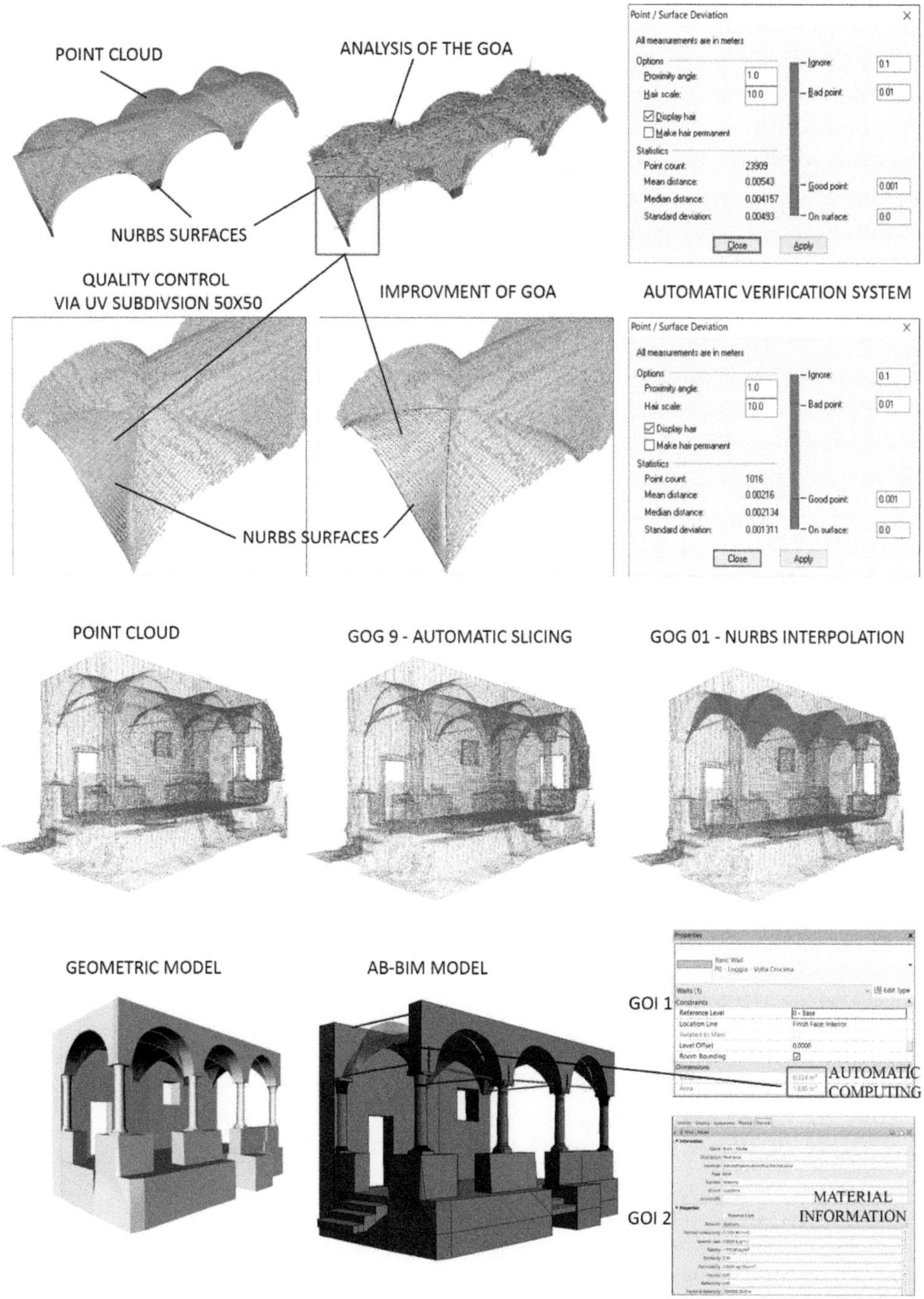

Figure 15.15. Automatic verification system (AVS) for AB-BIM and the improvement of the model quality.

Finally, thanks to the increasing need to verify the model by various experts involved in the project, different subsequent analyses have been possible to draft new BIM protocols to support BIM operators not very expert in 3D modeling.

5 CONCLUSION

In the last decade, BIM has reached an incredible interest in several scientific disciplines. BIM allows accurate reconstruction of buildings and a link with an enormous amount of information, improving results for various types of analysis. The new paradigm of complexity achieved in recent BIM projects is also considered in new national and international regulations to create new LOD definition for complex objects in the built heritage field, to document important structures, and historic buildings. The level of automation (LOA) of the generative process for as-built BIM turns out to be one of the most significant challenges facing the AEC sector. The use of innovative data acquisition and processing algorithms ensures the harmonization of scientific competencies involved in the 3D data capture, processing and data digitalization, as well as the 3D generative process of BIM, architectural and engineering computing, design, BIM-based analysis and information management.

One of the aims of this research was to improve the generation and management of complex BIM for the existing structures. The proposed integrated modelling and management information system (MMIS) provides a holistic view, in which different disciplines contribute to the achievement of various results. This goal implies new knowledge and methods such as those briefly described in this chapter.

The conducted literature study presented the state-of-the-art implementation and research with a focus on the benefits of SCAN to BIM. Automation of digitalization will provide new techniques for existing buildings linked to emerging technologies like augmented and virtual reality, cloud computing, real-time and dynamic modeling software, building performance, integrated project delivery (IPD) and additive manufacturing (AM).

Future developments should focus on this actual need and try to insert and implement the management of NURBS interpolation in BIM application to assure the morphological (grade of generation, or GOG) and typological aspects (grade of Information, or GOI) of buildings with all their unique peculiarities and values. The characteristic of irregular architectural elements and their correspondence to the constructive logic of the structure is useful in several projects, including structural analysis and monitoring of complex infrastructures. MMIS has implemented the management information over time by accurate AB-BIMs, preserving and maintaining the complexity of built heritage. Laser scanning datasets processed with new modeling techniques based on BIM technology are opening new opportunities for the architecture and construction industry. At the same time, information can be encapsulated in the BIM database to provide not only a detailed geometric reconstruction but also an enriched model (the "I" in BIM).

REFERENCES

Azhar, S. (2011) Building Information Modeling (BIM): Trends, benefits, risks, and challenges for the AEC industry. *Leadership and Management in Engineering*, 11(3), 241–252.

Banfi, F. (2016) Building information modelling–A novel parametric modeling approach based on 3D surveys of historic architecture. *Euro-Mediterranean Conference*. Springer, Cham.

Banfi, F. (2017) BIM orientation: Grades of generation and information for the different type of analysis and management process. *International Archives of the Photogrammetry, Remote Sensing and Spatial Information Sciences*, 42(2/W5), 57–64. https://doi.org/10.5194/isprs-archives-XLII-2-W5-57-2017.

Banfi, F., Fai, S. & Brumana, R. (2017) Bim automation: Advanced modeling generative process for complex structures. *26th International CIPA Symposium on Digital Workflows for Heritage Conservation 2017*. Copernicus GmbH.

Barazzetti, L. & Banfi, F. (2017) Historic BIM for mobile VR/AR applications. In: *Mixed Reality and Gamification for Cultural Heritage*. Springer, Cham, Switzerland. pp.271–290.
Barazzetti, L., Scaioni, M. & Remondino, F. (2010) Orientation and 3D modelling from markerless terrestrial images: Combining accuracy with automation. *The Photogrammetric Record*, 25(132), 356–381.
Bosché, F., Ahmed, M., Turkan, Y., Haas, C.T. & Haas, R. (2015) The value of integrating Scan-to-BIM and Scan-vs-BIM techniques for construction monitoring using laser scanning and BIM: The case of cylindrical MEP components. *Automation in Construction*, 49, 201–213.
Brumana, R., Dellatorre, S., Oreni, D., Previtali, M., Cantini, L., Barazzetti, L., . . . Banfi, F. (2017) HBIM challenge among the paradigm of complexity, tools and preservation: The Basilica di Collemaggio 8 years after the earthquake (L'Aquila). *International Archives of the Photogrammetry, Remote Sensing and Spatial Information Sciences*, 42(2W5), 97–104.
Della Torre, S. (October 14, 2010). Learning and Unlearning in Heritage Enhancement Processes. *ESA Research Network Sociology of Culture Midterm Conference: Culture and the Making of Worlds*. Available at SSRN: https://ssrn.com/abstract=1692099 or http://dx.doi.org/10.2139/ssrn.1692099
Fai, S. & Sydor, M. (2013, December) Building information modelling and the documentation of architectural heritage: Between the 'typical'and the 'specific'. *2013 Digital Heritage International Congress (DigitalHeritage)*. Vol. 1. IEEE, pp.731–734.
Fedorik, F., Makkonen, T. & Heikkilä, R. (2016) Integration of BIM and FEA in automation of building and bridge engineering design. *ISARC. Proceedings of the International Symposium on Automation and Robotics in Construction*. Vol. 33. Vilnius Gediminas Technical University, Department of Construction Economics & Property.
Jackson, P. & Eynon, J. (2016) BIM and Infrastructure. In *Construction Manager's BIM Handbook*. John Wiley & Sons Ltd, pp.16–23.
Kim, H., Shen, Z., Kim, I., Kim, K., Stumpf, A. & Yu, J. (2016) BIM IFC information mapping to building energy analysis (BEA) model with manually extended material information. *Automation in Construction*, 68, 183–193.
Murphy, M., McGovern, E. & Pavia, S. (2013) Historic building information modelling–Adding intelligence to laser and image based surveys of European classical architecture. *ISPRS Journal of Photogrammetry and Remote Sensing*, 76, 89–102.
NBS National BIM Report. (2017) Available from: www.thenbs.com/knowledge/nbs-national-bim-report-2017.
Piegl, L. & Tiller, W. (1996) Algorithm for approximate NURBS skinning. *Computer-Aided Design*, 28(9), 699–706.
Piegl, L. & Tiller, W. (2012) *The NURBS Book*. Springer Science & Business Media.
Previtali, M., Scaioni, M., Barazzetti, L. & Brumana, R. (2014) A flexible methodology for outdoor/indoor building reconstruction from occluded point clouds. *ISPRS Annals of the Photogrammetry, Remote Sensing and Spatial Information Sciences*, 2(3), 119.
Sacks, R., Radosavljevic, M. & Barak, R. (2010) Requirements for building information modeling based lean production management systems for construction. *Automation in Construction*, 19(5), 641–655.
Volk, R., Stengel, J. & Schultmann, F. (2014) Building Information Modeling (BIM) for existing buildings: Literature review and future needs. *Automation in Construction*, 38, 109–127.

Index

ISPRS book series

1. Advances in Spatial Analysis and Decision Making (2004)
   Edited by Z. Li, Q. Zhou & W. Kainz
   ISBN: 978-90-5809-652-4 (HB)

2. Post-Launch Calibration of Satellite Sensors (2004)
   Stanley A. Morain & Amelia M. Budge
   ISBN: 978-90-5809-693-7 (HB)

3. Next Generation Geospatial Information: From Digital Image Analysis to Spatiotemporal Databases (2005)
   Peggy Agouris & Arie Croituru
   ISBN: 978-0-415-38049-2 (HB)

4. Advances in Mobile Mapping Technology (2007)
   Edited by C. Vincent Tao & Jonathan Li
   ISBN: 978-0-415-42723-4 (HB)
   ISBN: 978-0-203-96187-2 (E-book)

5. Advances in Spatio-Temporal Analysis (2007)
   Edited by Xinming Tang,Yaolin Liu, Jixian Zhang & Wolfgang Kainz
   ISBN: 978-0-415-40630-7 (HB)
   ISBN: 978-0-203-93755-6 (E-book)

6. Geospatial Information Technology for Emergency Response (2008)
   Edited by Sisi Zlatanova & Jonathan Li
   ISBN: 978-0-415-42247-5 (HB)
   ISBN: 978-0-203-92881-3 (E-book)

7. Advances in Photogrammetry, Remote Sensing and Spatial Information Science. Congress Book of the XXI Congress of the International Society for Photogrammetry and Remote Sensing, Beijing, China, 3–11 July 2008 (2008)
   Edited by Zhilin Li, Jun Chen & Manos Baltsavias
   ISBN: 978-0-415-47805-2 (HB)
   ISBN: 978-0-203-88844-5 (E-book)

8. Recent Advances in Remote Sensing and Geoinformation Processing for Land Degradation Assessment (2009)
   Edited by Achim Röder & Joachim Hill
   ISBN: 978-0-415-39769-8 (HB)
   ISBN: 978-0-203-87544-5 (E-book)

9. Advances inWeb-based GIS, Mapping Services and Applications (2011)
   Edited by Songnian Li, Suzana Dragicevic & Bert Veenendaal
   ISBN: 978-0-415-80483-7 (HB)
   ISBN: 978-0-203-80566-4 (E-book)

10. Advances in Geo-Spatial Information Science (2012)
    Edited byWenzhong Shi, Michael F. Goodchild, Brian Lees & Yee Leung
    ISBN: 978-0-415-62093-2 (HB)
    ISBN: 978-0-203-12578-6 (E-book)

11. Environmental Tracking for Public Health Surveillance (2012)
    Edited by Stanley A. Morain & Amelia M. Budge
    ISBN: 978-0-415-58471-5 (HB)
    ISBN: 978-0-203-09327-6 (E-book)

12. Spatial Context: An Introduction to Fundamental Computer Algorithms for Spatial Analysis (2016)
    Christopher M. Gold
    ISBN: 978-1-138-02963-7 (HB)
    ISBN: 978-1-4987-7910-4 (E-book)

13. Planetary Remote Sensing and Mapping (2018)
    Edited by Bo Wu, Kaichang Di, Jürgen Oberst & Irina Karachevtseva
    ISBN: 978-1-138-58415-0 (HB)
    ISBN: 978-0-429-50599-7 (E-book)